I0028892

Systems Engineering

Systems Engineering

Holistic Life Cycle Architecture, Modeling,
and Design with Real-World Applications

Sandra L. Furterer

CRC Press
Taylor & Francis Group
Boca Raton London New York

CRC Press is an imprint of the
Taylor & Francis Group, an **informa** business

First edition published 2022
by CRC Press
6000 Broken Sound Parkway NW, Suite 300, Boca Raton, FL 33487-2742

and by CRC Press
2 Park Square, Milton Park, Abingdon, Oxon, OX14 4RN

© 2022 Taylor & Francis Group, LLC

CRC Press is an imprint of Taylor & Francis Group, LLC

Reasonable efforts have been made to publish reliable data and information, but the author and publisher cannot assume responsibility for the validity of all materials or the consequences of their use. The authors and publishers have attempted to trace the copyright holders of all material reproduced in this publication and apologize to copyright holders if permission to publish in this form has not been obtained. If any copyright material has not been acknowledged please write and let us know so we may rectify in any future reprint.

Except as permitted under U.S. Copyright Law, no part of this book may be reprinted, reproduced, transmitted, or utilized in any form by any electronic, mechanical, or other means, now known or hereafter invented, including photocopying, microfilming, and recording, or in any information storage or retrieval system, without written permission from the publishers.

For permission to photocopy or use material electronically from this work, access www.copyright.com or contact the Copyright Clearance Center, Inc. (CCC), 222 Rosewood Drive, Danvers, MA 01923, 978-750-8400. For works that are not available on CCC please contact mpkbookspermissions@tandf.co.uk

Trademark notice: Product or corporate names may be trademarks or registered trademarks and are used only for identification and explanation without intent to infringe.

ISBN: 978-0-367-53280-2 (hbk)
ISBN: 978-0-367-53281-9 (pbk)
ISBN: 978-1-003-08125-8 (ebk)

DOI: 10.1201/9781003081258

Typeset in Times
by SPi Technologies India Pvt Ltd (Straive)

*I dedicate this textbook on systems engineering
to my husband, Dan, and three amazing children, Kelly, Erik, and Zach,
from whom I continually derive the meaning of life*

– Sandra L. Furterer

Contents

Section 1   Systems Engineering Fundamentals

Section 2   Systems Design, Development, and Deployment Life Cycle – Service Applications and Tools

Preface

This book is a practical application-based guide for systems engineering modeling and design, providing methods, tools, and examples of how to design and implement systems engineering. The book takes the reader through the design life cycle with tools and application-based examples of how to design a system, focusing on incorporating systems principles and tools to ensure system integration. A service system example is provided that helps the reader understand the models, tools, and activities to be applied to design and implement a system.

The first section of the book includes the systems principles, models, and architecture for systems engineering. We discuss systems principles and concepts; life cycle models, including the Vee model, which is the life cycle applied in the book; and the systems architecture that lays the groundwork for Model-Based Systems Engineering (MBSE) and the integration of the tools throughout the book.

The second section includes the systems design, development, and deployment life cycle with applications and tools. We discuss the conceptual design phase, including how concepts are developed, assessed, and prioritized; the requirements and architecture phase, including how to generate requirements based on storyboarding and the development of process scenarios and class diagrams; the design phase applying the systems engineering design tools to ensure system integration; a discussion of the integration, verification, validation, and implementation phases; the Operations and Maintenance Phase; and the systems engineering planning techniques.

The last section provides advanced systems engineering topics including, discussion of MBSE modeling; a discussion of lean, iterative, and agile life cycles that can reduce life cycle design and development time lines; and a discussion of the people, communication, and change skills that are needed to successfully apply systems engineering tools and methods.

As systems, technology, and design become more complex, systems engineering design and modeling principles, tools, and methods become more important. This book provides a guide for learning and applying systems engineering tools and methods, with a focus on active learning strategies that enhance and promote higher levels of learning. Organizations are challenged with reducing the product and service design life cycles, and this book provides real-world examples that can engage the learner to learn and apply systems engineering design tools and methods.

One of the examples in the book describes how a hospital applied systems engineering tools and methods to extract service and process requirements, design, test, and implement systems and processes that met the patients' needs for quick appointment and service turnaround times, in a women's healthcare center. Another example describes how a service-based system, developing a food system to reduce food insecurity, could be used to apply the systems engineering life cycle, tools, and methods, to ensure integrated requirements, design specifications, and verification and validation of test cases.

The ancient Chinese philosopher Confucius expressed his belief in the importance of learning from experience when he wrote: "I hear and I forget/I see and I remember/I do and I understand[.]" Confucius related the acquisition of understanding and knowledge directly to living and experiencing. We will be experiencing how to apply systems engineering principles and tools through examples of how my graduate students learned and applied the tools. These examples are not perfect, but they are real. They will give the learner an appreciation of how difficult it can be to understand these concepts and to apply them to a real-world problem where the answers are not in the back of the book. I encourage the teacher and the students to learn this material by also applying the tools to a real-world problem. It is extremely effective to have the entire class working on different parts of the same system, as the examples in the book demonstrate. Examples and pictures are worth a thousand words. I limited the verbiage and descriptions to the critical information needed to understand and apply the tools, incorporating examples that the students can follow to apply these tools and develop their system. Spend time on reviewing the examples, as they can help the learner understand how to apply the systems engineering tools and the type of content that is needed for each.

This book is for learning about systems thinking and how to design a system with the Vee model life cycle as the methodology and applying the tools incorporated into the book. It is focused on a senior level undergraduate or graduate course as an introduction to systems engineering. But it can also be used by more experienced practitioners to gain insights and ideas of how to better integrate and design their systems as well.

Acknowledgment

I would like to acknowledge all of my students that soaked up the knowledge and applied systems engineering on real-world systems in my Management of Engineering Systems classes at the University of Dayton in the Engineering Management, Systems and Technology department. In particular, I would like to acknowledge those students who contributed their examples of applications of the systems engineering tools to the Food Justice System that we designed as part of the course, as follows:

Student Teams

#	Teams	Missions	Contributors
1	Connecting organizations	Create website to encourage more people, connect organizations, find what people want, and get young people involved.	Hafeez Ur Rehman, Benjamin Boyd, Dillon Jones, Hansong Zhang
2	Creating food policy	Develop policies for the food justice system.	De Wang, Joshua Ebersbacher, Pratheeka Kancharla, Varija Goli
3	Kids growing food	Teach kids to grow microgreens and get microgreens to store to sell.	Erin Lisac, Jared Cebulski, and Fahad Alhumoud
4	Food storage	Create a centralized storage location for the community food.	Abdulrahman Faden, Gowtham Kondakkagari, Ganeshkrishna Valluru, and Fares Alotaibi
5	Food hub website	Create a website to network food residents, entrepreneurs, farmers, grocery stores (food hub).	Myles Fiascone, Taylor Zehring, and Abdulaziz Baroun
6	GEM City Market	Operationalize GEM City Market supermarket.	Omar Alraqqas, Bader Alturkait, Deepika Akarapu, and Kavin Nilavu
7	Table 33 Restaurant	Help make the Table 33 concept profitable, by becoming more efficient and better connecting local foods with the restaurant.	Abdulrahman Alrayyes, Ahmad Alkandari, and Mohammad Alansari

I would also like to acknowledge and thank our community partners for the dedication and commitment to solving the food insecurity problem in Dayton, Ohio, including those who worked with us on these projects: Citywide, Caitlyn Jacob; Hall Hunger, Etana Jacobi; Microgreens, Rob Fogg; Homeful, Mandy Knaul; Downtown Dayton, Al Fergusen; Gem City Market, Amah Salassi; and Table 33, Chris Harrison.

Thank you to all of the faculty and staff of UD that participated in the GEMnasium during this Spring semester. A special thanks to Dr. Kevin Hallinan, Professor of Mechanical and Aerospace Engineering who facilitated and organized the integration points across the classes learning within the GEMnasium and coordinated with stakeholders working to provide a just food system. A big thank you to the Institute of Applied Creativity for Transformation (IACT) team: Brian LaDuca, Executive Director, IACT; Adrienne Ausdenmoore, Director, IACT; and Mike Puckett, Place and Space Coordinator, IACT.

Last, but certainly not least, I'd like to thank my wonderful children, Kelly and Erik Furterer who reviewed this manuscript and provided wonderful ideas to enhance the concepts, verbiage, and examples and to my son Zach Furterer, who explored the Ancient Egypt exhibit at the Cincinnati Museum Center with me, where we discovered the picture of the West Wall Tomb of Paheri that provides a wonderful example of a functional decomposition from ancient Egypt.

Author Biography

Dr. Sandra L. Furterer is an Associate Professor and Department Chair at the University of Dayton, in the Department of Engineering Management, Systems and Technology. She has applied Lean Six Sigma, Systems Engineering, and Engineering Management tools in higher education, healthcare, banking, retail, and other service industries. She had achieved the level of Vice President in several banking institutions. She previously managed the Enterprise Performance Excellence center in a healthcare system.

Dr. Furterer received her Ph.D. in Industrial Engineering with a specialization in Quality Engineering from the University of Central Florida in 2004. She received an MBA from Xavier University and a Bachelor and Master of Science in Industrial and Systems Engineering from The Ohio State University.

Dr. Furterer is an author or co-author of several reference textbooks and handbooks on Lean Six Sigma, Design for Six Sigma, Lean Systems, Quality Management and Quality Improvement as well as several journal articles, conference proceedings, and more than 90 conference presentations.

Dr. Furterer has extensive experience in business process and quality improvements. She is an ASQ Certified Six Sigma Black Belt, an ASQ Certified Manager of Quality/Organizational Excellence, an ASQ Certified Quality Engineer, an ASQ fellow, and a certified Six Sigma Master Black Belt.

Dr. Furterer lives in Dayton with her husband Dan, rescue dogs Demi (Dr. Deming) and Lily (Lilian Gilbreth), and cat Louis. Her three wonderful kids are also close by.

Section 1

Systems Engineering Fundamentals

1 Systems Principles and Concepts

1.1 INTRODUCTION AND PURPOSE OF THE BOOK

There are many books on the subject of systems engineering. Why another book? As I was reviewing books to use for my graduate-level systems engineering course that I teach, I would start reading the book and think, "Oh, this is good, it describes a systems engineering life cycle." Then the two or three chapters would end, and I would think, "well, that's not enough to carry my students through the course to understand the entire systems engineering lifecycle, and where are the examples?" So, I did what any other crazy academic would do and set on my journey to write my own book that would help my students understand 1) the principles of systems engineering; 2) what a systems engineer does in practice; 3) how they apply the systems engineering lifecycle or methodology; and 4) how they apply the systems engineering tools and modeling to understand the systems elements, how they connect, and how a system can be designed, developed, integrated, tested, deployed, and even retired.

Why systems engineering? Systems engineering is more than a very popular buzzword. Systems engineering and systems thinking is a body of knowledge brought together from multiple disciplines to understand the patterns, elements, and connectivity to better design, identify potential risks and failures, and build a better system that meets the customers' requirements.

1.2 WHAT IS A SYSTEM?

Let's start this chapter by defining a system. As you can expect, there are many different definitions of a system. Let's start first with one of the more recent and fundamental definitions from the biology discipline.

A system is a set of elements in interaction (Bertalanffy 1968).

INCOSE (2015) defines a system as:

any set of related parts for which there is sufficient coherence between the parts to make viewing them as a whole useful

where as they define an engineered system as:

An engineered system defines a context containing both technology and social or natural elements, developed for a defined purpose by an engineering life cycle.

(INCOSE, 2015)

An engineered system can be made up of

- Hardware
- Software
- Facilities

- Policies
- Documents
- Processes
- People
- Etc.

There are many types of systems as follows:

Product and product systems are systems in which products are designed, developed, and delivered to the customer or consumer for their use or someone else's use. An example of a product system is shown in Figure 1.1.

Service and service systems are systems that provide outcomes or services through the performance of processes, people, and most times technology.

Enterprise and enterprise systems are where one or more organizations or individuals provide a product or service. In an enterprise system, there is a shared purpose or mission, coordination, communication, and interaction between the organization and parts of the system. A hospital or medical system is an example of an enterprise system, as shown in Figure 1.2. There are multiple organizations that are part of the medical system, including the hospital owner, the medical or physician groups that have privileges at the hospital, the vendors, suppliers, the linen company that cleans the linens, to name a few, that provide medical services to patients.

System of systems includes a number of enterprises, service, and product systems brought together to achieve a common purpose. The system of systems is usually interconnected systems of a broader scale, such as an airport that is made up of product systems (airplanes, equipment, facilities), service systems (airport security, ticketing, gate services), enterprises (airline companies, contractors). Another example could be a transportation system for a geographic area including highway system, rail system, air travel system, waterway system. An example is shown in Figure 1.3. However, it is important to note that one person's system may be another person's system of system. Some may call our product system, airplane example above, a system of system.

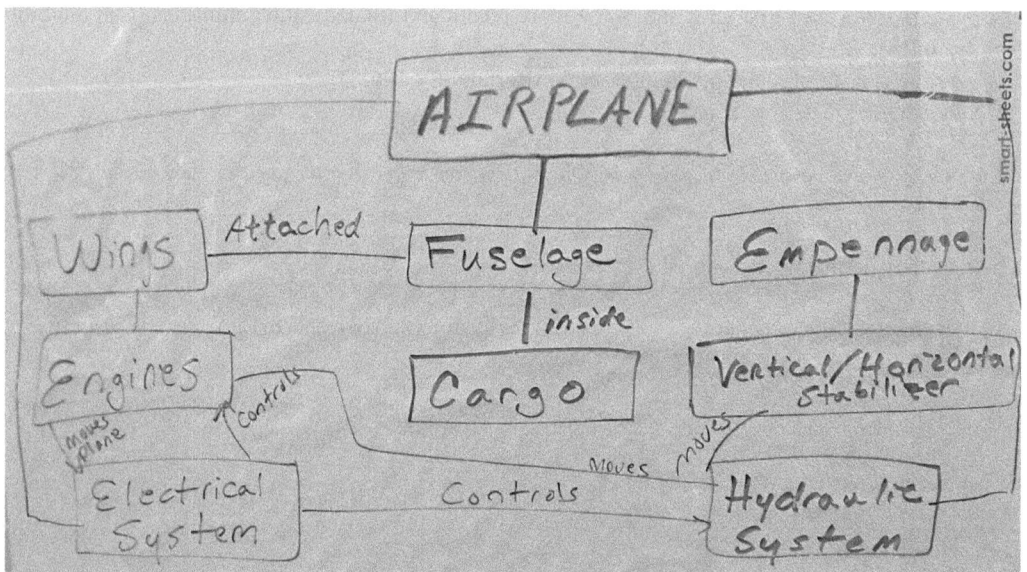

FIGURE 1.1   Product System – Airplane.

Source: Sandra L. Furterer course work, Management of Engineering Systems, 2018.

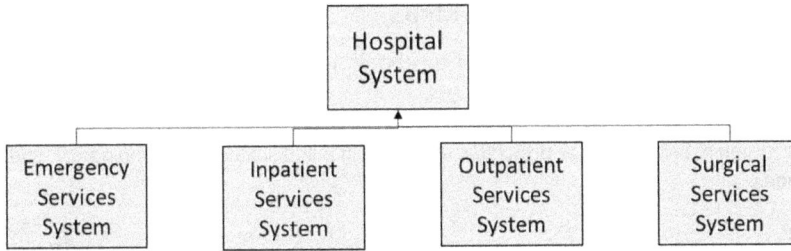

FIGURE 1.2 Example of an Enterprise System – a Hospital.
Source: S. Furterer course work, Management of Engineering Systems, 2018.

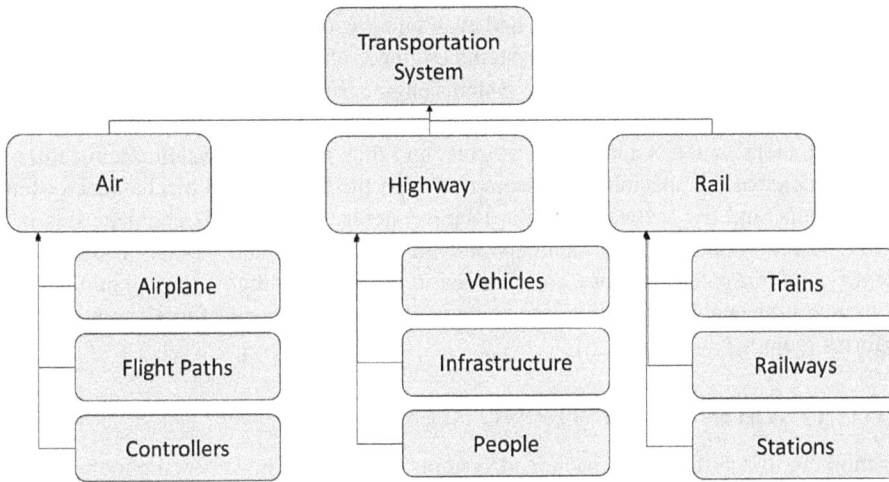

FIGURE 1.3 Example of a System of Systems.
Source: S. Furterer course work, Management of Engineering Systems, 2018.

Systems can be open systems, where they interact with their environment. An example of an open system is a rail system, which interacts with the environment, through the rails expanding and contracting with the weather and the people running their ticket through the ticket machine. The environment does not have to be the natural environment but can be the interaction through the boundary of the system, such as the traveler interacting with the ticketing machine, that is part of the railway access system.

A closed system is one that does not interact with their environment. Closed systems do not exist in nature. There are very few examples of a closed system. In thermodynamics, a closed system is one where mass is conserved within the boundaries of the system, but energy is allowed to freely enter or exit the system. In chemistry, a closed system is one in which neither reactants nor products can enter or escape. It can allow energy transfer of heat and light (Helmenstine, 2020).

There are natural systems that exist in nature, such as a wetland. There are man-made systems, such as a bridge or a highway. There are physical systems, that exist in a physical form, such as a bridge. There are conceptual systems, that exist as an idea, such as a drawing of a bridge or a blueprint of a building. There are also static and dynamic systems. A static system has structure, such as a building, but no behavior or activity. A dynamic system has structure and behavior or activity, such as the train arriving or leaving the rail station.

1.3   DEFINING SYSTEMS ENGINEERING

Systems engineering, which is the subject of this book, is defined as:

> A transdisciplinary and integrative approach to enable the successful realization, use, and retirement of engineered systems, using systems principles and concepts, and scientific, technological, and management methods.

(Fellows, 2019)

When we first start the semester in the Management of Engineering Systems course, my students, who come from a variety of engineering disciplines, including civil and environmental engineering, industrial engineering, chemical and materials engineering, mechanical engineering, and electrical and computer engineering, expect that they will be "designing" a system. As we delve further into the definition of systems engineering, and even into the application of the systems engineering tools, they begin to understand that the systems engineer and the systems engineering functions encompass the connective tissue across the systems engineering life cycle. Systems engineers don't design the system, per se, but they help to define the systems concepts, they facilitate the collection of the customers' and systems' requirements, and they model the system architecture. All of those activities enable the specialty engineers to design the product and mechanical systems, the electrical systems, and the software and hardware systems. The systems engineer aids in ensuring the integration of the systems elements and the verification and validation of the systems. They identify and mitigate risk. They are involved in the systems engineering planning functions. Depending upon the specific organization, some of these activities and functions may vary, as to who performs them.

1.4   WHY IS SYSTEMS ENGINEERING NECESSARY

Systems thinking, the systems approach, and systems engineering have become more important in the world today. Complexity of our systems increases due to the speed of technology change and shorter life cycles. As more software applications guide our evolving systems, to achieve more autonomy and intelligence, the need to understand and test the interconnectivity between the hardware, the software, and the customers' requirements becomes more important. Our world is increasingly becoming more connected and reliant on each other through global supply chains. Enhanced competition in the marketplace contributes to pressures to design, develop, test, and deploy systems more quickly. Many organizations still lack basic infrastructure related to the maturity of design and operational processes, making systems engineering processes precarious and risky, but ever more important (Blanchard and Blyler, 2016).

Let's now step back in time to understand the history of systems thinking.

1.5   SYSTEMS THINKING HISTORY

A very early view of a system is from the Tomb of Paheri, constructed around 1500 BC. Paheri was a property owner and governor. One of the figures on the west wall shows Paheri's role of supervising farm work: harvesting, hunting, fishing, and loading of the boats. Figure 1.4 illustrates a systems model, a functional decomposition of the activities that were critical to life in ancient Egypt and the cyclical and year-round nature of these agricultural functions (e_pahery_01).

Between 385 and 323 BC, the philosopher, Aristotle was credited with saying "The whole is more than the sum of its parts" (Von Bertalanffy 1972, p. 407), which addresses the principle of holism to be discussed later in the chapter.

In more modern times, shortly after the start of the industrial revolution, Dr. W. Edwards Deming, the great quality guru, developed his systems thinking and management philosophies starting in

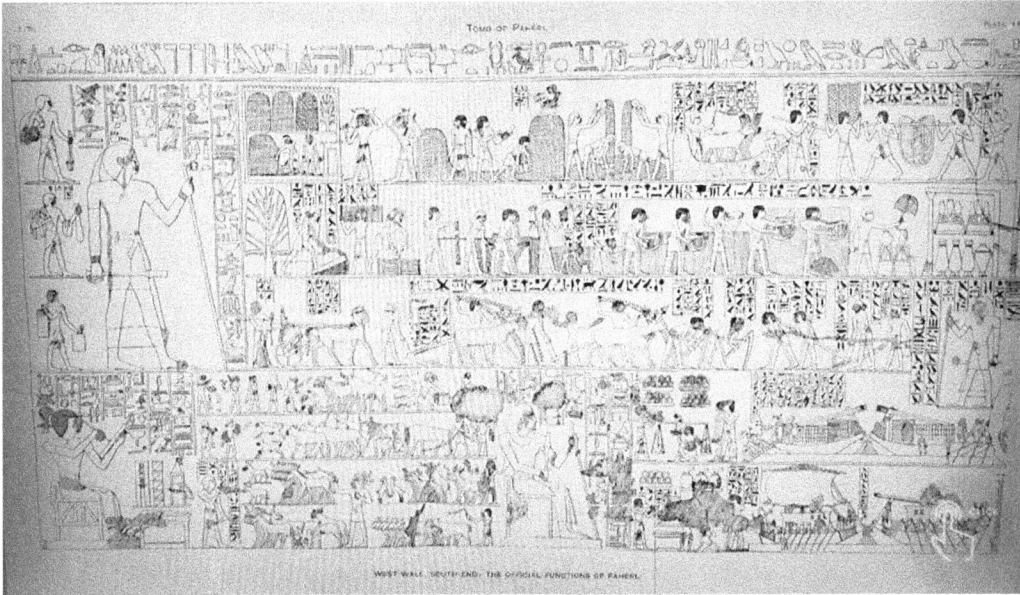

FIGURE 1.4 West Wall Tomb of Paheri.
Source: Zachary Furterer, Photo from the Ancient Egypt exhibit at the Cincinnati Museum Center, July 2019.

the 1920s to his death in 1993. He said, "Management of a system, cooperation between components, not competition. Management of people" was the secret to Japan's success in becoming an economic power after World War II (Deming and Orsini, 2013, Chapter 1). In Deming's System of Profound Knowledge, he synthesized his management philosophies from his 14 points, into four main elements: 1) appreciation for a system, 2) understanding variation, 3) theory of knowledge, and 4) psychology. A system is defined in this context as "a set of functions or activities within an organization that work together for the aim of the organization" (Evans and Lindsay, 2018, p. 55).

Deming defined a system as:

A system is an interconnected complex of functionally related components, divisions, teams, platforms, whatever you call them, that work together to try to accomplish the aim of the system. A system must have an aim. Without an aim there is no system. The aim of the system must be clear to everyone in the system. The aim includes plans for the future.

(Deming and Orsini, 2013, chapter 5)

Deming defined a view of his production system from a manufacturing system perspective but believed that quality and a system perspective would eventually be used in services as well, including hospitals, government, and the like (Deming, 1953). A food system can be illustrated through Deming's lens of a production system, as shown in Figure 1.5.

In the late 1920s, von Bertalanffy wrote:

Since the fundamental character of the living thing is its organization, the customary investigation of the single parts and processes cannot provide a complete explanation of the vital phenomena. This investigation gives us no information about the coordination of parts and processes. Thus, the chief task of biology must be to discover the laws of biological systems (at all levels of organization). We believe that the attempts to find a foundation for theoretical biology point at a fundamental change in the world picture. This view, considered as a method of investigation, we shall call "organismio biotogy" and, as an attempt at an explanation, "f/ie system theory of the organism"

[7, pp. 64 ff., 190, 46, condensed].

Food System

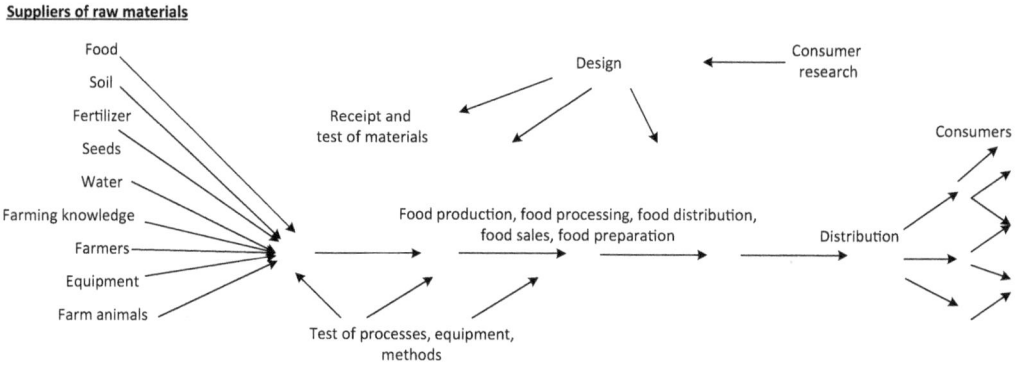

FIGURE 1.5 A food system through deming's lens of a production system.
Source: Sandra L. Furterer

This was recognized as a new concept in the biological literature and became widely accepted. If you replace the term "organism" above with "organized entities," the basis of systems theory applies. According to von Bertalanffy, he defined these organized entities as social groups, personality, or technology devices (von Bertalanffy, 1972).

Bertalanffy defined a system as "a complex of elements in mutual interaction" (Hammond, 2019, p. 1).

Bringing us to the 1990s, Peter Senge, in his book, *The Fifth Discipline*, described systems thinking as *"a discipline for seeing wholes. It is a framework for seeing inter-relationships rather than things, for seeing patterns of change rather than static snapshots"* (Senge, 1990, p. 68). In the next section, we will discuss the principles that provide the foundational understanding of systems and their behavior.

1.6 SYSTEMS PRINCIPLES

This section will describe the systems principles and concepts that are important to understand as the foundation of systems thinking and system development.

1.6.1 COMPLEXITY

The principle of complexity "is a measure of how difficult it is to understand how a system will behave or to predict the consequences of changing it" (Sheard and Mostashari 2009).

There are several types of system complexity: (SEBok)

1) Structural complexity looks at the structure of the system elements and the relationships and combinations of these elements.
2) Dynamic complexity of systems relates to the behavior of systems.
3) Sociopolitical complexity focuses on the effect of individuals or groups on complexity.

1.6.2 EMERGENCE

The principle of emergence relates to the concept that whole entities exhibit properties that are meaningful only when attributed to the whole, the interaction of the elements with each other (Checkland 1999). You can have parts or components of a car engine but until the different parts are connected and work together, they do not make a working system that exhibits the type of behavior that we want, propelling a car through the streets. The emergence principle relates to the concept of wholeness of the system and the interaction of the system elements to achieve the wholeness of the system.

1.6.3 HIERARCHY

The evolution of complex systems is facilitated by their hierarchical structure and the relationship of the system elements in a decomposition of elements to each other (Pattee 1973; Bertalanffy 1968; Simon 1996).

1.6.4 BOUNDARY

A boundary or membrane separates the system from the external world. It serves to concentrate interactions inside the system while allowing exchange with external systems (Hoagland, Dodson, and Mauck 2001).

1.6.5 SYSTEM DYNAMICS (BEHAVIOR, SYSTEM ELEMENTS)

The behavior of the system elements impacts the way a system operates to perform the desired functions.

1.6.6 SYSTEMS THINKING

Systems thinking considers patterns of behavior, the connectivity of system elements, and the impact of the interactions of systems elements impacting other parts of the system. Systems thinking considers the boundaries or scope of the system, the dependencies of the elements, and their relationships to each other. The goal of systems thinking is to simplify the understanding of a system, so that systems can be engineered and improved.

1.6.7 STATE AND BEHAVIOR

The state of a system is defined by the attribute, quality, or properties that it possesses at a point in time. A static system is one that remains in a single state, with no events, such as a building standing in place. A dynamic state is one that has multiple possible stable states but can move from one to another, such as a car is parked and a car can be moving. A homeostatic system is static but its elements are dynamic. The system maintains its state by internal adjustments such as a heater controlled by a thermostat.

1.6.8 FUNCTION

A function is defined as either a task or an activity that must be performed to achieve a desired outcome or as a transformation of inputs to outputs (Hitchins 2007). A system will perform and possess desired functions.

1.6.9 ABSTRACTION

Abstraction is *"A view of an object that focuses on the information relevant to a particular purpose and ignores the remainder of the information"* (ISO/IEC 2010). This focuses us on essential characteristics important to design the system, while ignoring the non-essential information. Abstraction is used in many of the system tools that describe the system and their elements, while perhaps ignoring the inner working of the particular elements. It allows us to focus on the interfacing or interacting elements (Sci-Tech Encyclopedia 2009; SearchCIO 2012; Pearce 2012).

1.6.10   System Elements

Elements of the system are the parts or functions that make up the system.

1.6.11   Relations

A system is characterized by its relations: the interconnections between the elements (Odum, 1994).

1.6.12   Interactions

Interactions provide an understanding of the behavior of a system based on the interactions of the elements (Hitchins 2009 p. 60).

1.6.13   View

A view allows us to focus on a particular property or behavior of the system to understand a complex system (Edson, 2008; Hybertson, 2009).

1.6.14   Parsimony

The principles of parsimony help us to choose the simplest explanation of a phenomenon, the one that requires the fewest assumptions (Cybernetics 2012).

1.6.15   Networks

The network provides the understanding of the system, through the relationships of the elements and the behavior of the elements (Lawson, 2010; Martin et al., 2004; Sillitto, 2010).

1.6.16   Dualism

Recognize dualities and consider how they are, or can be, harmonized in the context of a larger whole (Hybertson 2009). Dualism can also mean, division into two parts of the system.

1.6.17   Holism

A system should be considered as a single entity, a whole, not just as a set of parts (Ackoff, 1979; Klir, 2001).

1.6.18   Similarity/Differences

Both the similarities and differences in systems should be recognized and accepted for what they are (Bertalanffy 1975 p. 75; Hybertson 2009). Avoid forcing one size fits all and avoid treating everything as entirely unique.

1.6.19   Separation of Concerns

A larger problem is more effectively solved when decomposed into a set of smaller problems or concerns (Erl, 2012; Greer, 2008).

1.6.20   Modularity

Unrelated parts of the system should be separated, and related parts of the system should be grouped together (Griswold, 1995; Wikipedia, 2012a).

1.6.21 Encapsulation (Hide Internal Workings of System)

As a process, encapsulation means the act of enclosing one or more items within a (physical or logical) container (Parnas, 1972). Encapsulation is also a software development technique that consists of isolating a system function or a set of data and operations on that data within a module and providing precise specifications for the module (IEEE 1990). Encapsulation hides internal parts and their interactions from the external environment (Klerer, 1993; IEEE, 1990).

1.6.22 Change

Change is necessary for growth and adaptation and should be accepted and planned for as part of the natural order of things rather than something to be ignored, avoided, or prohibited (Bertalanffy, 1968; Hybertson, 2009). Things change at different rates, and entities or concepts at the stable end of the spectrum can and should be used to provide a guiding context for rapidly changing entities at the volatile end of the spectrum (Hybertson, 2009). The study of complex adaptive systems can give guidance to system behavior and design in changing environments (Holland, 1992).

Table 1.1 shows some examples of systems and systems thinking through the illustration of the principles of emergence or holism, the system name and elements, the possible states, and functions of the system.

Table 1.2 shows a description of a system with examples of the different properties and characteristics for a university system.

Now that we have a fundamental understanding of what a system is and some of the systems principles that enable us to design, build, and understand a system, we will move into the next chapter to understand the system life cycle model.

TABLE 1.1
Examples of Systems and Systems Thinking

System Name	System Behavior of Emergence or Holism	Elements	Interactions	State	Functions
Light Rail	Without rail, trains, people and tickets, software, network, the system wouldn't work	Rail, trains, website, software, tickets, people	Trains travel on rail, software runs the trains, need network to sell tickets on website	Train moving or not	Sell tickets, Trains move people
Computer	Computer won't work without its parts	Drives, keyboard, programming/software, CPU, hard drive	Hardware is dependent upon power supply, software runs drives, keyboards	Homeostatic, elements are performing calculations	Update data, open an application, perform calculations
App	Remove any components in the code, the app won't work	Hardware, software, coding	Power supply and interaction between human and computer	Code is static (for a point in time); App is dynamic – screens, information	Open an app, perform app functions
Cookies	If don't have all of the ingredients, no cookies	Flour, sugar, eggs, butter, chocolate chips	Stirring and putting together	Cooking – changes the ingredients	Eat the cookies, bake the cookies

(Continued)

TABLE 1.1 (Continued)

System Name	System Behavior of Emergence or Holism	Elements	Interactions	State	Functions
Guitar	Without pickups in electric guitar, can't hear the strings	Body, neck, pickups, strings, person	Body houses pickups, neck tightens string, strings vibrate, person	Strings move, strings being still	Make music
Ecosystem	Can't live without sunlight	Plants, people, sun, oxygen	Plants absorb sunlight, oxygen	Movement of sun, plants absorbing sun	Absorb water, absorb sun
Rocket	Need fuel, fuel pump, oxidizer or it won't go up	Pump, software, oxidizer, fuel, tank	Software engages pumps, puts fuel and oxidizer into tank	Flying, burning fuel	Fly, land
Sound system	If don't have all parts won't hear sound	Speaker, mic, person	Person provides input to mic, mic transfers signal to speakers	Mic not on, projecting sound	Pick up sound

Source: S Furterer course work, Management of Engineering Systems, 2018.

TABLE 1.2
System Description of a University System

System Definition	Your Description
Name of your system	University system.
Type of system: Product? Service? Enterprise? System of system?	Service system.
Purpose of the system	Sharing and learning knowledge, studying, and solving specific problem.
Elements, resources, or parts of your system: People: Hardware: Software: Facilities: Policies: Information: Documents: Etc.	Teachers, professors, students. Books, computers, tables, chairs. Software office, MATLAB. Classroom, library, gym. Academic integrity. Nature, science, and other journal articles or textbooks
Functions, behaviors, or services does the system provide	Libraries, labs, and classes for students to get knowledge and do experiments.

(Continued)

TABLE 1.2 (Continued)

Performance, how would you measure the system performance, or its ability to meet critical requirements?	Measure graduates of the universities salary, employment rate, and student satisfaction.
Complexity: Is it complex? And how?	Yes, has a lot of interacting departments, functions, information systems, policies, processes, and procedures.
Hierarchy: Draw a picture of the system hierarchy on the back of this page	Not included.
Open or closed system: Open? Closed? And why?	Open system as it interacts with the environment = the community and the physical environment.
Static or dynamic and why	Dynamic. In class, it needs teachers, students, computers, software, and data. In experiment, it needs workers, instructors, students, equipment, data, and software.
Emergence: Describe the emergence properties of your system	You can have students, teachers, and teaching assistants, but until they interact to learn, no learning occurs.

Source: S Furterer course work, Management of Engineering Systems, 2018.

1.7 SUMMARY

In this chapter we discussed the systems terminology and principles behind systems thinking. In the next chapter, we will cover the most common systems engineering life cycles.

1.8 ACTIVE LEARNING

1) Describe one of the system principles with examples for a system of your choice.
2) Create Table 1.2 with an example of a system of your choice.

BIBLIOGRAPHY

Bertalanffy, L. von. (1968). *General System Theory: Foundations, Development, Applications*, rev. ed. New York: Braziller.

Bertalanffy, L. von. (1972). The History and Status of General Systems Theory. *Academy of Management Journal, 15*(4), 407–426. https://doi-org.libproxy.udayton.edu/10.2307/255139

Blanchard, B, and Blyler, J. (2016). *Systems Engineering Management*, 5th ed., Wiley, Hoboken, NJ.

Checkland, P. B. (1999). *Systems Thinking, Systems Practice*. Chichester, UK: John Wiley & Sons Ltd.

Deming, W. E., & Orsini, J. N. (2013). *The Essential Deming. [electronic resource] : Leadership Principles from the Father of Total Quality Management*. New York, NY: McGraw-Hill.

Deming, W. E. (1953). Statistical Techniques and International Trade. *Journal of Marketing, 17*(4), 428–433. https://doi-org.libproxy.udayton.edu/10.2307/1247021

Fellows. (2019). *INCOSE Fellows Briefing to INCOSE Board of Directors*. January 2019.

Hammond, D. (2019). The legacy of Ludwig von Bertalanffy and its relevance for our time. *Systems Research & Behavioral Science, 36*(3), 301–307. https://doi-org.libproxy.udayton.edu/10.1002/sres.2598

Helmenstine, A. M. *Scientific Definition of a Closed System in Thermodynamics*. ThoughtCo, Feb. 11, 2020, thoughtco.com/definition-of-closed-system-604929.

Hitchins, D. (2007). *Systems Engineering: A 21st Century Systems Methodology.* Hoboken, NJ, USA: John Wiley & Sons.

IEEE. (1990). *Standard Glossary of Software Engineering.* Washington, DC, USA: Institute of Electrical and Electronics Engineers (IEEE), IEE 610.12-1990.

INCOSE. (2015). *Systems Engineering Handbook: A Guide for System Life Cycle Processes and Activities*, version 4.0. San Diego, CA, USA: International Council on Systems Engineering (INCOSE), INCOSE-TP-2003-002-03.2.2.

Parnas, D. (1972). On the Criteria To Be Used in Decomposing Systems Into Modules, *Communications of the ACM*, 5:(12) (December 1972) 1053–1058.

Senge, P. (1990). *Thee Fifth Discipline.* New York: Doubleday p. 68.

Sheard, S.A. and A. Mostashari. 2009. *"Principles of Complex Systems for Systems Engineering".* Systems Engineering, 12(4): 295–311.

Osirisnet.net, Tomb of Ancient Egypt. (n.d.). https://www.osirisnet.net/tombes/el_kab/pahery/e_pahery_01.htm, webmaster, Thierry Benderitter. Accessed: 8/11/2021.

2 Life Cycle Models for Systems Engineering

2.1 PURPOSE

This chapter describes life cycle models, including the Vee model, which is the life cycle applied in the book. We first start with a definition of a methodology, process and tools, and a life cycle. We then describe the Vee model which is the life cycle applied throughout this book. We then describe iterative and agile life cycle models.

2.2 METHODOLOGY, PROCESS, AND TOOLS

A methodology is *"a collection of related processes, methods, and tools. A methodology is essentially a 'recipe' and can be thought of as the application of related processes, methods, and tools to a class of problems that all have something in common"* (Bloomberg and Schmelzer, 2006). A process is a logical sequence of tasks designed to achieve a purpose or perform work. A method consists of techniques to perform a task, and it provides how to perform each task. A tool is an instrument that can enhance the efficiency, understanding, and documentation of a method and it should be applied or operated by someone using appropriate skills and training.

A life cycle, which is taken from the natural sciences, is used to describe the changes or stages that a system goes through from inception to its retirement and disposal. The Generic Life Cycle Model (ISO 15288) is shown in Figure 2.1.

In the exploratory stage, the feasibility of the potential system is explored. An example would be to develop a lift truck with enhanced automation, sensors, and artificial intelligence to detect potential safety hazards. In the concept stage, different concepts are generated, prioritized, and selected, for example, different lift truck concepts, such as incorporating automation that senses objects in the way of the lift truck and automatically slows or stops the truck. The development stage elicits requirements, develops the system architecture, and designs the system. In our example, requirements are elicited from existing lift truck customers in focus groups to generate customer requirements for the new lift truck concept. The systems engineering team develops system technical requirements that would satisfy the customer requirements. The design engineers and specialty engineers will design the lift truck and the manufacturing, assembly, and testing processes in the development stage. In the production stage the system is produced. For the lift truck, the parts and components are procured, manufactured, assembled, and tested. In the utilization and support stage, the product is rolled out to the customers and serviced, when needed. Eventually, the system is retired and disposed of or migrated to a new generation of the system. For the lift truck, the lift truck is replaced with a new design with the next generation of automation.

2.3 LIFE CYCLE METHODOLOGIES

There are three main life cycle categories of methodologies: (1) waterfall, which is also called linear or predictive, (2) evolutionary or incremental, and (3) iterative or agile. Many projects traditionally were performed using a waterfall, also called a linear or predictive project management approach. The project followed a linear method starting with requirements, design, implementation, verification, and maintenance, as shown in Figure 2.2. All the customers' and stakeholders' requirements would be developed for the entire product, system, or process once in the requirements analysis

DOI: 10.1201/9781003081258-3

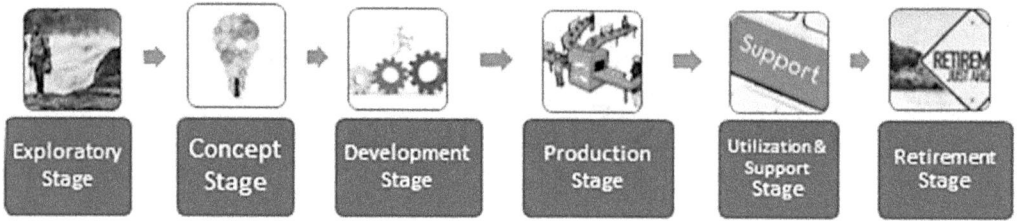

FIGURE 2.1   Generic life cycle model.
Source: Sandra L. Furterer, adapted from ISO15288 and SEBoK.

phase, then designed, developed, tested, and implemented. The scope, project plan, and resources are developed early in the project. The scope is closely managed throughout the project (Furterer and Wood, 2021).

In the evolutionary or incremental method, the product is produced through a series of iterations that add functionality within a specific time frame. The functionality is considered complete only after the final iteration. The spiral method is used to build increments of functionality. There are multiple spirals each consisting of requirements, analysis, and solutions being addressed concurrently instead of sequentially. It is critical that risk analysis is performed within each spiral. If the level of risk is acceptable the team moves forward into the next spiral, and so forth, where the requirements, solutions, and risk analysis are repeated, as shown in Figure 2.3.

In the iterative method, the scope is determined early in the project, but the time and cost estimates are routinely modified based on the team's increasing knowledge of the project parameters and requirements. Iterations develop the product, system, process, or service in a series of iterations or sequential cycles. Agile using Scrum is an iterative method where sprints of requirements and solutions are implemented quickly, with intense customer involvement in creating user stories to develop requirements. More about Agile is discussed in Chapter 14.

There are advantages and disadvantages to the different life cycle models, as shown in Table 2.1 (SEBok, 2020), (Furterer and Wood, 2021).

2.3.1   Vee Life Cycle Model

The Vee Life Cycle Model SEBoK Editorial Board, 2020, provides a methodology for performing the systems engineering activities to design, develop, implement, test, and operationalize the system, as shown in Figure 2.4.

The author developed the Systems Engineering Methodology, Activities & Tools Framework, shown in Table 2.2. The framework aligns to the Vee model and is used throughout this book to apply

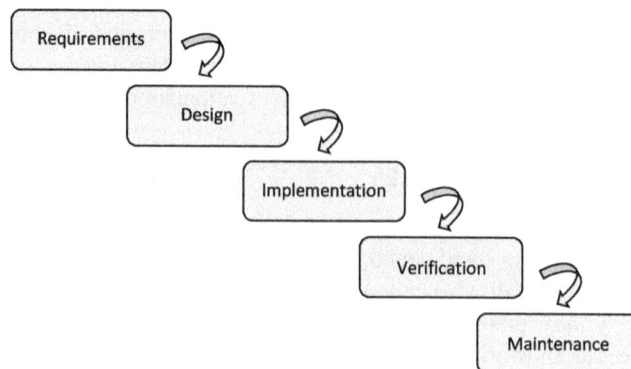

FIGURE 2.2   Waterfall life cycle methodology.

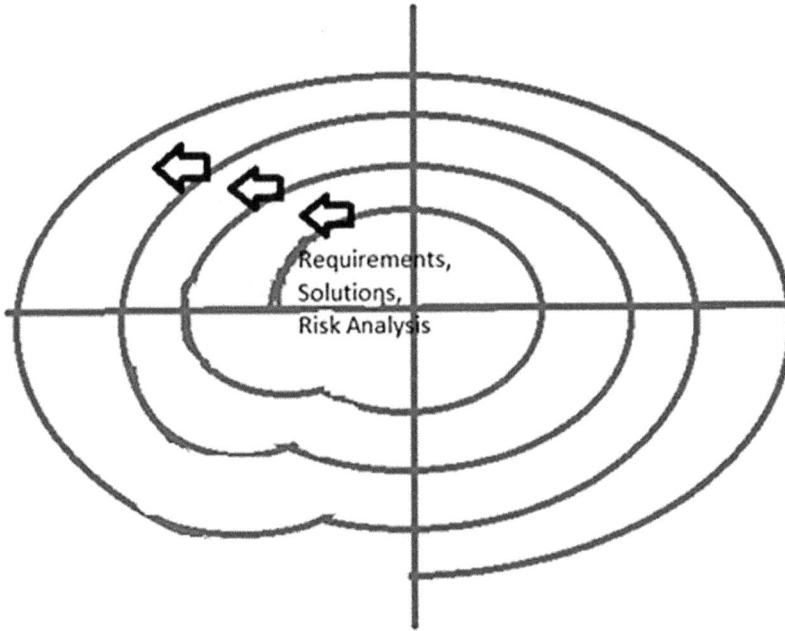

FIGURE 2.3 Spiral increment approach.
Source: Sandra L. Furterer adapted from SEBoK, 2020.

TABLE 2.1
Advantages and Disadvantages of Life Cycle Methodologies

Methodology Category	Advantages	Disadvantages
Waterfall, linear, predictive	• Can be easier to manage • Enables a more detailed plan early in the project • More predictable estimates of costs and schedules if requirements are stable • No partial solutions are given to the customer	• Can be rigid and inflexible • Requirements changes can cause increases in schedule and delivery times • Testing and feedback can be late in the project plan
Incremental	• Responsive to changes in the technology and requirements • Good method if requirements are not clearly defined early in the project • Incremental functionality can be delivered earlier in the project • Can be better for software development • Can be successful for service offerings and incrementally enhancing services	• More expensive because each new increment is a new project • Typically, can take much longer to deliver a finished product • Difficult to predict final costs and timelines • Active customer involvement is critical throughout the project • Can be more difficult to deliver increments of products or product systems
Iterative	• Can deliver interim functionality more quickly • Can be better for software development • Can be successful for delivery select service offerings iteratively and enhancing them in the next iteration	• More expensive because each new iteration is a new project • Can be much longer to produce the final delivered product • Need active customers engaged in defining requirements • Can be more difficult to deliver increments of products or product systems

Source: Sandra L. Furterer.

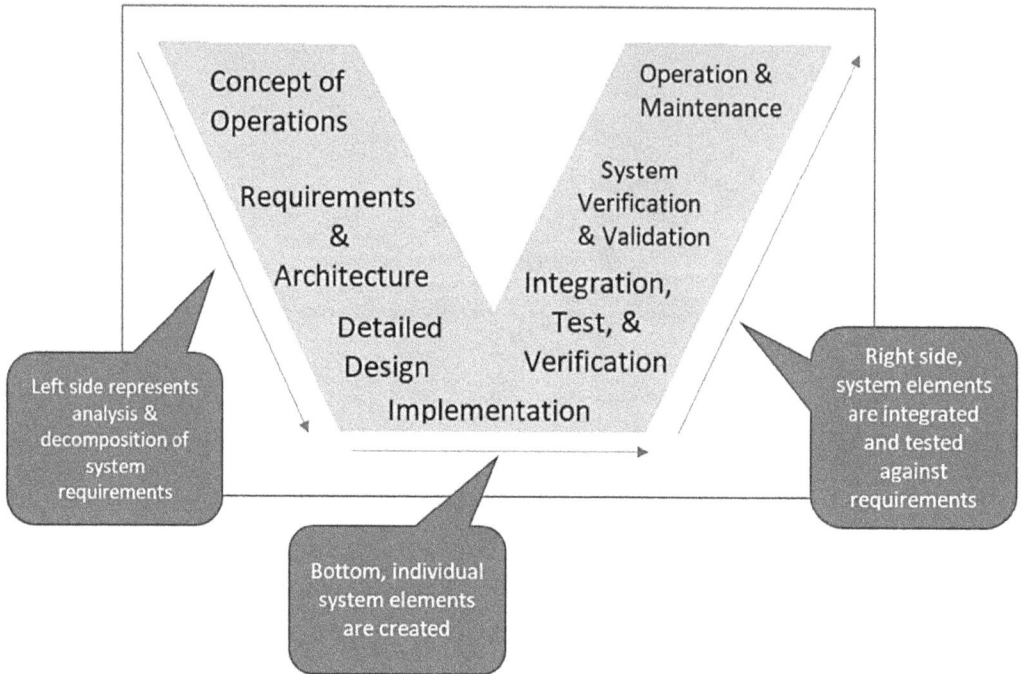

FIGURE 2.4 Vee life cycle model.

the systems engineering tools to teach the readers (students) how to apply the systems engineering tools. This provides a summary of the activities performed, the tools applied, and the principles that are needed as a basis for being able to perform each phase of the life cycle. The framework aligns to the Vee life cycle model's phases and then breaks each phase into more detailed activities that are performed within each phase. The tools applied to achieve the purpose of each phase are identified. The principles that should first be understood to best achieve the phase activities and outcomes are listed. The principles may be applied in later phases as well but are not repeated for each phase within the framework. A quick summary of the Vee model is described next, and then the details related to each element within the framework is described in detail within Chapters 4 through 12.

The Vee model can be used in either a sequential or an iterative method. However, typically, the entire Vee model is run through in a sequential manner, and then another iteration can occur with the next set of requirements, until the system is considered complete.

The left side of the Vee model represents analysis and decomposition of the requirements. The Vee model starts with the Concept of Operations phase, illustrated in the Vee model graphic on the top left, where the external and internal analyses are performed to assess the concept of the system within the marketplace and the ability of the organization to achieve the system. The next phase is the requirements and architecture phase where customer requirements are elicited, a system architecture is developed, and the customer requirements are translated into the system requirements. The detailed design phase is where the system is designed by the specialty engineering discipline areas, while the systems engineers focus on the integration and interface points to ensure connectivity and an integrated system view. At the bottom of the Vee model, the implementation of the system occurs where individual system elements are created. The system is built but not yet integrated and tested. Going up the right side of the Vee model represents that the system elements are integrated and tested and ultimately delivered to the customers. The integration, test, and verification phase integrates the system, tests, and verifies it with respect to the requirements and design. The System Verification and Validation phase validates the entire system from the perspective of the end user.

TABLE 2.2

Systems Engineering Methodology, Activities & Tools Framework

Vee Phase	Activities	Tools	Principles
Phase 1: Concept of Operations	Define Strategic Mission Goals: 1. Define mission strategy 2. Define goals of mission 3. Develop concept of operations 4. Perform external, internal, and SWOT analyses Define Mission Concepts: 1. Develop project charters 2. Perform stakeholder analysis 3. Collect Voice of Customer 4. Perform conceptual selection 5. Perform risk analysis	Strategic mission goals tools: • Mission Strategy; Mission goals; Concept of Operations, Concept; External analysis; Internal analysis; SWOT analysis Mission concepts tools: • Project charter with risk analysis • Stakeholder analysis • Critical to Satisfaction criteria • Pugh Concept Selection Technique • Risk Assessment, Risk Cube	• Complexity • Emergence (whole > sum of parts) • System of system • Hierarchy • Boundary
Phase 2: Requirements & Architecture	Develop Logical Architecture: 1. Develop Logical Architecture Elicit Requirements 2. Elicit customer requirements 3. Derive requirements • Develop system requirements • Develop specialty engineering requirements 4. Develop measurement plan 5. Develop an information model 6. Develop a quality management plan	Logical Architecture: • SIPOC • Value Chain & Functional Decomposition • Use Case Diagrams Requirements: • Process Scenarios via Process Architecture map • Use Cases • Customer requirements • System requirements • Specialty engineering requirements: software engineering; environmental engineering; safety and security engineering; Human Systems Integration • Data Collection Plan, with operational definitions • Class Diagram with information model and hierarchy model • Quality management plan	• System dynamics (behavior, system elements) • Cybernetics (information flow) • Systems thinking • Abstraction • Views

(Continued)

TABLE 2.2 (Continued)

Vee Phase	Activities	Tools	Principles
Phase 3: Detailed Design	1. Perform detailed design 2. Perform systems analysis	Detailed Design: • Physical Architecture Hierarchy Model • Physical Architecture Model: Physical Block Diagram; and/or SysML Block Definition Diagram; and/or SysML Internal Block Diagram • System Elements • QFD • Simulation (not required) Systems Analysis: • Selection criteria • Trade off analysis; and/or Effectiveness analysis; and/or Cost analysis; and/or technical risk analysis • Justification report	• Systems analysis • Wholeness and interactions
Phase 4: Implementation, Integration, Test & Verification	1. Perform system implementation 2. Perform system integration	• Integration constraints • Implementation strategy • System elements supplied • Initial operator training • Verification criteria • Verification test cases and results • N-squared diagram	• System elements • Modularity • Interactions • Networks • Relationships • Behavior
Phase 5: System Verification and Validation	1. Perform system verification 2. Perform system validation	• Verification and Validation criteria • Verification and validation test cases and results	• Synthesis

| Phase 6: Operation & Maintenance | 1. Deploy System
2. Perform training
3. Perform certification
4. Perform risk assessment
5. Perform improvement and maintenance planning
6. Perform disposal and retirement activities | • Training plan and materials
• Certification plan and materials
• Operations Manuals
• Performance reports
• Improvement and Maintenance plans; FMEA
• Disposal and retirement plan | • Control behavior and feedback
• Encapsulation (hide internal workings of system)
• Stability and Change |
| PLANNING (throughout all phases) | 1. Develop Systems Engineering Management Plan and associated plans | • Systems Engineering Management Plan: Assessment and Control Plan; Configuration Management Plan; Contractor Management Plan; Deployment Plan, Disposal and Retirement Plan; Information Management Plan; Interface Management Plan; Maintainability Program Plan; Measurement Plan; Quality Management Plan; Risk Management Plan; Specialty Engineering Plan; System Development Plan; System Integration Plan; System Integration Plan | |

Source: Sandra L. Furterer

The inner workings of the system are verified in the verification testing, and the external customer-facing aspects of the system are validated in this phase. Finally, the system is delivered, installed, and tested again on-site. The system is now turned over to the customers and is operational. This phase represents the longest phase of the system as the system is operated and maintained. This last phase also includes disposal and retirement of the system or transition to a new system. The Vee model then could be iterated again after the system is operational to enhance the system with new requirements, and the Vee model phases would be repeated until the new requirements are operational. The Vee model also represents starting from a high-level broad concept and moving into more detail down the left side of the Vee model, implementing individual system elements, and then integrating them going up the left side to a more integrated system.

2.4   SUMMARY

This chapter discussed the different life cycle methodologies, finishing with a high-level description of the Vee model that we use in the rest of the book. The next chapter covers the meta models that represent the system architecture of the information developed to develop, design, test, and operationalize the system. Then we move into the core of the book with chapters describing the methods and tools that can be used to create our system, with real world examples from students like yourselves.

2.5   ACTIVE LEARNING

1) Search for an article for one of the life cycle methodology categories, waterfall, incremental or iterative. Discuss how the methodology was applied in the article. Could one of the other methodologies have been equally successful, describe how it could have been approached with this alternate methodology. Share with the class.
2) Describe the Vee model. Use a reference article from the literature, such as the one listed here, to learn more about the Vee model (Vaneman, n.d.).
3) For the following systems in Table 2.3, which System Life Cycle Methodology would you use and why?
   • Waterfall
   • Spiral
   • Vee
   • Double Vee
   • Others?

TABLE 2.3
System Life Cycle Methodology by System Example Active Learning

System	System Life Cycle Methodology	Why?
Airplane		
Light rail system		
Traveling space station		
High-tech refrigerator		
Warehouse distribution system		

BIBLIOGRAPHY

Bloomberg, J. and Schmelzer, R. (2006). *Service Orient or Be Doomed!*. John Wiley & Sons: Hoboken, New Jersey.

Furterer, S. and Wood, D. (2021). *The ASQ Certified Manager of Quality/Organizational Excellence Handbook*, Fifth Edition (E-Book), ASQ Quality Press, Milwaukee, WI.

SEBoK Editorial Board, The Guide to the Systems Engineering Body of Knowledge (SEBoK), vol. 2.2, ed. R. J. Cloutier (Hoboken, NJ: Trustees of the Stevens Institute of Technology, 2020), accessed May 25, 2020, www.sebokwiki.org. BKCASE is managed and maintained by the Stevens Institute of Technology Systems Engineering Research Center, the International Council on Systems Engineering, and the Institute of Electrical and Electronics Engineers Computer Society.

Vaneman, W. (n.d.). The System of Systems Engineering and Integration, "Vee" Model

3 Systems Architecture for Model-Based Systems Engineering

3.1 PURPOSE

Chapter 3 covers the systems architecture that lays the groundwork for Model-Based Systems Engineering (MBSE). Seven different models are described along with a description of their elements, relationships, and purpose of each model. The meta models are used to identify the information to be assessed, formulated, and developed within each model to design, develop, and deploy the system. The models are 1) Mission Analysis model, 2) System Architecture and Requirements model, 3) Information and Application Model, 4) System Design model, 5) System Verification and Validation model, 6) Systems Operations model, and 7) Systems Engineering Management Plan model.

3.2 SYSTEMS MODELING OVERVIEW

"A model is an abstraction of something for the purpose of understanding it before building it" (Rumbaugh, et al., 1991). The model can be used in many ways, such as to build a bridge or a building, develop information systems applications, develop a business, design a system, or manufacture a car. In this book, we focus on systems modeling, where the goal is to design, develop, test, operationalize, and retire a system throughout its life cycle. We've earlier defined the different types of systems that can be built, and these conceptual models can be used for all of these system types.

Our conceptual or meta models will provide a common understanding to the users and the viewers of these systems models. These models convey knowledge in both a visual and a written way. To convey the knowledge in a consistent and unambiguous manner, we need to develop an ontology.

A definition of an ontology, from the philosophy discipline is "An explicit formal specification of how to represent the objects, concepts and other entities that are assumed to exist in some area of interest and the relationships that hold among them." Another definition is "a set of concepts and categories in a subject area or domain that shows their properties and the relations between them" (dictionary.com). The ontology helps us to describe a system in a precise way to convey the same meaning to the reader. It provides both an understanding of the model elements and the relationships of these elements to each other. An example of an ontology is shown in Figure 3.3, where the elements of the system mission model is shown, along with how they relate to each other. The mission consists of the mission strategy, goals, and operational concept. The external, internal, and Strengths, Weaknesses, Opportunities, and Threats (SWOT) analyses inform the system's mission. The project charter defines the mission project. The stakeholder analysis defines the system's stakeholders, while the systems concept selection technique helps us to select the mission's concepts. This chapter describes the system's ontology that can be used to apply MBSE.

We will use the Unified Modeling Language (UML) to provide a formal definition of the elements of the language. When designing a system, the designers will use the conceptual or meta models to ensure that they are describing the necessary systems elements and understanding their relationships to each other. This helps us to ensure that the appropriate connectivity, integration, and interfaces are designed, tested, and operationalized. The designers will develop instances or examples of their models that help them to realize their actual system application. For example, the conceptual model

DOI: 10.1201/9781003081258-4

in the Concept of Operations phase includes a system's mission. So, the systems engineer knows that they must define and communicate the system's mission. Every system should have a mission. However, the instantiation or instance of the mission is specific to the system's domain, the actual system being designed. An example would be "To develop an industrial lift truck that enhances the lift truck company's competitive advantage in the marketplace."

Many different conceptual modeling techniques have been used which include different levels of rigor and notation. We will draw from many of them as we design across the phases of the Vee model life cycle. Some of these are the Entity Relationship Models (Chen, 1976), UML (Object Management Group, 2010), SysML (Object Management Group, 2010), Business Process Model and Notation (bpmn.org), Fundamental Modeling Concepts (Fundamental Modeling Concepts, 2010), ArchiMate (The Open Group, 2010), Business Motivation Model (Business Rules Group, 2010), Service-Oriented Modeling and Architecture (Arsanjani, 2004), and Service-Oriented Modeling Framework (Bell, 2008) to name a few. Other less formalized modeling techniques are Six Sigma, Porter's Value Chain Analysis (Porter, 1985), Component Business Models (IBM, 2010), Variation Oriented Analysis (Arsanjani, 1999–2000), TOGAF (The Open Group, 2010), Zachman (Zachman, 2010), Information Architecture (Information Architecture, 2010).

3.3 CLASSIFICATION

3.3.1 AN OBJECT AND A CLASS

An object is a thing or categories of a thing that are all of the same kind. In structured analysis, we typically focused on the rules and logic of the customers' requirements. In object-oriented analysis, we focus on the objects or things and their attributes, then also define the methods or rules that the object experiences. A class is a collection of the objects, demonstrating the blueprint from which the individual objects are created. An example of an object is a patient, book, or a car. There may be many types of patients, books, or cars, but we can name one of them as a grouping of the types of patients or the types of books or cars.

3.3.2 STRUCTURE AND BEHAVIOR

Objects share two characteristics, state and behavior. A car has attributes or states such as type, color, and model. The car can have behavior such as speed, direction, and engine running. As you look at various objects around your room, you'll notice that objects vary in complexity. A book may be open or closed, where as a car can be speeding at several different speeds (1 to 100 and all of the speeds in-between). Systems will also vary in type and complexity.

3.3.3 CLASS DIAGRAM

The class diagram shows how the different entities (people, things, and data) relate to each other; in other words, it shows the static structures of the system. A class diagram can be used to display logical classes, which are typically the kinds of things the users or customers of the systems talk about – patients, charts, bills, registration, tests, and results. A class is depicted on the class diagram as a rectangle with three horizontal sections, as shown in Figure 3.1. The upper section shows the class name. The middle section contains the class' attributes. The lower section contains the class' operations (or "methods") (The Java Tutorials, 2011). At a conceptual or meta model level, the upper section describes the class and the middle level conveys the model's name. The conceptual model is a grouping of elements and their relationships.

A class diagram shows the relationships between the various classes. There can be inheritance and associations. Classes can inherit characteristics from other classes. A patient may be an inpatient or an outpatient. The inpatients and outpatients can have similar characteristics as a patient, such as a registration, tests, and physician. But an inpatient would have an inpatient room and bed, and an

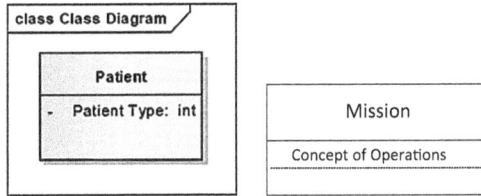

FIGURE 3.1 Example of a class at an instance level and a conceptual or meta level.
Source: S. Furterer.

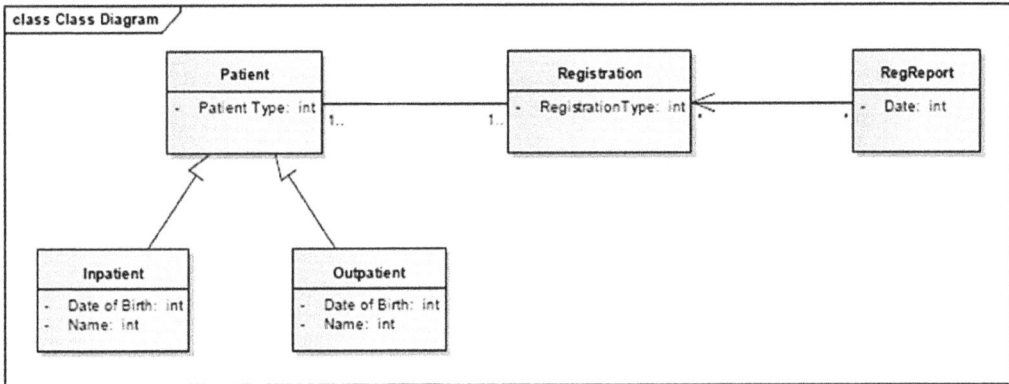

FIGURE 3.2 Class diagram example.
Source: S. Furterer.

outpatient would not. An inheritance is shown as an open triangle, pointing to the super class that the other classes are inheriting from. A solid line represents an association if both classes are aware of each other. A line with an open arrowhead represents an association known by only one of the classes. The class diagram is shown in Figure 3.2 (DeveloperWorks, 2011).

In Figure 3.2, the patient is the super class and the inpatient and outpatient inherit from the patient class. A patient has an association with a registration, and they know about each other. A registration report is associated with many patients, but the patient doesn't know anything about the registration report that registration uses to call the patients when they are registering. The numbers and asterisks associated with the classes describe the multiplicity of the instances of the classes to each other. The notation 1, * represents a relationship of 1 to many, such as one patient may have many registrations. The notation of 1, 1 represents a relationship of one to one, such as a patient has an account, there should only be one account for each patient. The notation *, * represents a many to many relationship, such as a patient may be on many reports, and a report can contain many patients.

3.4 MODEL DESCRIPTION

The systems architecture models provide the conceptual elements and their relationships that are required to design and operationalize our system. Understanding the elements that need to be considered helps us to ensure that the elements are identified and integrated to each other. For each of the models we provide:

- Title: Title of the Systems Architecture model
- Value & Purpose: The value and purpose of the model
- Definition: A definition of the elements
- Relationships: The relationships of the elements to each other

TABLE 3.1
Model-Based Systems Engineering Systems Architecture Meta Models

Meta Model	Vee Phase Used In	Purpose
1) Mission Analysis model	Phase 1: Concept of Operations	Define, assess, and scope the system's mission
2) System Architecture and Requirements model	Phase 2: System Architecture and Requirements	Develop the system's conceptual architecture, customer, and system requirements
3) Information and Application Model		Identify the information and applications that are part of the system
4) System Design model	Phase 3: Design	Design the system
5) System Verification and Validation model	Phase 4: System Verification Phase 5: System Validation	Develop and deploy systems testing and to validate the customer and system requirements
6) Systems Operations model	Phase 6: Operations and Maintenance	Plan, deploy, operate, maintain, and retire the system
7) Systems Engineering Management Plan model	All life cycle phase	Perform systems engineering management planning throughout the life cycle

3.4.1 Systems Architecture Models

In this chapter, we provide the conceptual meta systems architecture models (Table 3.1) for modeling systems aligned to the phases of the Vee life cycle model. There are seven different models applied with different phases of the Vee life cycle model.

3.4.2 Mission Analysis Model within Phase 1, Concept of Operations

In Phase 1, the Concept of Operations phase, the main purpose is to identify the mission of the systems design project, to develop project charters, perform a stakeholder analysis, and perform multiple analyses that help us to understand the external and internal influencers that impact the potential success of the system projects.

The Mission Analysis Model is shown in Figure 3.3. Next follows a description of the Mission Analysis Model.

TITLE: MISSION ANALYSIS MODEL

Value & Purpose:

The value and purpose of the mission analysis model is to define the system's mission, develop the system's project charters and respective scopes of each project, understand who needs to be involved in the project, and to perform an analysis to understand how the organization who is developing the system is positioned in the marketplace and within the company to achieve the mission.

For each model, the definition of each element and the relationships of these elements to other elements will be described for the model.

Mission Analysis Model Element Definitions:

Mission: The system's mission identifies the purpose of the system. It describes what the system will be able to achieve. The mission is made up of the mission strategy, mission goals, and operational concept.

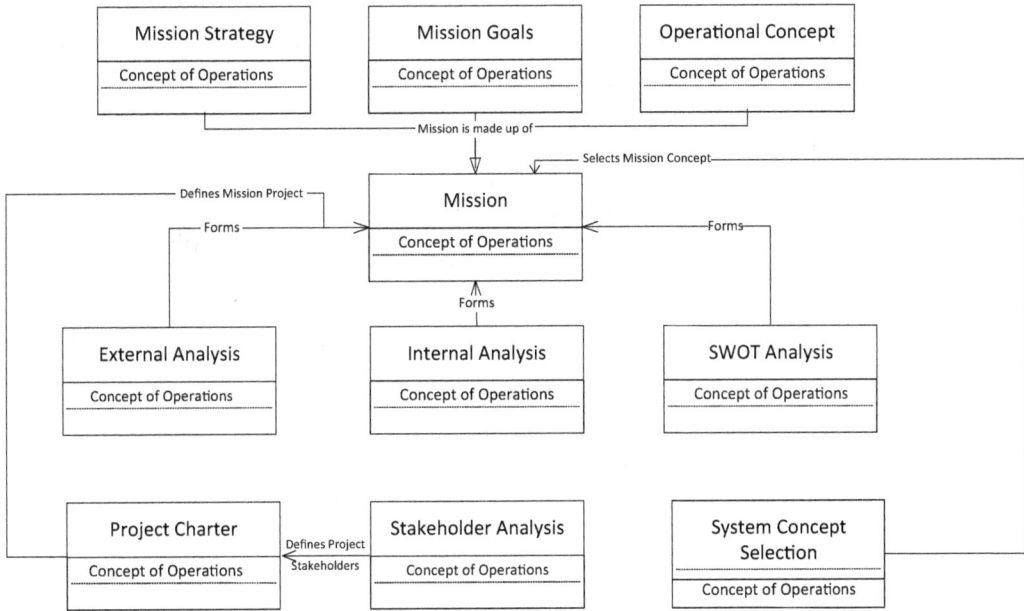

FIGURE 3.3 Mission analysis model.
Source: S. Furterer.

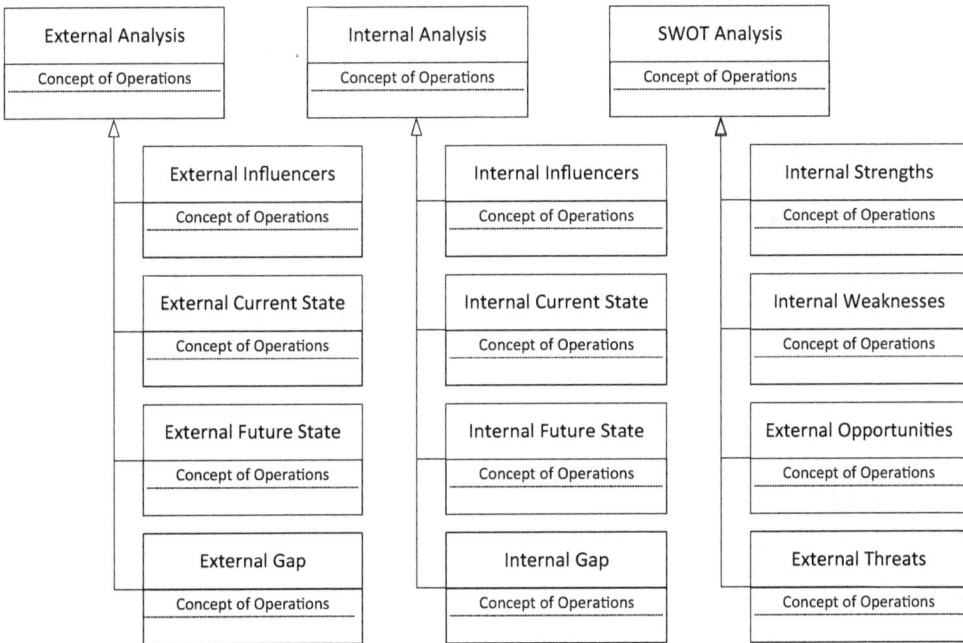

FIGURE 3.4 Mission analysis elements.
Source: S. Furterer.

Mission Strategy: The mission strategy encapsulates the long-term vision that the mission will achieve. It is a plan or a method to achieve a desired result, such as realizing the ultimate system.

Mission Goal: A goal is a qualitative desire that will help us to achieve the mission strategy. Its range is typically long-term and therefore relatively stable. Its focus should be narrow enough that project objectives can be described for it.

Operational Concept: The operational concept describes how the organization will be organized to design, develop, and deploy the system, including how the organization may leverage other organizations (for example, consulting firms, suppliers, etc.) to achieve the system's mission.

External, internal, and SWOT assessments evaluate the organization and the environment in which it operates to identify factors that will impact the system.

External Analysis: An analysis that is performed to scan and understand the environmental factors external to the organization that will design and deploy the system. Types of external factors to be investigated include political, economic, social, technological, environmental, and legal/regulatory.

- External Influencer: External is of or pertaining to the outside. So, an external influencer is something that has compelling force that is outside of the organization.
- External Current State: External factors of the current time frame.
- External Future State: The future vision of external factors if the system mission is achieved.
- External Gap: A difference between the current and future state that the system can attempt to fill related to external factors.

Internal Analysis: An analysis that is performed to understand the internal organization and the factors that could impact the successful design and deployment of the system. Internal factors include value activities and capabilities that the organization is able to deliver, financial, learning and growth opportunities, process/operational, and customer-related factors.

- Internal Influencer: The capacity to have the power to be a compelling force. Internal is of or pertaining to what is inside. An internal influencer is something that has compelling force that is internal, inside the organization.
- Internal Current State: The way the organization exists in the current time frame.
- Internal Future State: The future vision of what the organization could be based on achieving the mission.
- Internal Gap: A difference between the current and future state that the system can attempt to fill.

SWOT Analysis:

- Internal Strengths: The qualities of being strong that indicate the capacity to perform.
- Internal Weaknesses: An inadequate or defective quality.
- External Opportunities: A condition favorable for attaining a goal.
- External Threats: An indication of probable trouble.

Project Charter: To design, develop, and deploy a system, the work is typically divided into smaller projects. The project charter helps to clearly document the goals and scope of the projects.

Stakeholder Analysis: The stakeholder analysis helps to identify the stakeholders impacted and involved with the systems projects.

System Concept Selection: The system concept selection helps the team to generate systems concepts and choose desired concepts to move into the Requirements and Architecture phase.

System Element Relationships:
The mission is made up of the mission strategy, mission goals, and operational concept. The external, internal, and SWOT analyses provide assessments of the external and internal factors that may impact the successful implementation of the system. The project charter defines the mission projects. The stakeholder analysis defines the stakeholders. The system concept selection helps to select the system mission.

3.4.3 SYSTEMS ARCHITECTURE AND REQUIREMENTS MODEL WITHIN PHASE 2, SYSTEM ARCHITECTURE AND REQUIREMENTS

In Phase 2, the System Architecture and Requirements phase, the main purpose is to develop the system architecture as the foundation of the system and to elicit requirements from the customers and stakeholders.

The Systems Architecture and Requirements Model is shown in Figure 3.5. Next is a description of the model.

TITLE: SYSTEM ARCHITECTURE AND REQUIREMENTS MODEL

Value & Purpose:
The system architecture and requirements model provides the elements that enable the architecting of the system from a conceptual perspective to allow the foundation for the elicitation of the customer and system requirements. This phase remains at a conceptual, not a physical or design level.

System Architecture and Requirements Model Element Definitions:
System: A system is an interacting combination of system elements to accomplish a defined objective(s) (ISO/IEC/IEEE. 2015).

System elements: System elements are those parts of the system that will be defined, designed, developed, tested, implemented, and used. System elements can be physical products, parts or components, information, hardware, software, firmware, deliverables, facilities, functions, and people in roles that make up the system.

System types:

- Product system: A product system consists of a product or multiple products to achieve the system mission.
- Service system: A service system consists of processes, technology, and people to provide services to customers.
- Enterprise system: An enterprise system is a system working across multiple organizations to achieve the system mission. An enterprise system could include product and service systems.
- System of systems: System of systems are large and complex systems that consist of multiple systems that integrate and come together to achieve a system mission.

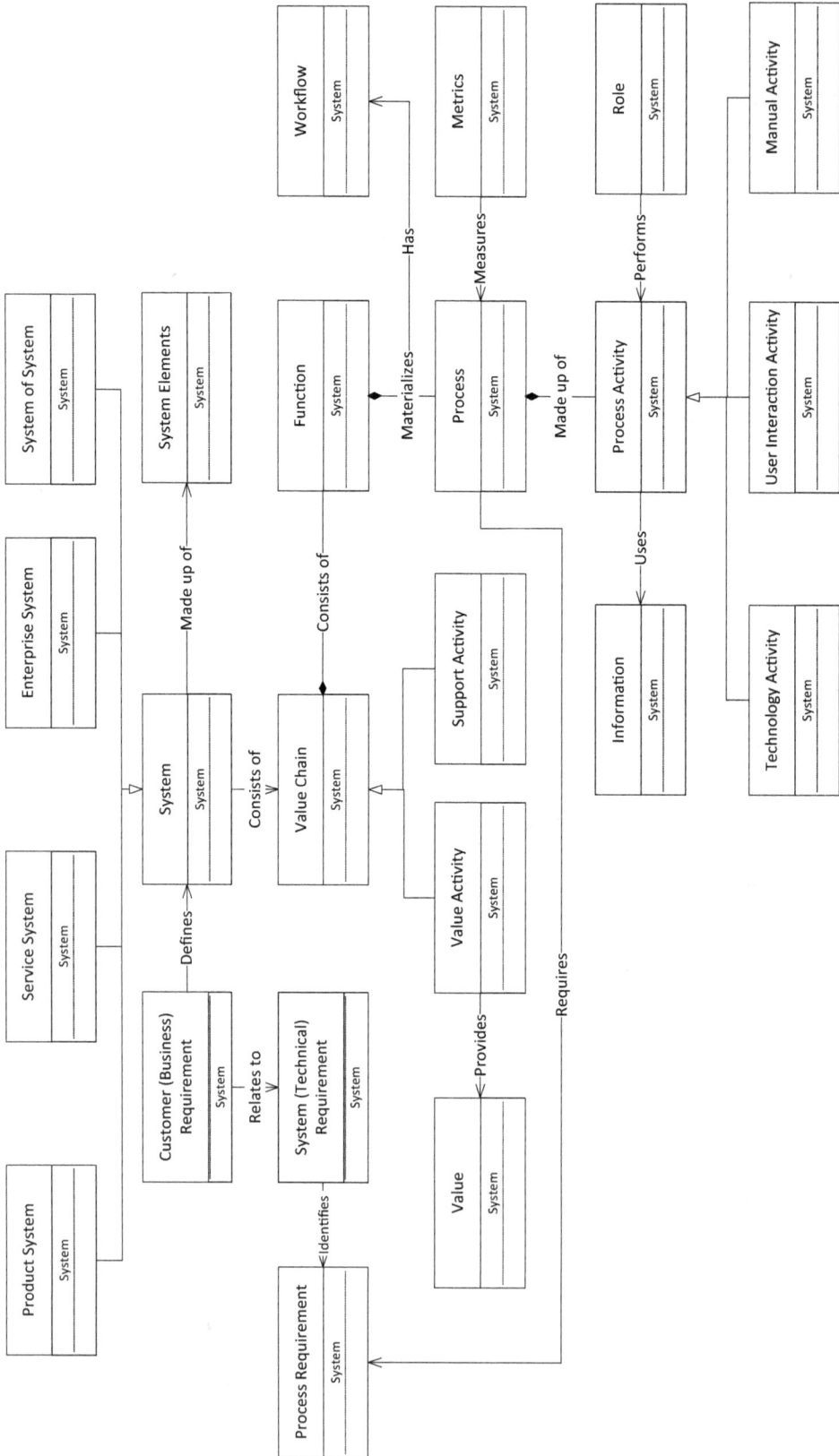

FIGURE 3.5 System architecture and requirements model.
Source: S. Furterer.

Customer requirements: The customer requirements are the statements elicited from the customers that define the needs of the system.

System (technical) requirements: System or technical requirements are statements that define the system, related to the customer requirements.

Process requirements: Process requirements are derived from system or technical requirements that are required to achieve the processes and the system.

Value chains: Value chains are groups of value and support activities that when performed create the system value.

Value activities: Value activities are those activities that are of value to the customers, that they would be willing to pay for to use the system.

Support activities: Support activities support the value activities.

Value: Value is defined by the customer and are derived from value activities within the system's value chains.

Functions: Functions are groups of activities that are required by the system to be performed.

Processes: Processes are activities performed in a specified sequence to achieve the purpose of work. Inputs are transformed into outputs either manually, with technology and people, or fully automated by technology. People are grouped into categories of roles that are trained to perform the specified work activities or tasks. Information is used in automated or manual formats to perform the processes.

Process Activities: Process activities are those tasks performed to achieve work, to transform inputs into outputs.

- Technology: Technology activities are those performed without manual intervention or are automated.
- User interaction activities: User interaction activities are those performed with the use of automation but can contain a manual component.
- Manual activities: Manual activities are those performed without the use of automation.
- Roles: Roles are categories of workers who perform parts of the process.
- Information: Information is data that provides knowledge to a process so that the activities can be performed and work can be achieved by the process. Information can be in an electronic automated format or a physical hard copy.

System Element Relationships:
A system is made up of system elements. The system consists of several system types, including a product system, service system, enterprise system, and system of systems. These types were described and defined in Chapter 1. The system design analysis, performed in the design phase, consists of trade-off analyses, cost analysis, effectiveness analysis, and a system risk analysis. The customer requirements define the system and relate to system or technical requirements. The technical requirements could require process requirements. A system consists of value chains, made up of value activities and support activities. The value activities provide value to the customers. The value chain consists of functions, which materialize processes. Processes are made up of process activities which consist of technology activities, user interaction activities, and manual activities. Process activities are performed by roles and use information.

3.4.4 Information and Application Model within Phase 2, System Architecture and Requirements

TITLE: INFORMATION AND APPLICATION MODEL WITHIN PHASE 2

Value & Purpose:

The information and application model (Figure 3.6), also called a domain model, provides a conceptual view of the information and software application technology used in the system. A conceptual model provides an abstract view of the classes being modeled. A class is a construct or idea, usually expressed as a noun, such as a person, place, or thing. It represents a concept and describes the state through attributes and behavior through methods. An instance of a class is an object. The conceptual model is independent of the design or implementation concerns that address efficiency of processing, concurrency, or data storage. The value of the information model is to understand and define the concepts used in the system and to describe the relationships of the concepts to each other.

Information and Application Model Element Definitions:

Business Domain: An area of the business that is of interest to be focused on to improve or build information systems within. The domain includes the concepts, their associations, attributes, and the appropriate constraints.

Business Vocabulary: The words of the business. The business vocabulary is the language of the business, of how the organization defines the concepts of the processes.

Concept: An idea or construct formed by mentally combining all of its characteristics.

Information Schema: The information schema is the diagram of how the information is related to each other and also describes the attributes of the information.

Application: A computer program that is developed and used for a specific need or use.

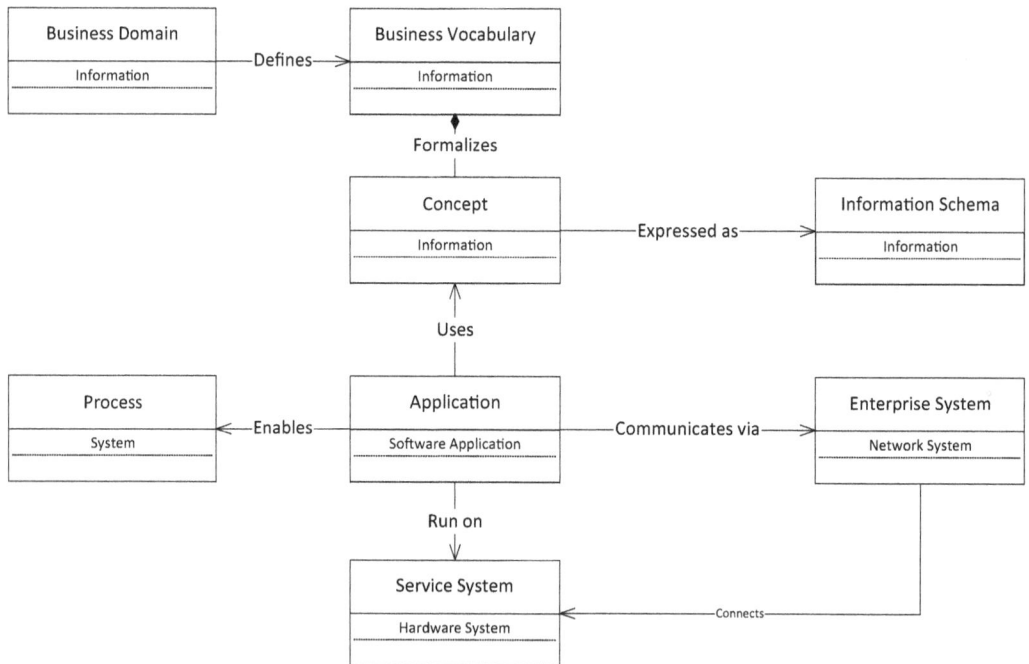

FIGURE 3.6 Information and application model.

source: S. Furterer.

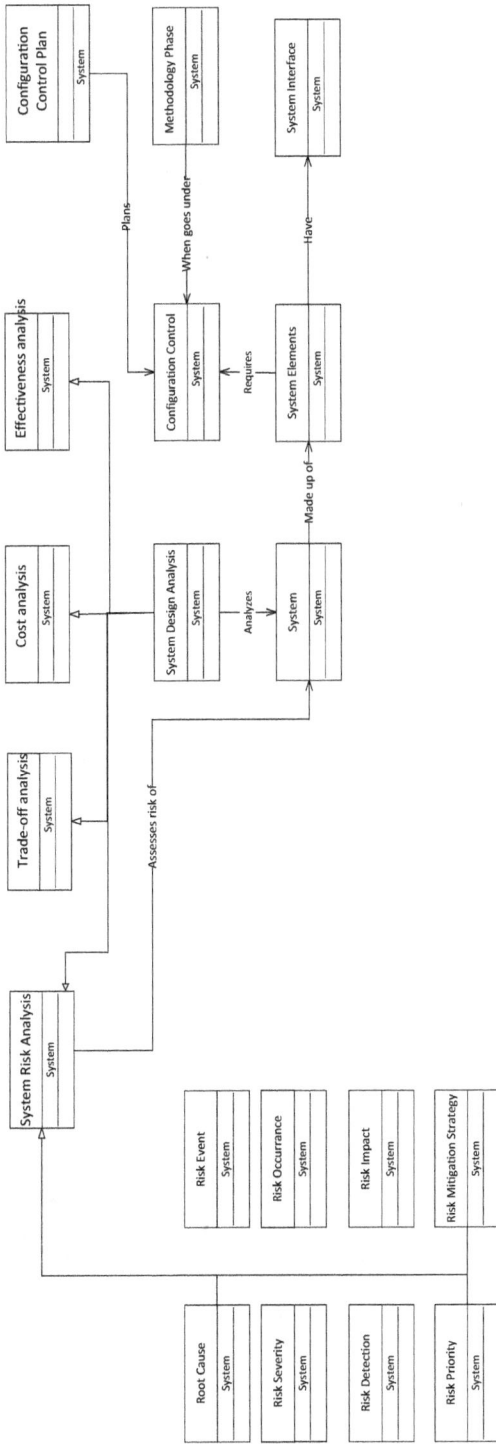

FIGURE 3.7 System design model.
Source: S. Furterer.

Enterprise System: The network that links applications and allows them to communicate to each other.

Service System: The hardware is connected through the enterprise system or network.

System Element Relationships:

The information model contains the business domain, which is the area of interest, that is the focus of the concepts and information systems applications. The business domain defines the business vocabulary. The business vocabulary contains the concepts, which are ideas expressed as nouns, people, places, or things. The concepts are the entities, which have roles in performing the processes. They are the inputs and outputs of process activities. The concepts have associations or relationships to each other. The concept is expressed as information schema. The applications use the information. The application communicates via the enterprise (network) system. The application enables the process. The applications run on service systems (hardware). The enterprise (network) system connects the service systems (hardware).

The concepts in the conceptual information model are expressed as information schema in the application model. These concepts are used to develop the optimized applications.

3.4.5 SYSTEMS DESIGN MODEL WITHIN PHASE 3, SYSTEM DESIGN

Value & Purpose: The System Design Model's (Figure 3.7) purpose is to enable the design of the system that identifies the system's elements and their interfaces, through the analyses performed.

System Design Model Element Definitions:

System design analysis: System design analysis includes the following analyses that help to define and design the system.

Trade-off analysis: Trade-off analysis consists of comparison of different alternative solutions that define system design elements.

Cost analysis: Cost analysis is a financial assessment of different system elements and system designs to make design decisions.

Effectiveness analysis: Effectiveness analysis provides an assessment of system and process effectiveness, to ensure an effective and efficient system and related processes to achieve the system.

System risk analysis: System risk analysis is an assessment of system risks, their impacts, mitigations, and prioritization of these risks.

- Risk events: the potential events that cause risk to the system, design, development, or operation
- Risk occurrence: the probability that the risk event could occur
- Risk impact: the effect of the risk on the system
- Risk mitigation strategy: the approach taken to reduce the potential risk from occurring
- Root cause: the ultimate reason for the risk event
- Risk severity: the level of impact or severity of the risk if it occurs
- Risk detection: the ability to detect the risk event
- Risk priority: the importance of mitigating the risk

Configuration control: Configuration control is a method for controlling deliverables consisting of information and system elements, so that when they change, the appropriate approvals are achieved and people are notified that they may have a deliverable that is affected by the change.

Methodology phase (goes under configuration control): The methodology phase is identified when each system element is placed under configuration control, typically when it is first baselined.

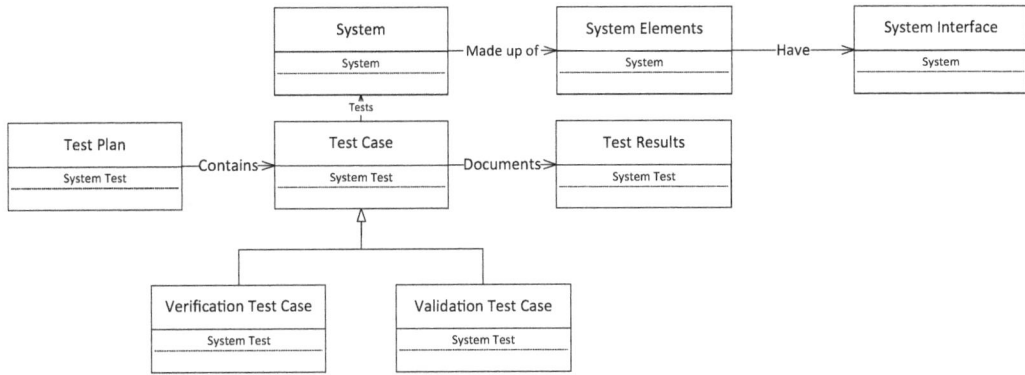

FIGURE 3.8   System verification and validation model.
Source: S. Furterer.

Configuration control plan: The configuration control plan describes the approach, activities, timelines to place deliverables, documents, system elements, processes, technology, etc., under configuration control.

System interfaces: System interfaces are points where different system elements connect to each other, either physically, via information flowing between them, or by people connecting the elements.

System Element Relationships:
The system design analysis is made up of different types of analysis, including trade-off analysis, cost analysis, effectiveness analysis, and risk analysis. System elements have system interfaces. System elements require configuration control. The configuration control has a methodology phase under which the elements go under configuration control. The configuration control plan is used to plan the configuration control for the system elements. The system risk analysis consists of risk events, risk occurrence, risk impact, risk mitigation strategy, root cause, risk severity, risk detection, and the risk priority.

3.4.6   SYSTEMS VERIFICATION AND VALIDATION MODEL WITHIN PHASE 4, SYSTEM VERIFICATION AND PHASE 5, SYSTEM VALIDATION

Value & Purpose:
The purpose of the Verification and Validation model (Figure 3.8) is to ensure that the system functions align according to the customer and system requirements defined in the system architecture and requirements phase.

System Verification and Validation Model Element Definitions:
Test Case: Test cases describe the test scenarios that ensure that the system achieves the customers' and system requirements. There are two different types of system test cases:

- Verification Test Case: Test cases that test all internal and external functions, customer, and system requirements.
- Validation Test Case: Test cases that test from the perspective of the system user (customer). They are the user/customer facing testing scenarios.

Test Plan: The test plan is a description of the approach to testing, with a listing of all of the verification and validation test scenarios.

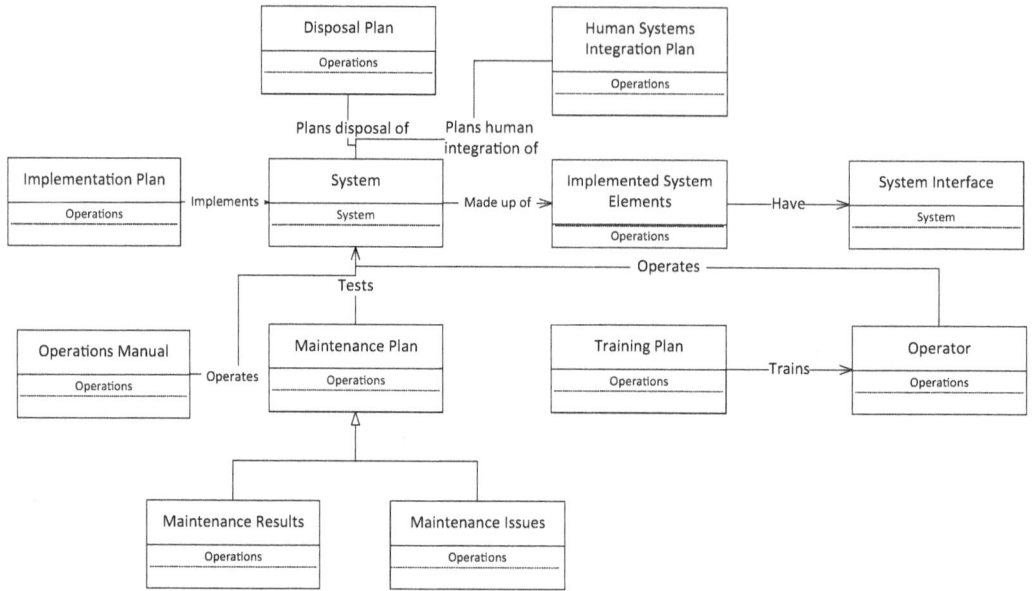

FIGURE 3.9   System operations model.
Source: S. Furterer.

Test Results: The test results are a compilation of the outcomes of each test case, identifying whether each test case meets the expected results of the test cases.

System Element Relationships:
The test plan contains the test cases that are used to test the system. The test cases document the test results. There are two types of test cases, verification and validation test cases. The system elements make up the system.

3.4.7   System Operations Model within Phase 6, Operations and Maintenance

Value & Purpose:
The purpose of the System Operations Model (Figure 3.9) is to describe how the system operates, is maintained, and is ultimately disposed. The model includes implementation, operations, maintenance, human integration, and training plans.

Systems Operations Model Element Definitions:

Implementation Plan: The implementation plan describes the strategy for implementing the system.

Implemented System Elements: The implemented system elements provides a list of the system elements that are ultimately implemented and made operational.

Operations Manual: The operations manual provides a detailed description of how the operators operate the system.

Maintenance Plan: The maintenance plan describes how to maintain and service the system.

Training Plan: The training plan describes how to train the system operators.

Operator: The operator is the person(s) who operate the system.

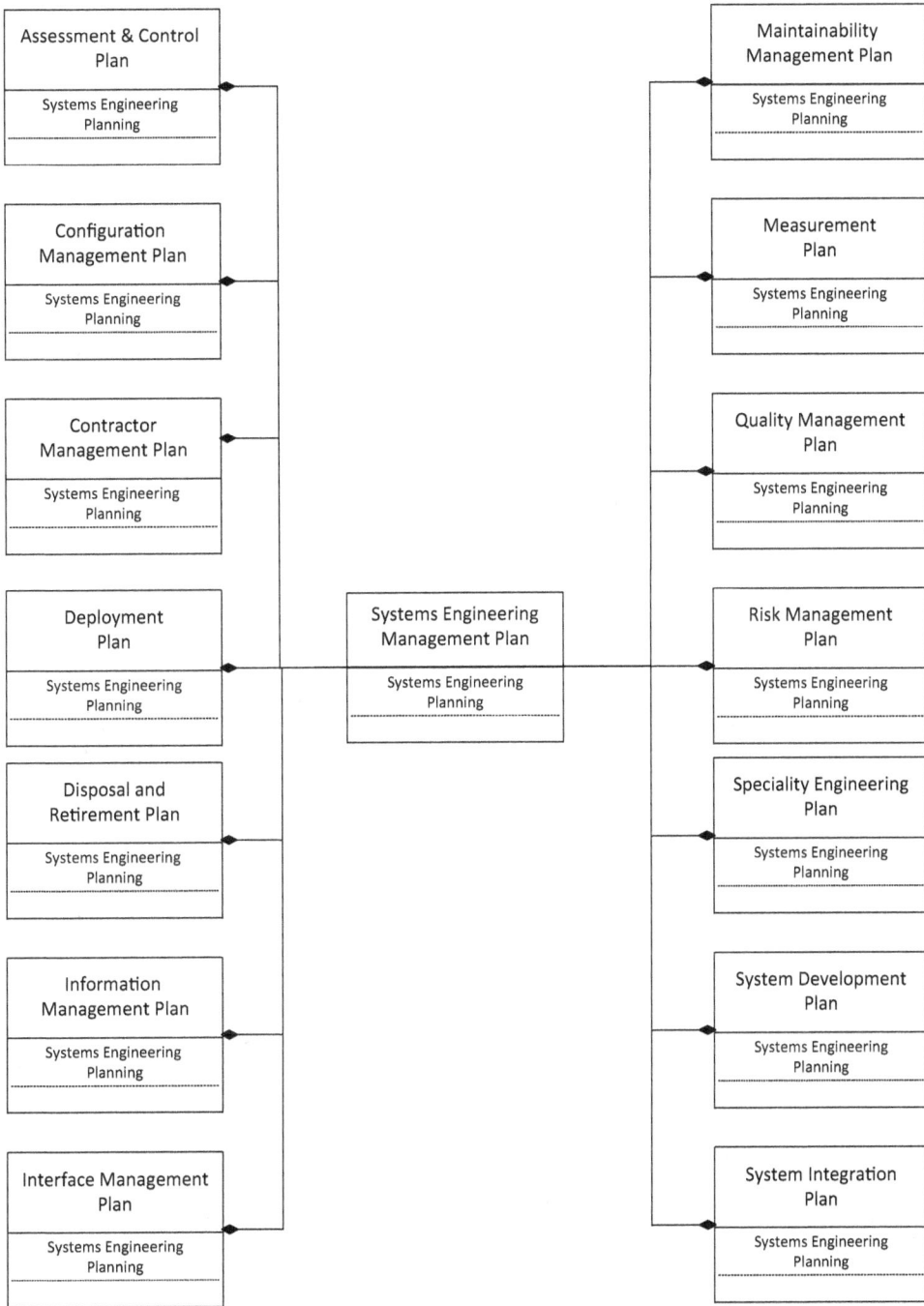

FIGURE 3.10 System engineering management planning model.
Source: S. Furterer.

System Element Relationships:

The system is made up of the implemented system elements. The implementation plan describes how to implement the system. The maintenance plan describes how to maintain the system. The maintenance plan includes the maintenance results and any maintenance issues. The training plans describe how the operators will be trained to operate the system. The operations manual describes how to operate the system. The human systems integration plan is used to plan how humans integrate and use the system. The disposal plan helps to plan the disposal of the system.

3.4.8 Systems Engineering Planning

Value & Purpose:
The purpose of the Systems Engineering Planning Model (Figure 3.10) is to describe the elements that make up the Systems Engineering Management Plan. The model includes the Systems Engineering Management Plan consisting of the plans identified under the Systems Engineering Management Planning Element Definitions.

Systems Engineering Management Planning Element Definitions:

1. Assessment and Control Plan: The Assessment and Control Plan describes how the key artifacts will be assessed and controlled throughout the systems engineering life cycle.
2. Configuration Management Plan: The Configuration Management Plan is used to plan how the processes and data will be configured and changed.
3. Contractor Management Plan: The Contractor Management Plan is used to describe how the contractors will be managed in relationship to the government or system owner organization.
4. Deployment Plan: The Deployment Plan will describe how the system will be deployed within the Phase 6 Operations and Maintenance.
5. Disposal and Retirement Plan: The Disposal and Retirement Plan is used to safely dispose of the system elements when the system is retired.
6. Information Management Plan: The Information Management Plan is used to manage the technical information for the development, design, and deployment of the system.
7. Interface Management Plan: The Interface Management Plan is used to manage the system interfaces, to ensure appropriation verification testing.
8. Maintainability Program Plan: The Maintainability Program Plan is used to maintain the system in the operations phase.
9. Measurement Plan: The Measurement Plan is used to design and manage metrics for measuring the key performance indicators for the system.
10. Quality Management Plan: The Quality Management Plan is used to plan and implement a quality management system.
11. Risk Management Plan: The Risk Management Plan is used to identify, assess, monitor, mitigate, and control system risks.
12. Specialty Engineering Plan: The Specialty Engineering Plan is used to plan the work of the required specialty engineering disciplines.
13. System Development Plan: The System Development Plan is used to plan the development of the system in the Phase 3 Detailed Design.
14. System Integration Plan: The System Integration Plan is used to plan how the system will be integrated to ensure verification and validation of the system.

System Element Relationships:
All of the plans identified in the model are part of the systems engineering management plan. They are developed at different points in the systems engineering life cycle.

Value & Purpose:
The purpose of the System Engineering Management Plan (SEMP) Model (Figure 3.10) is to do the planning that is part of the technical aspects of the system. There is other planning related to the systems engineering program and projects, which are not part of the SEMP Model.

3.5 SUMMARY

This chapter provided a description of the systems meta models that can be used as a basis for the Model-Based Systems Engineering. Understanding the models' elements and their relationships to each other help the systems engineer to design a well-integrated and functioning system. There were seven different meta models described 1) Mission Analysis model, 2) System Architecture and Requirements model, 3) Information and Application Model, 4) System Design model, 5) System Verification and Validation model, 6) Systems Operations model, and 7) Systems Engineering Management Plan model. More detailed descriptions of the activities and tools that help the systems engineer develop the information in each of the models are described in Section 2 of the book.

3.6 ACTIVE LEARNING EXERCISES

1) Find an article on systems engineering models and modeling. Are the meta models similar to ours? What are the similarities and differences?

BIBLIOGRAPHY

Archimate, *The Open Group*, http://www.archimate.org/en/about_archimate/, 2010.

Arsanjani, Ali, "Service-Oriented Modeling & Architecture". IBM Online article, 09 Nov 2004.

Arsanjani, Ali, "Principles of Advanced Software Engineering: Variation-oriented Analysis, Design and Implementation", 1999–2000.

Bell, Donald, UML basics: An introduction to the Unified Modeling Language,

Bell, Michael, *Service-Oriented Modeling: Service Analysis, Design, and Architecture*. Wiley, 2008.

Business Process Management Notation, Object Management Group/Business Process Management Initiative http://www.bpmn.org/, 2010.

Business Rules Group, http://www.businessrulesgroup.org/bmm.shtml, 2010.

Business Rules Group, Semantics of Business Vocabulary and Business Rules (SBVR), http://www.business-rulesgroup.org/sbvr.shtml, 2004.

Component Business Model, IBM, http://www-935.ibm.com/services/uk/igs/html/cbm-bizmodel.html, 2010.

DeveloperWorks, http://www.ibm.com/developerworks/rational/library/769.html#fig3

IBM Corp., Accessed: 8/11/2021.

Dictionary.com, Ontology. Accessed: 2011.

Fundamental Modeling Concepts, http://fmc-modeling.org/, 2010.

Information Architecture, http://www.iainstitute.org/, 2010.

ISO/IEC/IEEE. 2015. *Systems and Software Engineering - System Life Cycle Processes*. Geneva, Switzerland: International Organization for Standardization (ISO)/International Electrotechnical Commission (IEC), Institute of Electrical and Electronics Engineers. ISO/IEC/IEEE 15288:2015. The second definition is an expanded version of the ISO/IEC/IEEE version.

Kotiadis, K. & Robinson, S. *Conceptual Modeling: Knowledge Acquisition and Model Abstraction. Proceedings of the 2008 Winter Simulation Conference*. S. J. Mason, R. R. Hill, L. Mönch, O. Rose, T. Jefferson, J. W. Fowler eds., 2008.

Object Management Group, UML, http://www.omg.org/technology/documents/modeling_spec_catalog.htm#UML, 2010a.

Object Management Group, Product Rule Representation, http://www.omg.org/spec/PRR/1.0/ 2010b.

Chen, P., Pin-Shan, P. (March 1976). "The Entity-Relationship Model - Toward a Unified View of Data". *ACM Transactions on Database Systems* 1 (1): 9–36. doi:10.1145/320434.320440.

Porter, M. *Competitive Advantage*, The Free Press, A Division of Simon and Schuster, New York, NY, 1985.

RuleSpeak, http://www.rulespeak.com/en/, 2010.

Rumbaugh, J., Blaha, M., Premerlani, W., Eddy, F., Lorensen, W., *Object-Oriented Modeling and Design*, Prentice Hall, Englewood Cliffs, New Jersey, 1991.

The Java Tutorials, http://download.oracle.com/javase/tutorial/java/concepts/object.html, 2011.

TOGAF, The Open Group, http://www.opengroup.org/togaf/, 2010.

Watson, A., Visual Modeling: past, present and future, Object Management Group, 2010, Vice-President and Technical Director Object Management GroupTM.

Zachman, J., Zachman Framework, http://www.zachmaninternational.us/index.php/, 2010.

Section 2

Systems Design, Development, and Deployment Life Cycle

Service Applications and Tools

S.1   SECTION AND CHAPTER DESCRIPTIONS

The second section includes the systems design, development and deployment lifecycle with applications and tools. Chapters 4 and 5 will describe the Concept of Operations phase tools and activities. Chapter 4 will include the development of the strategic goals related to the system's mission. It includes the analyses performed to understand the system's landscape and the organizations that are part of the design, development, and deployment of the system. Chapter 5 will include the description of the tools and activities for developing the system's mission. Chapter 6 will cover the development of the system's architecture and chapter 7 will cover the requirements elicitation and chapter 8 will cover the requirements derivation. Chapter 9 will describe the detailed design phase. Chapter 10 will cover phases 4 and 5 of the Vee model, with phase 4 being the implementation, integration, test, and verification, and then phase 5 being the system verification and validation. Chapter 11 will cover phase 6, operations and maintenance. The section will end with chapter 12 that will discuss systems planning activities and tools.

The author developed the Systems Engineering Methodology, Activities & Tools Framework, shown in Table 1. This provides a summary of the activities performed, the tools applied and the principles that are needed as a basis for being able to perform each phase of the lifecycle. The framework aligns to the Vee lifecycle model's phases, and then breaks each phase into more detailed activities that are performed within each phase. The tools applied to achieve the purpose of each phase are identified. The principles that should first be understood to best achieve the phase activities and outcomes are listed. The principles may be applied in later phases as well but are not repeated for each phase within the framework. The details related to each element within the framework will be described in detail within chapters 4 through 12.

S.2   CASE STUDY DESCRIPTION

This section will provide two real-world community-based case studies. The first is developing the system and processes for a Women's Healthcare Center developed with the author, and the second designing a food system to reduce food insecurity that was performed by the author's graduate and advanced undergraduate students in a Systems Engineering course. The first case study is used in the Management of Engineering Systems course as a guide for the students to understand how the systems engineering models and tools are applied to a service system. Then the students design a system. For the particular semester, the students designed a food system to help to reduce food insecurity in the Dayton, Ohio area.

S.3   WOMEN'S HEALTHCARE CENTER

Many outpatient facilities are focusing on providing comprehensive services to women in a comfortable setting. In a qualitative study of women who had received a mammogram in the prior three years, without a history of cancer, satisfaction was related to the entire experience, not just the actual mammogram procedure. The authors of the described study found seven satisfaction themes from the focus groups: (1) appointment scheduling, (2) facility, (3) general exam, (4) embarrassment, (5) exam discomfort/pain, (6) treatment by the technologist, and (7) reporting results (Engelman, Cizik and Ellerbeck, 2005). This supports the focus of designing a seamless experience for women in the Women's Center through applying Systems Engineering methodology, tools, and principles.

S.4   FOOD SYSTEM TO REDUCE FOOD INSECURITY

Forty-two million Americans have trouble consistently accessing appropriate levels of affordable, nutritious food. Major contributing factors to food access is people residing in low income

DOI: 10.1201/9781003081258-5

neighborhoods, where food deserts exist. A food desert is an area where it is difficult to obtain food, due to lack of full service grocery stores. The purpose of chapters 4 through 12 is to demonstrate the application of systems engineering tools to model a food justice system in an Engineering Management course. These models help to develop connectivity and integration between disparate parts of the food justice system. As part of an engineering management graduate course, Management of Engineering Systems, the students worked in a transdisciplinary collaborative environment, called the GEMnasium, part of the Institute of Applied Creativity and Transformation at the university. The Systems Engineering Vee design lifecycle methodology was used as the framework to enable the systems engineering modeling of the food justice system. Seven teams of students designed different value chain activities to connect and integrate different parts of the system. Application of the systems engineering tools enabled a systems view of the food justice system to identify integration points and requirements that developed potential solutions to improve access to food, providing a more just food justice system. The instructor's and students' assessments demonstrated that the framework was impactful in helping students to view the food justice system as an integrated system of stakeholders, activities and goals. The instructor assessment graded the students' reports at a 92% average grade, with a 5% standard deviation. The percentage of the students' positive responses (top two positive rating categories) for the eight questions assessing the effectiveness of the experiential learning experience ranged from 88% to 100%. The results demonstrated the successful application of the systems engineering methodology and tools, and the collaborative interactive sessions between the stakeholders and the students in understanding and modeling the food justice problem. This case study will provide a real-world example of a community-based systems engineering project to help the reader better understand how to apply the six phases of the Vee lifecycle model. The case study and the tools were developed by graduate and advanced undergraduate students in the Management of Engineering Systems course. They are not perfect, but will help the reader understand how to apply the tools based on these real examples.

TABLE S.1
Systems Engineering Methodology, Activities & Tools Framework

Vee Phase	Activities	Tools	Principles
Phase 1: Concept of Operations	Define Strategic Mission Goals: 1. Define mission strategy 2. Define goals of mission 3. Develop concept of operations 4. Perform external, internal and SWOT analyses Define Mission Concepts 1. Develop project charters 2. Perform stakeholder analysis 3. Collect Voice of Customer 4. Perform conceptual selection 5. Perform risk analysis	Strategic mission goals tools: • Mission Strategy; Mission goals; Concept of Operation, Concept; External analysis; Internal analysis; SWOT analysis Mission concepts tools: • Project charter with risk analysis • Stakeholder analysis • Critical to Satisfaction criteria • Pugh Concept Selection Technique • Risk Assessment, Risk Cube	• Complexity • Emergence (whole > sum of parts) • System of system • Hierarchy • Boundary

(Continued)

TABLE S.1 (Continued)

Vee Phase	Activities	Tools	Principles
Phase 2: Requirements & Architecture	Develop Logical Architecture 1. Develop Logical Architecture Elicit Requirements 2. Elicit customer requirements 3. Derive requirements • Develop system requirements • Develop specialty engineering requirements 4. Develop measurement plan 5. Develop an information models 6. Develop a quality management plan	Logical Architecture: • SIPOC • Value Chain & Functional Decomposition • Use Case Diagrams Requirements: • Process Scenarios via Process Architecture map • Use Cases • Customer requirements • System requirements • Specialty engineering requirements: software engineering; environmental engineering; safety and security engineering; Human Systems Integration • Data Collection Plan, with operational definitions • Class Diagram with information model, and hierarchy model • Quality management plan	• System dynamics (behavior, system elements) • Cybernetics (information flow) • Systems thinking • Abstraction • Views
Phase 3: Detailed Design	1. Perform detailed design 2. Perform systems analysis	Detailed Design: • Physical Architecture Hierarchy Model • Physical Architecture Model: Physical Block Diagram; and/or SysML Block Definition Diagram; and/or SysML Internal Block Diagram • System Elements • QFD • Simulation (not required) Systems Analysis: • Selection criteria • Trade off analysis; and/or Effectiveness analysis; and/or Cost analysis; and/or technical risk analysis • Justification report	• Systems analysis • Wholeness and interactions
Phase 4: Implementation, Integration, Test & Verification	1. Perform system implementation 2. Perform system integration	• Integration constraints • Implementation strategy • System elements supplied • Initial operator training • Verification criteria • Verification test cases and results • N-squared diagram	• System elements • Modularity • Interactions • Networks • Relationships • Behavior

(Continued)

TABLE S.1 (Continued)

Vee Phase	Activities	Tools	Principles
Phase 5: System Verification and Validation	1. Perform system verification 2. Perform system validation	• Verification and Validation criteria • Verification and Validation test cases and results	• Synthesis
Phase 6: Operation & Maintenance	1. Deploy System 2. Perform training 3. Perform certification 4. Perform risk assessment 5. Perform improvement and maintenance planning 6. Perform disposal and retirement activities	• Training plan and materials • Certification plan and materials • Operations Manuals • Performance reports • Improvement and Maintenance plans; FMEA • Disposal and retirement plan	• Control behavior and feedback • Encapsulation (hide internal workings of system) • Stability and Change
PLANNING (throughout all phases)	1. Develop Systems Engineering Management Plan and associated plans.	• Systems Engineering Management Plan: Assessment and Control Plan; Configuration Management Plan; Contractor Management Plan; Deployment Plan, Disposal and Retirement Plan; Information Management Plan; Interface Management Plan; Maintainability Program Plan; Measurement Plan; Quality Management Plan; Risk Management Plan; Specialty Engineering Plan; System Development Plan; System Integration Plan; System Integration Plan	

Source: Sandra L. Furterer

4 Concept of Operations Phase 1
Define Strategic Mission Goals

In this chapter, we discuss the Concept of Operations phase analyses that are performed to inform the strategic goals for the mission.

4.1 PURPOSE

The Concept of Operations Phase is the first phase in the Systems Engineering Vee Life Cycle Model. The purpose of the phase is to perform an analysis of the mission and define its strategic goals. Within this phase, the systems engineering team also develops the project charters to break the system into manageable projects. A stakeholder analysis is performed to identify the stakeholders of the system and their impacts and concerns related to the system. Additionally, the system's concepts are generated, prioritized, and selected in this phase.

4.2 ACTIVITIES

The activities performed and the tools applied in the Concept of Operations phase are shown in Table 4.1

In this chapter, we cover Define Strategic Mission Goals and the related tools.

The systems architecture model for the Concept of Operations phase, discussed in Chapter 3, is shown in Figures 4.1.

We now describe the first activity within the concept of operations phase, 1) Define mission strategy.

4.3 DEFINE MISSION STRATEGY

A mission strategy is a pattern or plan that focuses the organization on the major strategic goals of the mission. A well-formulated strategy helps to provide a roadmap to achieve the mission.

For example, for an existing lift truck manufacturer, the mission strategy may be to capture additional market share by incorporating autonomous features into the next generation of their lift trucks. For a community whose citizens face food insecurity, the strategy may be to develop technology and processes that can connect citizens, food producers, processors, distributors, preparers, and consumers to reduce the prevalence of food deserts.

Typically, an organization develops the mission strategy within their strategic planning process, which typically has a three- to five-year planning horizon. The senior leadership should share the strategies that apply to the particular system mission with the systems engineering team. In the case of the food justice system, the transdisciplinary challenge was defined collaboratively with the course instructors and the community-partner stakeholders as "To grow and sustain a just and resilient community food system" [1]. The ultimate system sponsor will typically have an idea of the system's mission that they are trying to build. The system sponsor will define and communicate the system's mission. As a part of the systems planning process, the program manager will develop the

DOI: 10.1201/9781003081258-6

TABLE 4.1

Activities and Tools in the Concept of Operations Phase

Vee Phase	Activities	Tools	Principles
Phase 1: Concept of Operations	Define Strategic Mission Goals: 1. Define mission strategy 2. Define goals of mission 3. Develop concept of operations 4. Perform external, internal, and SWOT analyses Define Mission Concepts 1. Develop project charters 2. Perform stakeholder analysis 3. Collect Voice of Customer 4. Perform conceptual selection 5. Perform risk analysis	Strategic mission goals tools: • Mission Strategy; Mission goals; Concept of Operation, Concept; External analysis; Internal analysis; SWOT analysis Mission concepts tools: • Project charter with risk analysis • Stakeholder analysis • Critical to Satisfaction criteria • Pugh Concept Selection Technique • Risk Assessment, Risk Cube	• Complexity • Emergence (whole > sum of parts) • System of system • Hierarchy • Boundary

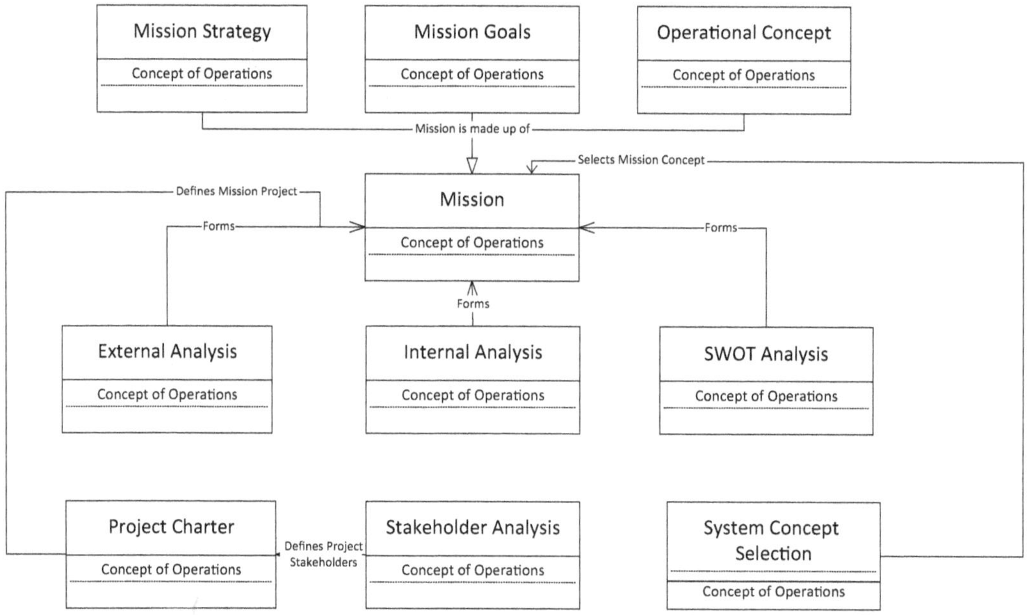

FIGURE 4.1   Mission analysis model.

Source: S. Furterer

organizational structure of those designing and developing the system. The activities typically entail creating sub-system teams aligned to a logical grouping of the value chain activities or functionality.

The mission strategy, strategic goals, and operational concept for the Women's Center Healthcare Center is provided in Figure 4.2. From the external analysis, we find that there could be an issue in providing additional physician resources to read the imaging results; that approximately 50% of the campaign for construction costs of a new facility have been raised from donors to the hospital; that the industry lacks robust business process management (BPM) technology that would provide process times; and performance measures to design and improve processes; and that contracts will be needed for supplier partners.

For the food system case study, the students were both graduate and advanced undergraduate students from across the School of Engineering, resulting in a mix of engineering disciplines, including engineering management, mechanical engineering, aerospace engineering, civil and environmental

Mission Strategy:	Develop a new women's healthcare center that provides women's services.
Strategic goals:	1) improve the turnaround time to provide imaging results for women's screening and diagnostic mammograms 2) Reduce the long delays in getting an appointment, and in registering when the patient comes into the center for their appointment. 3) Improve processes and technology to seamlessly connect the patients to the cancer center, when the need exists, and also to identify VIP patients. 4) There is also lack of sufficient spiritual care resources at the women's center, due to resource and staffing constraints.
Operational Concept	The goal of this system engineering project is to design the processes and metrics which result in optimal flow for the new Women's Center and align to the Voice of the Customer. There are many benefits that this system can provide, including patient satisfaction, increased capacity due to efficient workflow, and resultant revenue are potential financial benefits of this project.

FIGURE 4.2 Women's center healthcare center mission strategy, strategic goals and operational concept. *Source*: S. Furterer

engineering, electrical and computer engineering, and chemical and materials engineering. This enabled a wide variety of perspectives with the students being able to draw from their engineering disciplines. Seven teams, with three to four students were formed with a mix of the different disciplines. The initial system mission focus was defined based on the non-profit stakeholders' organizational missions within the food system.

The initial kickoff meeting with all of the students, faculty, and food system stakeholders was held at the beginning of the semester. A leader from each non-profit stakeholder organization described their organizations' missions and gave an excellent history of how food insecurity and food deserts developed and the causes of food insecurity. The non-profit organizational stakeholders focused on the outcomes and strategies of the Community Food Projects Competitive Grant Program (CFPCGP). The CFPCGP helps to fight food insecurity through developing community food projects to help promote self-sufficiency in low-income communities [2]. These organizations help to bring food security, defined as:

Community Food Security is a condition in which all community residents obtain a safe, culturally appropriate, nutritionally sound diet through an economically and environmentally sustainable food system that promotes community self-reliance and social justice.

(Mike Hamm and Anne Bellows).[3]

The desired outcomes of the CFPCGP are to [2]:

- Meet the needs of low-income people by increasing their access to fresher, more nutritious food supplies.
- Increase the self-reliance of communities in providing for their own food needs.
- Promote comprehensive responses to local food, farm, and nutrition issues.
- Meet specific state, local, or neighborhood food and agricultural needs for infrastructure improvement and development.
- Plan for long-term solutions.
- Create innovative marketing activities that mutually benefit agricultural producers and low-income consumers.

The following strategies can help to achieve the desired outcomes of the CFPCGP [2]:

- Create community food systems
- Build collaboratives/networks/partnerships
- Build organizational and individual capacity
- Promote public education and outreach
- Advocate for policies to change local food infrastructure

The mission strategy for the systems engineering course was defined to identify the stakeholders, integration points among them, and potential solutions that will enhance food security within the Dayton community, shown in Figure 4.3. Each team then developed their own mission strategy, strategic goals, and operational concepts to support the overarching strategy.

There are a total of seven teams consisting of three to four students per team, focused on the following food system mission strategies, as shown in Table 4.2.

Mission Strategy:	Model and design a food justice system that will identify current and future resources, integration points amongst stakeholders, and potential solutions that will enhance food security in the Dayton community.

FIGURE 4.3   Systems Engineering Overarching Mission Strategy.

TABLE 4.2
Concept of Operations Mission Strategies by Team

#	Teams	Missions
1	Connecting organizations	Create website to encourage more people, connect organizations, find what people want, and get young people involved.
2	Creating food policy	Develop policies for the food justice system.
3	Kids growing food	Teach kids to grow microgreens and get microgreens to store to sell.
4	Food storage	Create a centralized storage location for the community food.
5	Food hub website	Create a website to network food residents, entrepreneurs, farmers, grocery stores (food hub).
6	GEM City Market	Operationalize GEM City Market supermarket.
7	Table 33 Restaurant	Help make the Table 33 concept profitable, by becoming more efficient, and better connecting local foods with the restaurant.

4.4 DEFINE STRATEGIC GOALS OF MISSION AND CONCEPT OF OPERATIONS

The strategic goals are the actionable end results that the mission strategy is trying to achieve. The operational concepts are how each team and the related stakeholders are organized to be able to achieve the mission strategy and strategic goals. The operational concepts ensure that the teams are resourced and organized sufficiently to move forward into the concept generation, requirements analysis, design, and deployment phases.

Table 4.3 shows the mission strategies, strategic goals and operational concepts for the seven teams, based upon the information provided in the initial kickoff and information sharing meeting with the stakeholders and the students.

4.5 PERFORM EXTERNAL, INTERNAL, AND SWOT ANALYSES

The teams then performed the external, internal, and Strengths, Weaknesses, Opportunities, Threats (SWOT) analyses that formed the potential systems' concepts.

4.5.1 EXTERNAL ANALYSIS

The external analysis includes investigating and documenting the external influencers that impact the system. The current state related to the external influencer, the proposed future state, and the gap between the two are investigated and documented. The external analysis influencers should be identified first for the current state related to the system of interest. The future state for each influencer should then be identified, typically through discussions with the stakeholders of the system. The gap between the current state and the future state should be identified by the systems engineering team based upon the difference between the current and future state for each of the external influencers. This helps to begin extraction of the potential concepts for this phase (Figure 4.4).

The external analysis is shown in Figure 4.5 for the Women's Healthcare Center. From the external analysis, we find that there could be an issue in providing additional physician resources to read the imaging results that approximately 50% of the campaign for construction costs of a new facility have been raised from donors to the hospital; that the industry lacks robust business process management (BPM) technology that would provide process times and performance measures to design and improve processes and that contracts will be needed with supplier partners.

As part of the conceptual analysis, the students met with the community partners as an entire class and then had team sessions with the stakeholders that aligned to their chosen food system activities. The external analyses are shown in Table 4.4 through 4.10. The session topics included: 1) Each community organization shared their mission and the services that they provide; 2) Each community organization shared a story of how their organization has helped the community and described their stakeholders' challenges; 3) Students interviewed the assigned community partners to identify elements related to the external and internal analyses, including their capabilities and limitations; 4) Each community partner then shared a story that describes a dream future state. The external analyses included the following influencers categories: political, economic, social, technological, environmental, and legal. The analyses identified external influencers specific to each team's mission, but also elements that could influence the entire food justice system. The external analysis elements can be identified through interviews with knowledgeable stakeholders, other subject matter experts, and the literature including scholarly articles, books, trade journals, and reputable internet sources. Governmental sources are valuable external references for the external analysis as well.

TABLE 4.3

Mission Strategies, Strategic Goals, and Operational Concepts

#	Team	Mission Strategies	Strategic Goals	Operational Concepts
1	Connecting organizations	Building a website which can provide an information exchange and food ordering platform to improve food security in the westside of Dayton.	1) Design a form containing questions about basic information (e.g. gender, age), foods' need or preference, and the user experience collecting the data from community members for further analysis which will be done in the next step. 2) Address the data and arrange them by priority. The information will be given to Citywide and let them prioritize suggested goods to farmers. 3) Improve communication between community members, farmers, and Citywide using digital and physical advertisement to purchase products and host events to sell food. 4) Create a forum to exchange their information and discuss topics pertaining to food sales, needs, events, etc. Citywide can administer the forum to get the information from both the farmer and the community member.	The goal of this system engineering project is to design a website connecting community members, farmers, community partners, and Citywide to address and work to improve food accessibility in the west side of Dayton. The benefits of this website include providing more chances for farmers and community partners to advertise and sell their products; making fresh and diverse amounts of foods more accessible to community members; providing a direct and effective method for the communication among farmers, community partners, and community members.
2	Creating food policy	The team is working to create a wide encompassing system to show all the current and future resources that are working to alleviate the food justice issue seen in the Dayton area	1) Creating a working document of current food justice systems. 2) Reduce the confusion surrounding which systems can be utilized for different food justice related issues. 3) Continuously improve overarching system to incorporate future systems.	The goal of this project is to create a structure of the current resources available that are addressing the food justice issue in the Dayton area. Currently, the resources that are being put to use are not effectively being used because of the lack of structure of a total system. By building such a system, the food justice problem will be able to be handled swiftly and properly.
3	Kids growing food	Food justice in Dayton, Ohio	Healthy food, affordable food, culturally pertinent food, ease of availability of food, and knowledge of nutrition for all citizens of Dayton.	Food production, processing, distribution, sales, preparation, and consumption acting in accordance with the strategic goals of food justice in Dayton.

#	Name	Description	Objectives	Goal
4	Food storage	Centralized storage location for community food.	1) The idea of centralized storage locations is to increase the shelf life of food while keeping it safe at a central location which would make distribution convenient. 2) There is a lack of food supply during certain periods of the year, therefore this ultimately ensures its supply throughout the year even during scarcity. 3) Selecting the storage place nearby to a redline region. 4) Reducing the percentage of food loss by storing respective vegetables at its best temperature. 5) Improve the storage system by providing a new technology that meets the requirements of efficient storage	We aim to improve the previous condition of the storage facilities of the community food and make distribution more convenient by introducing and following smarter and low-cost quality storage methods. This could be attained by encouraging local organizations, activities, and interactions among the project. Enhance and improve the processes for developing and maintaining the storage system. Creating appropriate conditions where food can be stored for 30 days or longer which is an important condition of winter marketing.
5	Food hub website	Use the current working model in the Downtown Dayton Partnership as a start and expand into other geographic portions of the city.	Leverage current relationships with agriculture and restaurateurs, attract food entrepreneurs, and use those to invest in underprivileged areas of the city.	Network the current relationships with local farmers and restaurant owners with food entrepreneurs willing to invest in underprivileged neighborhoods in need of a greater food system.
6	GEM City Market	To develop a Food Sales system to Operationalize Gem City Market – Supermarket	1) To provide access to healthy food resources. 2) Obtain healthy food resource from local community farms. 3) To maintain the quality and quantity of the food resources both from the community farms and distributors. 4) To promote hybrid pricing in the store. 5) Analyze the marketing and sales statistics 6) Improve according to demand in the supermarket.	Develop a food sales system to operationalize the Gem City Supermarket by providing the community with healthy food options, access to a variety of food resources and make it a worker-community owned one.
7	Table 33 Restaurant	Provide healthy food with affordable prices, helping local farmers.	1) Reduce the cost of ingredients by purchasing them from local farmers. 2) Increase profits of Table 33.	Reduce poverty and the food desert in the area, increase Table 33 sales.

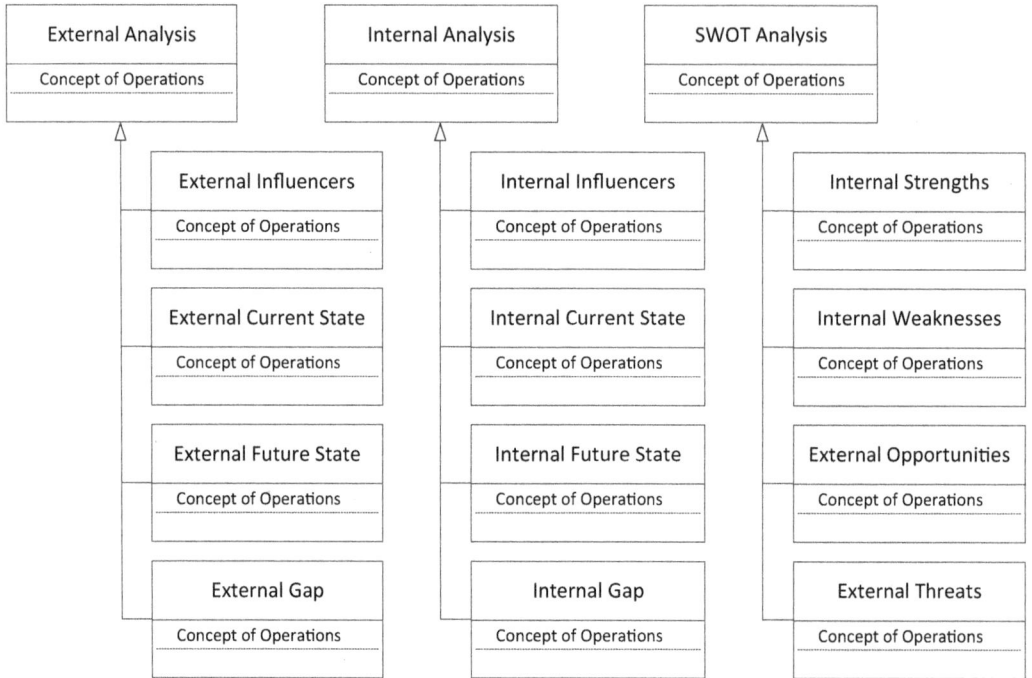

FIGURE 4.4 Mission analysis elements.

Team 1 identified several external gaps related to connecting organizations to provide food for those in need, including little control of public funding by community members, lack of grocery locations on the west side of Dayton, lack of partners willing to be involved with a program for a local food market, lack of knowledge of how farmers can be self-sustaining, lack of funds to achieve the mission, young people not wanting to attend meetings to volunteer and lack of social networks to engage volunteers across various age groups, lack of information regarding organic food, lack of food delivery systems, and lack of marketing for food.

Team 2 identified several gaps related to creating food policy, including lack of resources to help locally, lack of programs to bring money into the city for food programs, and areas that have the knowledge to help, lack the interest.

Team 3 developed external influencers' gaps related to kids growing food, including lack of ongoing funding, lack of stable home life for children, lack of available nutritional food, need for equipment and materials for growing food, schools not being onboard to help, and unresponsive social workers.

Team 4 developed their external analysis related to food storage gaps, including setting up a new farmers' market from scratch, the need to encourage more volunteers to work on the problem, need for increasing health awareness of nutritional food, need for growing sponsors and growers, need to improve the number of markets.

Team 5 developed their external analysis related to the food hub website, including lack of neighborhood investment, lack of entrepreneurs, lack of support for the food system, lack of equity and inclusion, lack of software applications for delivery, lack of farms, antiquated regulations, and lack of city involvement in planning a market.

External Analysis			
External Influencers	**Current State**	**Industry or Future State**	**Gap**
Political	Moderate physician support	Strong physician support	Physician resource availability
	Buy-in of city council	Buy-in of city council	None
Economic	Met 50% of need through campaign	100% of need met	50% of campaign
	No monetary resources for additional radiologist for same day results	Additional radiology resources	Gap of additional radiologist resource
	Lack of dedicated women's center resources, not connected to hospital	Dedicated women's center resources	Lack of dedicated resources for women's center
Social	Social responsibility plan in place	Social responsibility plan in place	None
	Excellent donor network, friends of the hospital	Excellent donor network, friends of the hospital	None
Technological	No Business Process Management (BPM) information technology	State of art BPM information technology	Lack of Business Process Management (BPM) information technology
	No technology for VIP alerts	Technology for VIP alerts	Lack of alert technology
Environmental	Construction delays, and disposal of materials	Construction complete, environmental plan deployed	Lack of environmental plan, construction and traffic delays
	No disposal of materials plan	Disposal of material plan in place	Lack of disposal plan
Legal	Lack of established contracts with all providers	All providers with agreed to contracts	Need contracts

FIGURE 4.5 External analysis for the women's healthcare center.

TABLE 4.4
Team 1 – Connecting Organizations – External Analyses for Food System

External Analysis			
External Influencers	**Current State**	**Industry or Future State**	**Gap**
Political	Higher administration controls investment of money in the community	Public has more of a say where money is invested	Little control of public funding by community members
	About two corner food stores in five to six cities in west side of Dayton	Multiple locations in west side of Dayton where affordable food can be obtained	Lack of locations where fair priced food is available
	SNAP requires federal funding but funding has been unstable lately	Funding that is stable	Lack of sustainable federal funding for SNAP

(Continued)

TABLE 4.4   (Continued)

External Analysis

External Influencers	Current State	Industry or Future State	Gap
Economic	Cost of labeling foods and meats for a store can be thousands of dollars	Cost of labeling foods and meats for a store can be thousands of dollars	None
	SNAP receives discounts on organic meats from Brooksville Farmstand	SNAP receives discounts on organic meats from Brooksville Farmstand	None
	Food at food stand is slightly more than at Kroger and Walmart but less than at the local corner stores	At least 10 food stands functioning in different places at different times	Lack of partners willing to be involved in program
	Profit margin is low due to high indirect costs	Have food stands and farms become self-sustaining	Lack of knowledge of how farmers can become self-sustaining
	Little or no funds (government + private)	Trying to engage them by telling them about food justice importance	Lack of federal, state, and local funds to support the mission
Social	People in Public housing tend to have high unemployment rate or low income	Have funding for the community to invest in greater opportunities to lower unemployment and raise income	Funding is too high a risk for traditional loans
	Different thinking habits between generations	More of an online presence/ have more events rather than have meetings only	Young people do not care to go to meetings
	More retired folk volunteer than younger people		Lack of social networks involving and accommodating all generations
	Younger generation struggles with time availability and desire to get involved		Cannot control people's free time and priorities
	Mission-based farm in Brookville working with Citywide	Multiple farms involved with Citywide	A lot of small farms do not have the equity to do this
	Less awareness about this big issue	Doing seminars and public gatherings	Lack of information about organic foods/what it tastes like
Technological	Credit card and SNAP software applications are separate.	Have one software available where both credit card and SNAP can be accepted	Vendors may not have access to both payment methods
	A good portion of the community may have little access to social media platforms/internet.	Be able to reach more people through phone messaging and social media	Lack of communication via more modern methods of advertising and communication (e.g. websites, phone apps)
	No sophisticated delivery system	Should be able to deliver home to home	Lack of food delivery system
	No organized marketing (word of mouth)	Use social media platform	Lack of organization of how farmers market their foods and information to the consumers
	Certain pickup areas in communities where people come to pick up the food	Permanent area where a farmer's market can exist	Lack of established areas in the west side of Dayton where farmers can set up regularly

(Continued)

TABLE 4.4 (Continued)

Environmental	Weather could cause food and crop prices at food stands/stores to rise	Weather could cause food and crop prices at food stands/ stores to rise	None
	Loss of livestock/crop due to deep freeze in winter	Loss of livestock/crop due to deep freeze in winter	None
Legal	Must have authorization available to accept SNAP	Must have authorization available to accept SNAP	None
	Foods and meats must be labeled in a store	Foods and meats must be labeled in a store	None

TABLE 4.5
Team 2 – Creating Food Policy – External Analysis for Food System

External Analysis			
External Influencers	Current State	Industry or Future State	Gap
Political	Moderate Support from the city of Dayton	Strong support from the city and the University of Dayton	Resources are not currently well fleshed out to the point that local leaders can help in a meaningful way
Economic	New policy can only be implemented if it provides a beneficial economic impact	All future policies will provide Dayton with economic incentives	Not enough time is being spent developing programs that can bring money into the city
Social	Minimal support from those that are not directly afflicted by current poor policies	The whole Dayton area will work together to find a solution	Areas that have the highest economic impact on the subject matter (UD, Oakwood) are extremely disinterested in the problem

Team 6 developed their external analysis for the GEM City Market gaps including funding for the market, lack of healthy food options due to lack of grocery stores on the west side of Dayton, lack of interaction with the community, people having addictions, lack of opportunities, history of redlining, and a food crisis on the west side of Dayton.

Team 7 developed their external analysis for the Table 33 restaurant, including the need to increase local food growth, gap in food quality, need to increase funds, resources, and income, need to increase sustainability, and lack of growing teaching licenses.

TABLE 4.6

Team 3 – Kids Growing Food – External Analysis of Food System

External Analysis

External Influencers	Current State	Industry or Future State	Gap
Political	One time P&G Grant	Replenishable grant every year	P&G agreement to replenish every year
Economic	$20,000 grant will eventually run out	Refunded grant	P&G agreement
	Self-funded	No personal funds applied	Self-funds applied
Social	At-risk children	Stable children	Lack of stable home life
	70 kids, ages 15–19	70 kids or more, ages 15–19	None
	Opportunistic children	Responsible children	Recognition of responsibilities
	Forced to take care of younger siblings while parents busy with multiple jobs	Forced to take care of younger siblings while parents busy with multiple jobs	None
	Nutrition not a priority for children	Children prioritizing nutrition	Lack of available nutritional foods
	Unreliable attendance of children at school	Reliable attendance at school	Lack of accountability at home
Technological	Aquaponics using repurposed materials (solo cups)	Permanent materials in use	Permanent materials
	PVC pipe degradation	No PVC pipe degradation	Higher quality pipe
Environmental	Children experience toxic stress at home	Children not exposed to toxic stress	Toxic stress exists at home
	Lack of nutrition producing zombie-like state	Engaged children	Lack of available nutritional foods
	Upkeep of aquaponics in winter i.e. fish	All year aquaponic system	Lack of functioning shelter for system
Legal	School could be considered liable	School embracing liability	School not on board
	Unresponsive social workers	Responsive social workers	Responsiveness of social workers
	Dealing with minors	Dealing with minors	None
	Children often in trouble with the law	Children not legally in trouble and in attendance at school	Lack of proper guidance or role model at home

4.5.2 INTERNAL ANALYSIS

The internal analysis looks at internal influencers within the stakeholders' organizations to assess

TABLE 4.7
Team 4 – Food Storage – External Analysis of Food System

External Analysis			
External Influencers	**Current State**	**Industry or Future State**	**Gap**
Political	No farmers' market is around region, where there are other food shopping sources like Kroger's, instead being located around redline region	Setup farmers' market at redline region	Setting up a new farmers' market from scratch
	West Dayton is facing a high challenge regarding food insecurity	Mitigate the challenges and make it smooth running	Encourage more volunteers to research and work on solving the problem
	Public health is not the main concern for the government	Public health should be the main concern for the government than making profit of food industry	Government should come forward along with all of the communities to spread the health awareness
	Government highly subsidizes processed food ingredients	Government should support organic food	Government becoming concerned of the consumers health and subsidizing organic food
Economic	Limited number of partners	High number of partners and sponsors	Attracting and increasing sponsors and partners
	Minor part of community growing own food	Major part of community is growing own food	Increasing number of people growing own food and increase the number of gardens
	Major percent of West Dayton not buying the food grown by the community	Having major part of West Dayton buy community food	Increasing major number of people buying the community food
	Homefull is making a turnover less than $15000 a year	Make a turnover higher than $15000 a year	None
	No food bank method used	Food bank concept could be utilized	Introduce food bank concept to achieve central storage location
Social	The regional food is sold to a limited market area.	Sell food close by at the hospitals and schools	Increase the selling location and improving market
	Excellent supporter network, employees, and friends at the organization	Excellent supporter network, employees, and friends at the organization	None
	Diversity and multicultural	Increase diversity and multicultural	Diversity improvement
	Improves community in teaching methods of food processing and implementation	Involves community in teaching methods of food processing and implementation	None

(Continued)

TABLE 4.7   (Continued)

External Analysis

External Influencers	Current State	Industry or Future State	Gap
Technological	Storage space is minimal	Have a centralized and efficient storage space	Invest and develop efficient storage space
	No cooking space and crock-pot along with storage	Have a storage space along with cooking using crock-pot for cooking	Develop a storage space along with cooking space
	No mobile truck for food distribution simultaneously for storage	Usage of mobile truck for distribution and storage	Buying a truck with storage capacity
	No bikes with coolers used for distribution of food	Usage of bike with a cooler for distribution from the storage center	Introduce a bike distribution from central storage location
Environmental	Training people how to grow food	Training people how to grow food	None
	Offering expertise to people to grow niche products	Offering expertise people to grow niche products	None
	Homefull distributes food to about 1,000 people	Distribute food to all under privileged people	Reach and distribute food to all people in the area in need
	Fixing green house and garden which are broken down and wasted	Continue as the current	None
	Not all food growers are aware to cope with all of the weather conditions	All food growers being aware to cope with all the weather conditions	Educate all the food growers to cope with all weather conditions
Legal	Farmers cannot afford the legal help they need	Farmers should be able to afford the legal help they need	Try to fix the need
	Not all farmers are aware of food law and policies	Farmers must be aware of the food law and policy	Educate the farmers
	Following all of the laws regarding packaging, labeling, transportation, and storing	Following all of the laws regarding packaging, labeling, transportation, and storing	None

TABLE 4.8

Team 5 – Food Hub Website – External Analysis of Food System

External Analysis

External Influencers	Current State	Industry or Future State	Gap
Political	Neighborhood Investment	Economic Revival of Urban Areas	Lack of neighborhood investment
Economic	Need entrepreneurs for food market	Entrepreneurs for food market	Lack of entrepreneurs

(Continued)

TABLE 4.8 (Continued)

Social	Lack of support for food system	Support for food system	Generating support through jobs and locally sustained markets
	Systemic Racism	Equity and Equality for all types	Lack of equity and inclusion
	Business Safety	Secure business and investments	Increased security incentives
Technological	Individual food services like apps (Uber Eats, etc.)	Grocery Delivery via Apps	Lack of software applications for delivery
Environmental	Need of more farming to sustain food systems	Local Support from Local Farmers	Lack of farms
Legal	Food code is impeding	Modernize regulations	Antiquated regulations
	City Planning	City involvement in store working	Lack of city involvement in grocery store planning

TABLE 4.9
Team 6 – GEM City Market – External Analysis of Food System

External Analysis			
External Influencers	Current State	Industry or Future State	Gap
Political	Funds to develop the GEM City Supermarket	Revenue from the market	Food crisis due to redlining led to the development of the supermarket
	Government donations of $100,000	To develop websites and to establish the market	None
	Federal donations of $250,000	Getting important needs activated for the market	None
Economic	Hybrid pricing	Affordability of the products	None
	Fund raising for the market	Establishment and development of the supermarket	Due to redlining and food crisis
	Improve economic status of the communities	To make the store worker owned community (co-op)	Unaffordable pricing and lack of reach to healthy food
	Gentrification	Development of the community	Bias between communities
Social	Work opportunities	Increase in employment rate of the community	Due to differentiation between communities and people having other addictions
	Meeting community needs through interactions	Community organizations and meetings conducted by the community	Lack of interactions within the community
	Helping farmers to grow and sell their crops	Supply food resources to the market	Lack of opportunities

(Continued)

TABLE 4.9 (Continued)

External Analysis

External Influencers	Current State	Industry or Future State	Gap
Technological	Website	Awareness about the food desert	Redlining
	Mobile grocery store	Reachable to everyone	None
	Advertisement	Promotion of the supermarket	None
Environmental	Food sourcing	25% of food resources obtained from the community farms	Lack of opportunities
	Availability of health food	Needs are met	Food crisis
Legal	Land trust	Buying the land and renting it out	Avoiding legal issues
	Taxes	On time payments	None

TABLE 4.10
Team 7 – Table 33 Restaurant – External Analysis of Food System

External Analysis

External Influencers	Current State	Industry or Future State	Gap
Political	Living in urbanizednation	Create farms in our nation	Increase local food growth
	U.S. dietary standards do notmove beyond sufficiency	Focus beyond food sufficiency	Increase food quality
Economic	Low funds	Get more funds	Increase funds
	Limited resources	More resources	Increase resources
	Low daily income	Better daily income	Increase income
Social	No food growingknowledge	Teach people how to growtheir food	Have better growingknowledge for thecommunity
	Ignoring recommended U.S. consumption characteristics	Follow recommended U.S. consumption characteristics	Consume 1200 poundsof food yearly
Technological	No equipmentfor delivering food	Get catering equipment for delivering food	Increase delivery efficiency
Environmental	Turn farms to urbanized land	Create more farms	Increase environmental sustainability
Legal	Lack of growing teaching licenses	Get growing licenses	More growing teachers

their potential value activities that can be performed by the organization and capabilities that the organization possesses related to the system.

The systems engineer interviews the stakeholders to identify the current state, future or desired state, and the gaps between the current and future state. The internal influencers help the team to identify how prepared the organization is to design and deploy the system and what gaps exist that may identify needs for the system.

The internal analysis for the Women's Healthcare Center is provided in Table 4.11. From the internal analysis, we find that 50% of the campaign for funding for the construction for the new center is needed. We also find that there are no procedures for the work processes nor document control to control the procedures when they are created, and training is needed to ensure consistency in execution of the processes.

The internal analyses for the seven teams are shown in Tables 4.12 through 4.18.

TABLE 4.11
Internal Analysis for Women's Healthcare Center

	Internal Analysis		
Internal Influencers	**Current State**	**Future or Desired State**	**Gap**
Value Activities	Long delays in scheduling appointments	Ability to schedule service online within one week	Delays in scheduling of appointments
	Long wait times for service	No wait times for service	Long wait times for service
	Three days to provide results	Same day results	Two-day gap to provide results
	No hand-off to Cancer Center	Seamless hand-off to Cancer Center, when needed	No connectivity to cancer center
	No ability to identity VIP patients	Automated alert to identify VIP patients	No technology for VIP alerts
	No ability to perform surgery at women's center	Ability to perform surgery on demand in women's center	No surgical center within women's center
Capabilities	Schedule service within two weeks	Ability to schedule service online within one week	Gap of one week for appointments
	Patient registration when they arrive to center	Ability to pre-schedule before arrival to center, with simple check in	No ability to pre-schedule, long registration wait times
	Three days to provide results	Same day results	Gap of two days for providing results
	No ability to connect to Cancer Center	Seamless hand-off to Cancer Center, when needed	No technology or patient navigator
	No ability to provide on-demand spiritual care	Must connect and schedule spiritual care	Identified resources on-site for spiritual care
	No seamless connection to Cancer Center	Seamless handoff to Cancer Center, when needed	Patient navigator to connect to Cancer Center
	No ability to identify VIP patients	Identify VIP patients automatically alerted	Technology to alert for VIP patients' real-time

(Continued)

TABLE 4.11   (Continued)

Internal Analysis

Internal Influencers	Current State	Future or Desired State	Gap
Balanced Scorecard:			
Financial	Met 50% of need through campaign	100% of need met	50% of campaign
	Lack of spiritual care resources	Assigned spiritual care resources	Financials for dedicated spiritual care resources
	Lack of additional radiologist resource	Additional radiologist	Financials for additional radiologist
	Resources split between hospital and women's center	Dedicated women's center technician resources	Financials for dedicated technician resources
Learning & Growth	Lack of training, procedures, and documentation control	Controlled documents and procedures; available and accessible training	Need for training, procedures, and document control
	Lack of process design and improvement	Lean Six Sigma program and continuous improvement culture	Just beginning Lean Six Sigma program
	No continuous process improvement program or culture	Lean Six Sigma program and continuous improvement culture	Just beginning Lean Six Sigma program
Internal (Process)	Poorly designed processes	Excellent design of processes	Technical skills and expertise
	No Lean Six Sigma program	Lean Six Sigma program and continuous improvement culture	Just beginning Lean Six Sigma program
	No process measurement	Lean Six Sigma program and continuous improvement culture	Just beginning Lean Six Sigma program
	No technology to measure processes	Business Process Management System (BPMS) for process measurement and improvement	Lack of BPMS
Customer	Loyal customer base	Loyal customer base	None
	Measures patient satisfaction	Measures patient satisfaction, and identifies improvements	No improvement program based on patient satisfaction results
	Low patient satisfaction results	High patient satisfaction results, in top 1%	Low patient satisfaction results

Team 1's internal analysis identified critical gaps related to the lack of a system to understand the types of food that people eat and how they obtain their groceries. They also identified the need to recruit farmers to provide food to people in need and the lack of the community involvement to aid farmers.

TABLE 4.12

Team 1 – Connecting Organizations – Internal Analysis for Food System

Internal Analysis

Internal Influencers	Current State	Future or Desired State	Gap
Value Activities	Organizers possibly risk time and money to get the food stands and farms going	Bring in more volunteers to help with the food stand and the farming aspect to reduce time and increase productivity	Unknown solution to keep farms self-sustaining
	Unknown knowledge of how people currently purchase their foods	Understanding of how people in these areas purchase foods	Lack of system to see how people tend to obtain their groceries (i.e. corner stores, faraway box stores, markets, etc.)
	Unknown knowledge of what types of foods people in these communities tend to eat and what they want	Understand what they look for in product (use to dry packaged meats compared to fresh bloody meat)	Lack of system to see what people desire as their food and eating habits
	Family of HH farm helps them sometimes.	Volunteers help farmers with setup, picking, selling, etc.	Lack of community involvement to aid farmers
Capabilities	Help organizations get started and try to become sustainable	Bring in additional organizers, businesses to contribute to the cause	Lack of an effective way to recruit farmers
	Able to help promote the farm food stands in the impoverished areas in Dayton	Marketing method to promote local goods to west side of Dayton communities	Lack of effective marketing method to advertise local farm fresh foods to communities
	Cannot deliver fresh meat without having any preservatives	Able to deliver fresh meat home to home	Poor logistics in current system for home delivery of meats

Team 2 identified several gaps related to lack of planning regarding policy change, a poorly designed policy tracking system, and lack of communication between stakeholders.

Team 3 identified a lack of children being interested in growing their own foods, lack of processes and standards for growing micro greens, and lack of manpower and funds to do so.

Team 4 discovered several gaps related to food storage when moving food from farms to the markets, including difficulty to find people jobs, not having enough funds for vehicles, slow investment times, difficulty in reaching people to sell the food from local farms, the market not being developed, not having enough storage places available and difficulty in reaching people for training related to food storage.

Team 5 identified gaps related to the food hub website including the food hub being constrained within the west side of Dayton, the gap of commercial space for selling and distributing food, limited connection with local restaurants and stores, a limited farmer network, limited investors for a food hub, and lack of financing for connecting farmers to food sellers and consumers.

Team 6 identified several gaps related to the GEM City Market including lack of food production resources, no proper nutrition awareness, no past community store, and no resources to provide healthy food options on the west side of Dayton.

Team 7 identified several gaps related to the Table 33 Restaurant including the need to hire trainers in different fields, no hands-on full-time workers, the need to reduce food waste and the food desert, and the need to grow microgreens to save money.

TABLE 4.13

Team 2 – Creating Food Policy – Internal Analysis for Food System

	Internal Analysis		
Internal Influencers	**Current State**	**Future or Desired State**	**Gap**
Value Activities	None	Fully Fleshed out policy change map that makes creating change in the area somewhat bearable	Lack of planning regarding policy change
	Community leader organization meetings to help fix the food justice issue	None	None
	Capabilities	Lack of organized planning	Developed planning process
Process for developing plan of action		Poorly designed policy tracking system	Good tracking system
Information and design expertise	Lack of communication	Clearly communicated goals	Communication process

TABLE 4.14

Team 3 – Kids Growing Food – Internal Analysis of Food System

	Internal Analysis		
Internal Influencers	**Current State**	**Future or Desired State**	**Gap**
Value activities	Lack of continued participation from children	Ongoing participation of children	Lack of child responsibility and interest
	No defined processes	Defined, sustainable process	Lack of process
	No process measurement	A defined process which includes measurement capability	Lack of standard
Capabilities	Working aquaponic system during non-winter months	Sustainable aquaponic system	Lacking the land, manpower, funds
	Teacher funded food opportunities (smoothie, stew)	Non-teacher funded food opportunities	Lack of funds

TABLE 4.15

Team 4 – Food Storage – Internal Analysis of Food System

	Internal Analysis		
Internal Influencers	**Current State**	**Future or Desired State**	**Gap**
Value Activities	Helping people find jobs	Increase the number of jobs	Difficulty in reaching jobs
	Distribution food to people with bicycles	Have bigger vehicles	Not enough funds for vehicles
	Making decent amount of money	Make more money	Slow investment

(Continued)

TABLE 4.15   (Continued)

Capabilities	Reach about 1000 people	Ability to reach more people that will buy food	It is difficult to reach people to sell food
	Empowering people to sell their food	Provide a corner market to sell food	Market not yet built or open
	Need for storage of food	Centralize storage facilities	Not enough tools or storage places available
	Teaching people how to grow food	Training center to teach people how to grow food	Hard to reach people for providing the training, no training center

TABLE 4.16
Team 5 – Food Hub Website – Internal Analysis of Food System

	Internal Analysis		
Internal Influencers	Current State	Future or Desired State	Gap
Value Activities	Restricted within West Dayton area	Full use of city of Dayton	Constrained within West Dayton
	Commercial Space is limited	Fill more commercial space	Gap of commercial space for selling, distributing food
Capabilities	Work with limited number of local restaurants and stores	Need to expand system to broader array of local restaurants and stores	Gap of limited connections
	Work with limited number of local farmers	Need to expand number of local farmers	Gap of limited farmer network
	Additional entrepreneurship support for investors	Need for consistent investors	Gap of consistent investors available
	Develop Financing by the city of Dayton	Need to develop financing by city of Dayton	Lack of financing for connecting farmers to sellers and consumers

TABLE 4.17
Team 6 – GEM City Market – Internal Analysis of Food System

	Internal Analysis		
Internal Influencers	Current State	Future or Desired State	Gap
Value Activities	Lack of food resources	Providing food resources through community farms and distributor	No proper ways to produce food
	Lack of quality and healthy food	Meeting good standards of quality and quantity of food	No proper nutrition awareness
Capabilities	New supermarket	To make the store a community or worker owned one	No past community owned store
	No proper healthy food	Provide healthy food and meet the required quantity and quality	No resources to healthy food options

TABLE 4.18

Team 7 – Table 33 Restaurant – Internal Analysis of Food System

	Internal Analysis		
Internal Influencers	**Current State**	**Future or Desired State**	**Gap**
Value Activities	Not enough proper trainers	Provide experts and trainers	Hire trainers in different fields
	No hands-on full-time workers	Apply contracts with workers	Put regulation for turn overs
	Not much food events and operations	Participate in more events locally	Reduce food waste and food desert
Capabilities	No ability to grow microgreens	Grow microgreens	Grow microgreens and save money

4.5.3 SWOT ANALYSIS

The Strengths, Weaknesses, Opportunities and Threats (SWOT) analysis is used to provide the context in which to select the mission strategies and goals that the organization will seek to achieve. A SWOT analysis is used to assist the team in selecting the strategic goals that ensure the best fit between the internal strengths and weaknesses of the organization and the external opportunities and threats that face the organization.

The SWOT analysis is used to understand the internal strengths and weaknesses of the organization to achieve the potential system, while the external opportunities and threats in the marketplace or external to the organization are identified.

The information collected during the SWOT analysis can include the organization's strengths and weaknesses, competitive information, government requirements, and economic environmental factors.

The strengths include how the organization is positioned to capitalize on the opportunities. The weaknesses of the organization describe how the organization could be exposed and not be able to take advantage of the opportunities. The opportunities are those potential advantages that can be achieved in the marketplace by developing the system. The threats are the perils in the marketplace that could derail the system from being realized. The SWOT analysis combines both the internal aspects of the organization with the external aspects of the marketplace or connections between the stakeholders. There can be some similar elements across the external analyses and the opportunities and threats and the internal analyses and the strengths and weaknesses. The systems engineer may have a preference to do just the external and internal analyses or the SWOT analysis. However, they are all presented, so that the reader can decide which is most appropriate for their system design methodology and approach.

The SWOT analysis for the Women's Healthcare Center is shown in Figure 4.6. The SWOT analysis demonstrates that the organization's strengths include their reputation, location, ability to renovate an existing center, the skill of the clinical and non-clinical staff, and the loyal donors that contributed to the renovation campaign. The weaknesses of the organization related to the proposed women's center are the inefficient processes, poor process measures, lack of stakeholder support for designing and improving the processes, the cost of construction, and lacking technology to support the processes. The external environment provides many opportunities, including market availability

SWOT ANALYSIS			
Internal		External	
Strengths	Weaknesses	Opportunities	Threats
Reputation	Inefficient Processes	Market availability	Competition
Location	Poor measures	Loyal patients	Technology process and VIP alerts
Facility	Lack of support	Need in marketplace	Regulations required
Clinical & Non-Clinical Staff	Cost of construction	Market niche	Accreditation will be needed
Donors	Lack of process technology		Other healthcare organizations are building women's centers
Physicians	No Lean Six Sigma program		
State-of-art technology	No Business Process Management System		
	Scheduling of appointments delays		
	No same day results		
	No patient navigator to connect to cancer center		
	Long wait times and delays in providing services		

FIGURE 4.6 SWOT analysis for the Women's Healthcare Center.

and receptivity for needing a women's center, loyal patients of the hospital and the current women's center, the need for women specific healthcare services provided in one location, and a market niche for this type of center. The potential market threats include other competitors beginning to look at providing this type of center, the lack of robust technology in the market to support the processes, the need for meeting regulations, and meeting accreditation requirements, while we are designing, building, and running the women's center.

The SWOT analyses for the seven teams are shown in Tables 4.19 through 4.25.

TABLE 4.19
Team 1 – Connecting Organizations – SWOT Analysis for Food System

SWOT ANALYSIS

Internal		External	
Strengths	Weaknesses	Opportunities	Threats
Promotion of cause	Farm(s) involvement	Need of local food system in west side Dayton	Low profit margin
Advertise events	Volunteer involvement	Marketing ability	Unpredictable weather
Needy customers	Low management of operations	Consistent customer income	Regulations on purchases with SNAP/WIC
Local farms in area	Limited food items available	Create relations between farmer and customers	Non-sustainable farms
Mobile tents/stands/food market	Limited communication methods with community members	Multiple methods of communication	Low customer desirability

TABLE 4.20

Team 2 – Creating Food Policy – SWOT Analysis for Food System

SWOT ANALYSIS

Internal		External	
Strengths	**Weaknesses**	**Opportunities**	**Threats**
Knowledge of policy needs	Ability to make policy change happen quickly	Focus on food justice system in community	Resistance to policy change
Passion within stakeholder organizations	Ability to connect across different stakeholders in the city	Many power players: former congressman, academics, non-profit expertise, governmental expertise	Slow processes in local, state, and federal government to make policy change
Momentum for policy change		Passion exists to address the challenges	Lack of state funding availability
		Ability to make policies economically viable	

TABLE 4.21

Team 3 – Kids Growing Food – SWOT Analysis of Food System

SWOT ANALYSIS

Internal		External	
Strengths	**Weaknesses**	**Opportunities**	**Threats**
Passion for growing food within schools	Lack of nutrition	Balanced nutrition need	Lack of available funding to support food growing in schools
Passion for sharing nutrition knowledge	Lack of knowledge about nutrition or health	Ample knowledge in community of nutrition and health	Ability to connect to families and restaurants to sell grown food
	Lack of receptivity of students to grow and eat food grown within schools		

TABLE 4.22

Team 4 – Food Storage – SWOT Analysis of Food System

SWOT ANALYSIS

Internal		External	
Strengths	**Weaknesses**	**Opportunities**	**Threats**
Community workers with a very good intention of serving the public	Lacks good storage facilities	No competition within the community, as the goal is to serve the community	Wealth and influence of other processed food market
Diversity in the community	Process food producers have a varying market	Underserved markets for organic food products	Changing weather conditions
Growing niche products	Not enough distribution facilities	Few organic food growers	Improper methods could spoil the entire crop

(*Continued*)

TABLE 4.22 (Continued)

Skilled and knowledgeable staff who provide training to new community members	No proper distribution centers, need to collect food from different regions	Increase in health consciousness among public	Changing minds of consumers
		Emerging need for organic food products	
		Volunteers and school programs spreading awareness of food justice	
		Agriculture sector is a priority of the government	

TABLE 4.23
Team 5 – Food Hub Website – SWOT Analysis of Food System

SWOT ANALYSIS

Internal		External	
Strengths	Weaknesses	Opportunities	Threats
Connectivity with some investors in the market	Lack of ability to connect to broader range of investors and funding sources	Passion for food justice system	Lack of investors for funding
Knowledge of non-profit, food justice system, and investors	Lack of ability to connect with wider number of local farmers	Need for food hub connecting growers, distributors, consumers	Lack of connectivity within multiple levels of government
	Lack of technology to connect growers, distributors, consumers		

TABLE 4.24
Team 6 – GEM City Market – SWOT Analysis of Food System

SWOT ANALYSIS

Internal		External	
Strengths	Weaknesses	Opportunities	Threats
Effective store design	The capacity of money and space	Employment to the community people	Crossing the planned budget
Providing healthy food resource	Location and density around the market	Awareness to the community about healthy food	Recruitment of experienced store manager
Eliminating the traces of redlining	Inability to convince nearby markets to sell the land		Customer retention
Community organizations	Low-profit margin		Not meeting the 100% requirements of the community
	Long development time		

TABLE 4.25

Team 7 – Table 33 Restaurant – SWOT Analysis of Food System

SWOT ANALYSIS

Internal		External	
Strengths	**Weaknesses**	**Opportunities**	**Threats**
Passion to serve the community, hire former prison population and people in recovery	Not enough income for workers	Desire for more restaurants and expansion of existing restaurant	Lack of connectivity to growers
Restaurant has a great deal of community support, great reputation, and great food	Workforce with potential issues, due to hiring former prison population and people in recovery	Need in downtown Dayton for additional restaurants leveraging organic food and local farmers	High cost of organic food items
Owners well-connected in the non-profit community and food justice system	Not enough profits for the restaurant	Support of local community	Seasonality of food
	Business model allows patrons to pay what they think is appropriate, which may not cover costs of meal and staff		

4.6   SUMMARY

We move into the mission analysis, in the next chapter, which helps us to divide the system development into manageable projects, identify the stakeholders, select the system's concepts, and perform the first risk assessment.

4.7   ACTIVE LEARNING EXERCISES

1) Select a system to design. Develop the mission strategy for the system.
2) Develop the mission goals for the system.
3) Develop the operational concept for the system.
4) Interview stakeholders to understand the system problem to be solved.
5) Develop interview questions to develop an external analysis.
6) Develop interview questions to develop an internal analysis.
7) Develop interview questions to develop a SWOT analysis.
8) For your identified system, interview the stakeholders and create an external analysis.
9) For your identified system, interview the stakeholders and create an internal analysis.
10) For your identified system, interview the stakeholders an create a SWOT analysis.

REFERENCES

[1] No Author. (2019). "GEMnasium Spring 2019 One Sheeter", University of Dayton.
[2] No Author. (2010). *Community Food Projects Indicators of Success*, National Research Center, Inc., Boulder, CO, www.n-r-c.coom
[3] Hamm, M., & Bellows, A. (2007). Community Food Security Programs: What Do They Look Like? *Community Food Security Programs*.

5 Concept of Operations Phase 1

Define Mission Analysis

In this chapter, we discuss the Concept of Operations Phase mission analysis activities and tools that are performed to develop the program and project structure for the system development activities.

5.1 PURPOSE

The Concept of Operations Phase is the first phase in the Systems Engineering Vee Life Cycle Model. The purpose of the phase is to perform an analysis of the mission and define its strategic goals. Within this phase, the systems engineering team also develops the project charters to break the system into manageable projects. A stakeholder analysis is performed to identify the stakeholders of the system and their impacts and concerns related to the system. Additionally, the system's concepts are generated, prioritized, and selected in this phase.

5.2 MISSION ANALYSIS

5.2.1 ACTIVITIES

The activities performed and the tools applied in the Concept of Operations Phase are shown in Table 5.1. In this chapter, we cover the activities and related tools within the Define Mission Concepts.

TABLE 5.1
Activities and Tools in the Concept of Operations Phase

Vee Phase	Activities	Tools	Principles
Phase 1: Concept of Operations	Define Strategic Mission Goals: 1. Define mission strategy 2. Define goals of mission 3. Develop concept of operations 4. Perform external, internal, and SWOT analyses Define Mission Concepts 1. Develop project charters 2. Perform stakeholder analysis 3. Collect Voice of Customer 4. Perform conceptual selection 5. Perform risk analysis	Strategic mission goals tools: • Mission Strategy; Mission goals; Concept of Operation, Concept; External analysis; Internal analysis; SWOT analysis Mission concepts tools: • Project charter with risk analysis • Stakeholder analysis • Critical to Satisfaction criteria • Pugh Concept Selection Technique • Risk Assessment, Risk Cube	• Complexity • Emergence (whole > sum of parts) • System of system • Hierarchy • Boundary

DOI: 10.1201/9781003081258-7

5.2.2 DEFINE MISSION ANALYSIS

5.2.2.1 Develop Project Charters

The systems planning infrastructure is developed prior to the Concept of Operations Phase. The systems program manager or systems engineer develops the Systems Engineering Management Plan (SEMP) and the Systems Engineering Plan (SEP). This defines the team, program, and project's structure. The SEMP is discussed in Chapter 12 in more detail. Once the team structure is developed, the teams can then develop their project charters which defines the problem they are solving, the project goals, and scope. The seven case study teams' project charters are shown in Tables 5.2 through 5.8.

The important fields in a project charter include the project overview, problem statement, goals, and scope. The project charter includes the following fields that help to define the project. The project charter is a "living" document that should be revised and enhanced as the project progresses, additional information is collected, and the problem is better understood (Duffy and Furterer, 2020).

5.2.2.1.1 Project Charter Fields

Project overview: The project overview provides a summary of the project, with high-level information describing the purpose of the project. It sets the stage for the problem to be explored.

Problem statement: The problem statement provides a description of the problem to be solved and the system to be developed. The problem statement should be as descriptive as possible, with the information available when it is written. The problem statement can be further enhanced once additional information is collected as the project progresses.

Goals: The goals provide the desired end state of the project and system to be achieved. The goal(s) should be specific, measurable, attainable, realistic, and time-based (SMART).

Project scope: The project scope provides the boundaries of the project, including the high-level description of the system to be developed. A description of what is in scope and out of scope of the project, related to high-level concepts, functionality, time frames, and resources should be discussed in the project scope field.

There may be many other fields that are part of the project charter, including the team members, the budgets, the timelines, and a project risk analysis, but we focus here on just the critical fields discussed above and shown in the tables next as examples.

You may notice from the sample project charters that the length and detail supplied in each field varies, based on the amount of knowledge that the teams had collected at this point in the projects. This information can be further refined as the teams elicit information from and interview the subject matter experts. This information should also be reviewed with the project sponsor and the subject matter experts to validate the project charter.

The project charter for the Women's Healthcare Center is shown in Figure 5.1.

5.2.2.2 Perform Stakeholder analysis

The stakeholder analysis is a tool that can be used to define the people and their roles that are impacted by the systems project and identify who the team members should be in the team. The stakeholder analysis includes a listing of the stakeholders, type of stakeholder, primary roles that have a stake in the project, potential impact/concerns, and receptivity (Duffy and Furterer, 2020).

Stakeholder (Group Name): The stakeholder's group name describes the function that the stakeholder groups hold related to the project. The stakeholder's name should never be a specific person but the function that they perform. People change jobs or functions but the function usually persists.

Type: If there are a large number of stakeholders, they can be categorized as primary or secondary types. Primary stakeholders are those who are directly impacted by the system being developed and may be operating the system or benefit directly from the system. The secondary stakeholders are those who are impacted by the system but not as directly as the primary stakeholders.

Primary Role: The role of each stakeholder group with respect to the project is described.

TABLE 5.2
Team 1 Connecting Organizations Project Charter

Project Overview	Problem Statement	Project Goals	Project Scope
A large focus of Food Justice in Dayton has been to bring in more quality and fair priced foods to the west side. Citywide has been partnered with HH farms in Brookville, Ohio for several years, trying to do so with their home-grown organic foods. The current method of operation is having a mobile farm stand which travels to various locations depending on the week. Along with bringing quality foods into these communities, there is a desire to better connect the communities with the farmers. Currently, there is not a structured method in place to reach out to these community members. Reaching out means to have a network of communication from the farmer to the customer and vice versa. The increase of order is our advanced goal to achieve by this new method. This includes surveying what types of food the customers are looking for, feedback on the farm stands and food, providing advertisement of when and where the next farmers market will be. The website would improve and broaden advertisement to the communities. Additionally, it is a more modern form of communication the younger generation may have more interest in viewing. The website will be designed to invoke all age groups ranging from the younger generation who have access to the internet to the older generation. This supports the purpose of designing a means of communication to connect the farmers to the customers via an online website and paper handouts.	C.J. from Citywide described a lack of communication between Citywide and the people in the communities regarding the people in the communities regarding when and where the mobile farmers' market will be. Currently, the way they get this information out to the public in these low-income areas is by paper handouts. Additionally, there is a gap in knowledge as to what types of food people in these communities are looking for and have little access to. In low-income households with children living below the poverty line in 1993, 27% had no telephone service. This made it difficult to contact them via one mean of communication. The desire is to design and create a website where information can be centrally passed between the farmers, Citywide, and the people in the communities. Additionally, a paper-based system will be included to reach out to those with no internet access.	1. Design a website and paper-based method of communication between the farmers and the west side of Dayton customers. 2. Indirectly create relations between the farmer, the customer, and these communities. 3. Encourage and gain interest by local farmers to participate in the west side of Dayton mobile markets. 4. Gain community interest and younger generation in volunteering for event setup and help farmers. 5. Ergonomically design website to accommodate ease of navigation of the website.	The scope of this project will be focused on creating two methods of communication between the farmers and the customers. One method will be using internet access to a public-based website, where farmers can upload information about food and events. It will include the advertisement from the farmer of a schedule of the upcoming events and locations and what types of foods are being offered to the public. Customers would be looking at the information and providing feedback for future events. The website would be admin controlled by Citywide and can evaluate the overall satisfaction of local community members. The second method is paper-based handouts of similar information. These handouts will be delivered to community homes and provided at public gathering places (i.e. churches). Resources required will be the use of a computer to create the online website and the paper handout forms for surveys and advertisement. The project does not include method of delivery of these goods or the operations of each individual farmer. Additionally, it does not include the process of money exchange between the farmers/customers as well as the funding that sustain these farms.

TABLE 5.3
Team 2 Creating Food Policy Project Charter

Project Overview	Problem Statement	Project Goals	Project Scope
Food security is one where every resident gets access to nutritious food in a sustainable way. The food movement in the United States, around 2010, had focused attention on hunger and access to food and created a drive for local food. Many advocated that food should be a basic human right. While it seems that there is an endless supply of food in the United States, many people find themselves lacking access to food, particularly healthy food, since buying fresh, organic produce is beyond the reach of many. Food security eludes millions of Americans who suffer chronic underconsumption of adequate nutrients. The Good Food Purchasing Policy (GFPP) developed by the Los Angeles Food Policy Council in 2012 equally embraces five values – local, sustainable, fair, humane, and healthy – which together offer a proper structure for the justifiable food system.	There are numerous reasons and root causes for food insecurity, so could be the solutions. One can address the issue when the problem, challenges, and gaps within that are clearly understood. Hence, our motive is to understand the current Food Justice system and then design a Future Food Justice system based on the requirements identified as a part of the study.	To develop current and a future Food Justice policy development system model.	There are many stakeholders involved, like Community partners and other organizations working for the same cause, people affected, and needs amendments in the current processes and policies to come up with a desired future Food Justice system. The scope of the project is to understand the current Food Justice system and develop a future Food Justice system based on the gaps identified in the current system.

TABLE 5.4
Team 3 Kids Growing Food Project Charter

Project Overview	Problem Statement	Project Goals	Project Scope
R.F. is a high school biology teacher who works closely with at-risk students. He struggles to keep these students engaged every day due, in part, to their lack of steady nutrition but has applied a multitude of different strategies, including after-school programs, competitions, and small jobs, to help stimulate them in any way that works. This provides the foundation to hone in on one of RF's strategies by creating a process for an ongoing, sustainable program for children to grow microgreens at home for profit.	At-risk high school children are performing poorly in class due, in part, to lack of access to healthy, affordable food. Children are unsuccessful in recognizing and taking responsibility for their independent ability to acquire food and/or money to buy food. The short-term desire is for children to earn money by growing microgreens at home, but the long-term desire is for children to take control of their own well-being.	The goals of this project are to get children interested in growing microgreens or their own food, give children the ability to make money, give children the chance to learn and try to apply responsibility, good choices, and independence in a low-risk environment, allow children to see the benefits of healthy eating, change overall lifestyle of at-risk students, both physically in their health and mentally in their willingness to participate and take responsibility.	The purpose of the project scope is to define how far the system solution will reach. A well-defined scope ensures adequate, manageable solutions. The boundaries include the children, teacher, and consumer organization as well as the materials, supplies, training, and processes to enable the children to grow microgreens for a profit. The scope includes the production, processing, distribution, and sale of microgreens.

TABLE 5.5
Team 4 Food Storage Project Charter

Project Overview	Problem Statement	Project Goals	Project Scope
Food storage allows food to be eaten for some time (typically weeks to months) after harvest rather than immediately. It is both a traditional domestic skill and, in the form of food logistics, an important industrial and commercial activity. Food preservation, storage, and transport, including timely delivery to consumers, are important to food security, especially for the majority of people throughout the world who rely on others to produce their food. Food is stored by almost every human society and by many animals.	Safe food storage using refrigeration requires adhering to temperature guidelines.	Food storage is a measure of the availability of food and individuals' accessibility to it, where accessibility includes healthy food. The goals are to design a food storage solution to hold food between the farms and distribution at safe temperatures.	The project scope is to design a food storage solution to hold food between the production and distribution at safe temperatures.

TABLE 5.6
Team 5 Food Hub Website Project Charter

Project Overview	Problem Statement	Project Goals	Project Scope
Mapping the current system that exists within the Downtown Dayton Partnership, between farmers, restaurants, and local businesses and expanding that system to involve greater Dayton area (the food deserts).	The current greater Dayton area lacks proper suppliers for fresh food and distribution methods for produce. This lack of proper distributors means there is an absence of sustainable produce and healthy food options in lower-income neighborhoods. This lack of access to fresh food increases the likelihood of the cycle of poverty through poor nutrition, hygiene, and health.	Connect current existing farmer-restaurant economy in the downtown area with the current food entrepreneurs in the food deserts of Greater Dayton.	Geographically, the Downtown Dayton Partnership is only involved in a limited area of downtown Dayton. Limited commercial space, funds for support, farming connections, and systemic racist. Excludes the details of food growth, the interest of communities outside of Dayton (suburbs), and the socio-economic status of those living in the affected area. Includes: those within the detailed area, the distribution method, farmer–restaurant partnerships, and the supply–demand of the area.

TABLE 5.7
Team 6 GEM City Market Project Charter

Project Overview	Problem Statement	Project Goals	Project Scope
We are operationalizing the Gem City Supermarket to help eliminate food deserts and provide fresh and healthy varieties of food to the community. The location of the supermarket will be vital to the communities which are facing food shortages.	The community of West Dayton is facing a food desert as a consequence of redlining policies. There aren't grocery stores servicing the area, so individuals in the community must travel far in order to obtain their groceries. There needs to be an effective solution to help elevate the problem of healthy food shortage.	The goal of this system engineering project is to develop a food sales system to operationalize the Gem City Supermarket by providing the community with healthy food options, access to a variety of food resources, and make it a worker-community owned one.	Includes food production, community, food distribution, resources, materials, equipment, sales, and required system changes. The scope includes the new Gem City Market, community farming, food storage, and hybrid pricing.

TABLE 5.8
Team 7 Table 33 Project Charter

Project Overview	Problem Statement	Project Goals	Project Scope
Food desert is increasing in the area. There is lack of grocery stores that provide healthy food, which is affecting people's nutrition. Some people do not have access to food while others cannot afford it. So, Table 33 decided to help the community out by offering affordable healthy food by purchasing local ingredients from local farmers which will impact the prices in their menu.	Table 33 is a local restaurant that helps reduce food desert in the area. However, cost of food is not affordable for the majority of people that are living in the area. Therefore, a concept of reducing price of meals that are offered by Table 33 was generated. This concept is based on creating a relationship between Table 33 and local farmers, which will help get ingredients for a lower price. This will eventually reduce cost of meals and increase profits of Table 33. Also, this will benefit local farmers and allow more people to be able to dine in the restaurant.	Increase Table 33's profit and benefit local farmers.	Experience of faculty members, lack of local farmers, and Table 33's customers. It excludes food storing, food production, and sales.

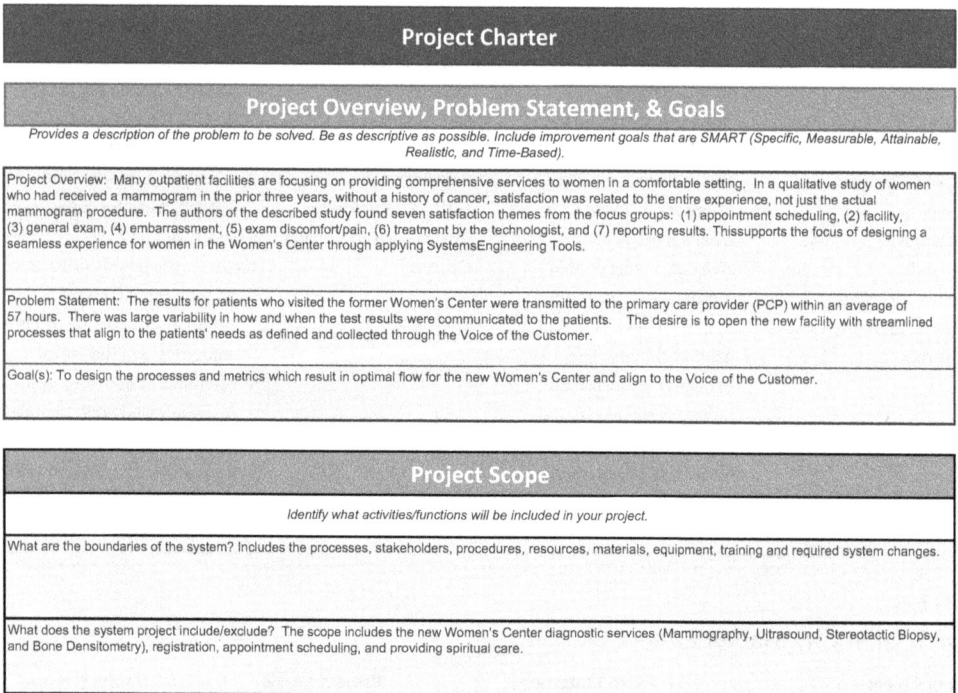

Project Charter

Project Overview, Problem Statement, & Goals

Provides a description of the problem to be solved. Be as descriptive as possible. Include improvement goals that are SMART (Specific, Measurable, Attainable, Realistic, and Time-Based).

Project Overview: Many outpatient facilities are focusing on providing comprehensive services to women in a comfortable setting. In a qualitative study of women who had received a mammogram in the prior three years, without a history of cancer, satisfaction was related to the entire experience, not just the actual mammogram procedure. The authors of the described study found seven satisfaction themes from the focus groups: (1) appointment scheduling, (2) facility, (3) general exam, (4) embarrassment, (5) exam discomfort/pain, (6) treatment by the technologist, and (7) reporting results. This supports the focus of designing a seamless experience for women in the Women's Center through applying Systems Engineering Tools.

Problem Statement: The results for patients who visited the former Women's Center were transmitted to the primary care provider (PCP) within an average of 57 hours. There was large variability in how and when the test results were communicated to the patients. The desire is to open the new facility with streamlined processes that align to the patients' needs as defined and collected through the Voice of the Customer.

Goal(s): To design the processes and metrics which result in optimal flow for the new Women's Center and align to the Voice of the Customer.

Project Scope

Identify what activities/functions will be included in your project.

What are the boundaries of the system? Includes the processes, stakeholders, procedures, resources, materials, equipment, training and required system changes.

What does the system project include/exclude? The scope includes the new Women's Center diagnostic services (Mammography, Ultrasound, Stereotactic Biopsy, and Bone Densitometry), registration, appointment scheduling, and providing spiritual care.

FIGURE 5.1 Women's healthcare center project charter.

Potential Impacts/Concerns: The project sponsor, team lead, or systems engineer should interview stakeholder representatives to understand their primary concerns regarding the project and/or how they are impacted by the project and system.

Initial and Future Receptivity: The project sponsor and team lead should assess how receptive the stakeholders are to the project as it is initiated. They should also assess how receptive the stakeholders need to be when the system is deployed for use. The receptivity is typically assessed based on a rating scale from strongly support, moderate support, neutral, moderate against, and strongly against. How does the project sponsor and the team leader assess the stakeholders' receptivity to the project? Assessing receptivity can be done in several ways: 1) talk to the stakeholders to assess their receptivity, 2) monitor body language and conversations of the stakeholders within and outside of meetings, and 3) talk with other people who know the stakeholders.

Examples of supportive and engaged stakeholders, team members, and leaders:

- Stakeholders attend meetings, are prompt, and engaged
- Have a positive outlook and willingness to participate
- Exude genuine interest in the project and system
- Share ideas in meetings and brainstorming sessions
- Performs tasks outside of meetings without being prompted
- Offer management perspective to help the team with improvement ideas

Examples of non-supportive and non-engaged stakeholders, team members, and leaders:

- Complain within and outside of meetings
- Tone in meetings is confrontational
- Most ideas are discouraged
- May not provide ideas in meetings and brainstorming sessions
- Out of box thinking is frowned upon and not encouraged

Some examples of non-supportive body language are arms crossed, roll eyes, always on phone, holding head in hands, putting hands out in front of body as if to stop someone from advancing.

The Women's Healthcare Center stakeholder analysis is shown in Table 5.9. The stakeholders that have a stake in the system being developed are identified. Their role with respect to the system is described. At least one representative of each stakeholder group should be interviewed to understand their potential impacts or concerns related to the system. The initial receptivity to the system being developed should be assessed during these discussions with the stakeholders. The systems engineer should assess the level of future receptivity that is needed by the time that the system is going into production for each stakeholder. A plan should be developed where there are gaps between the initial and future receptivity, to move the stakeholders to a higher level of engagement, and receptivity on the systems team.

The majority of the stakeholders are concerned with quality of care, customer service, patient, and physician satisfaction, meeting the volume of requested services. At the beginning of the project,

TABLE 5.9
Women's Healthcare Center Stakeholder Analysis

Stakeholder	Primary Role	Potential Impacts/ Concerns	Initial Receptivity	Future Receptivity
Patient (women)	Receive services in the center	• Customer Service • Quality • Efficiency	Neutral	Neutral
Referring physicians	Refer patients to the center and communicate results to the patient	• Quality of care	Moderate Support	Strong Support
Payers	Pay for services, such as managed care, Medicare employers, self-pay	• Quality • Cost effective care	Neutral	Neutral

(Continued)

TABLE 5.9 (Continued)

Stakeholder	Primary Role	Potential Impacts/ Concerns	Initial Receptivity	Future Receptivity
Donors	People who donate money to the center	• Quality of care • Meet patient requirements	Strong Support	Strong Support
Imaging Technologists	Work in the center and perform procedures	• Patient satisfactions • Improved/well-designed work environment • Associate satisfaction	Moderate Support	Strong Support
Radiologists	Read and provide results of imaging procedures	• Reduced volume and revenue • Physician Satisfaction	Moderate Support	Strong Support
Administration	Manage the hospital and center	• Volume • Revenue • Patient Satisfaction • Physician Satisfaction • Productivity	Strong Support	Strong Support
Patient scheduling	Registration who registers patients and perform insurance authorizations, centralized scheduling who make the patient appointments	• Volume • Productivity • Timeliness of processes	Moderate Support	Strong Support
Information Technology	Provide phone and computer systems	• Meet requirements • On time • On budget	Strong Support	Strong Support
Physicians	Provide women's services	• Reduced volume and revenue • Physician Satisfaction	Moderate Support	Strong Support
Marketing and Development	Perform business development, marketing and fund raising	• New business • Funds available • Able to reach customers	Strong Support	Strong Support

most of the stakeholders' initial receptivity is from neutral to strongly supporting this project. We need most stakeholders to be strongly supportive of our system when we implement our new processes and the system.

The Food System case study stakeholder analyses are shown in Figures 5.2 through 5.8.

5.2.2.3   Collect Voice of Customer

The Voice of the Customer (VOC) is a term that is used to "talk to the customer" to hear their needs and requirements or their "voice." The VOC represents the customers' requirements for the system, product, service, or process. For the systems engineering project in the Concept of Operations phase, the Critical to Satisfaction (CTS) criteria represents the beginning of the VOC. The CTS criteria are the elements of a system, product, service, or process that significantly affect the output. The CTS criteria are gathered by the systems engineering team through interactions with the customers or ultimate users of the system. There are many ways to collect the VOC, such as interviews, focus groups, surveys, customer complaint and warranty data, market research, competitive information, and customer buying patterns.

Stakeholder Analysis					
Identify the internal and external customer roles and concerns to ensure their expectations are addressed. Define roles, impacts, concerns, and receptivity.					
Stakeholder	Type	Primary Role	Potential Impacts/ Concerns	Initial Receptivity	Future Receptivity
Team	Primary	Owner	Amount of users; propaganda of website; connection with other partner or organization.	Strongly Support	Strongly Support
Farmer	Primary	Supplier/user	The accuracy of information; weather; price of production; market behind the website.	Neutral	Moderate Support
Citywide	Primary	Partner	Ability of analyzing; connection with community and website;the accurency of information; propaganda of website; the deal of orders	Moderate Support	Strongly Support
Community partner	Primary	Supplier/user	The accurracy of information; weather; price of production; market behind the website.	Neutral	Moderate Support
Community member	Primary	Customer/user	The quality of the food or production; the reflection of their needs or suggestion; the reliability of website	Neutral	Moderate Support

FIGURE 5.2 Team 1 connecting organizations stakeholder analysis.

Stakeholder Analysis					
Identify the internal and external customer roles and concerns to ensure their expectations are addressed. Define roles, impacts, concerns, and receptivity.					
Stakeholder	Type	Primary Role	Potential Impacts/ Concerns	Initial Receptivity	Future Receptivity
Land Developer	Secondary	Land levelling	Provide quality land for farming	Moderate Support	Moderate Support
Labor/Wrokers	Secondary	Help in farming	Improve the quality of crops by fertilizers/pesticides	Moderate Support	Moderate Support
Farmers	Primary	Grow Crops	Quality of crops produced	Strongly Support	Strongly Support
Food Processing Industry	Primary	Process the grown crops into consumable food	Quality of food processed. Meet the demand	Strongly Support	Strongly Support
Distributor	Secondary	Distributes the processed food to Retailers	Timliness, Maintaining quality of packaging	Strongly Support	Strongly Support
Retailer	Primary	Making food available at stores	Timliness, Quality of service	Strongly Support	Strongly Support
Customers	Primary	Consume/Buy food	Customer Service, Efficiency, Quality	Neutral	Neutral
Administration	Secondary	Manage the store	Improve customer and employee satisfaction, Revenue, Timliness	Strongly Support	Strongly Support

FIGURE 5.3 Team 2 creating food policy stakeholder analysis.

Stakeholder Analysis					
Identify the internal and external customer roles and concerns to ensure their expectations are addressed. Define roles, impacts, concerns, and receptivity.					
Stakeholder	Type	Primary Role	Potential Impacts/ Concerns	Initial Receptivity	Future Receptivity
Children	Primary	Grow and harvest microgreens	Fair payment for work and efforts, adequate training, quality of microgreens	Neutral	Strongly Support
Children's Guardian	Secondary	Support and enable child throughout program	Adequate training for children, no mess or loss of money	Neutral	Moderate Support
Teacher Representative	Primary	Buy microgreens from students and sell to participating businesses, as well as oversee program	Quality of children's microgreens, satisfied consumers, no loss of money	Strongly Support	Strongly Support
Individual Consumer	Secondary	Consume microgreens	Quality of microgreens	Moderate Support	Strongly Support
Participating Businesses	Primary	Purchase microgreens from teacher representative	Cost and quality of microgreens, ease of use of program's service, satisfied customers	Moderate Support	Strongly Support

FIGURE 5.4 Team 3 kids growing food stakeholder analysis.

Stakeholder Analysis					
Identify the internal and external customer roles and concerns to ensure their expectations are addressed. Define roles, impacts, concerns, and receptivity.					
Stakeholder	Type	Primary Role	Potential Impacts/ Concerns	Initial Receptivity	Future Receptivity
Downtown Restaurants	Primary	Economic stability	Growth of economy introduces more risk, especially with businesses not in economically stable areas	Moderate Support	Moderate Support
Farmers	Primary	Growing fresh produce	An increase in demand as a result of growing amount of business	Neutral	Strongly Support
Citizens of Dayton Area	Primary	Purchasing available food	Increase in economic health, property prices rise	Moderate Support	Strongly Support
Citizens of Dayton Suburbs	Secondary	Support the food system	Draw money back into the city and out of suburban economies	Moderate Support	Strongly Support
Downtown Dayton Partnership	Secondary	Bring knowledge from downtown food system to suburbs of Dayton	Responsibility to current interestes may be impacted with increase in business downtown. More interests = more responsibility and possibly less oversight	Neutral	Moderate Support
University of Dayton	Secondary	Invest/research/support the Dayton area food system	Potential investment/liability in the development of sustainable food systems in the Dayton area.	Moderate Support	Strongly Support

FIGURE 5.5   Team 4 food storage stakeholder analysis.

Stakeholder Analysis					
Identify the internal and external customer roles and concerns to ensure their expectations are addressed. Define roles, impacts, concerns, and receptivity.					
Stakeholder	Type	Primary Role	Potential Impacts/ Concerns	Initial Receptivity	Future Receptivity
Downtown Restaurants	Primary	Economic Stability	Growth of economy introduces more risk, especially with businesses not in economically stable areas	Moderate Support	Moderate Support
Farmers	Primary	Growing fresh produce	An increase in demand as a result of growing amount of business	Neutral	Strongly Support
Citizens of Dayton Area	Primary	Purchasing available food	Increase in economic health, property, prices rise	Moderate Support	Strongly Support
Citizens of Dayton Suburbs	Secondary	Support the food system	Draw money back into the city and out of suburban economies	Moderate Support	Strongly Support
Downtown Dayton Partnership	Secondary	Bring knowledge from downtown food system to suburbs of Dayton	Responsibility to current interests maybe impacted with increase in business downtown. More interests = more responsibilty and possible less oversight	Neutral	Moderate Support
University of Dayton	Secondary	Invest/research/support the Dayton area food system	Potential investment/liability in the development of sustainable food systems in the Dayton area	Moderate Support	Strongly Support

FIGURE 5.6   Team 5 food hub website stakeholder analysis.

Stakeholder Analysis					
Identify the internal and external customer roles and concerns to ensure their expectations are addressed. Define roles, impacts, concerns, and receptivity.					
Stakeholder	Type	Primary Role	Potential Impacts/ Concerns	Initial Receptivity	Future Receptivity
Community farmers	Primary	Provide a portion of products	Product quality, meeting needed demand	Strongly Support	Moderate Support
Supplier	Primary	Provide products to market	Product quality, meeting needed demand	Strongly Support	Strongly Support
Workers	Primary	Work and operate the market	Quality of service, timeliness, following work procedures	Strongly Support	Strongly Support
Customers (community residents)	Primary	Buy the products	Customer service, customer satisfaaction	Neutral	Neutral
Administration	Secondary	Manage the market	Quality control, customer & worker satisfaction, productivity, revenue	Strongly Support	Strongly Support
Distributor	Primary	Distributed needed products to the markets	Late deliveries, product quality, meeting demands	Strongly Support	Strongly Support
Marketers	Secondary	Make the products appealing	Ability to reach customers	Strongly Support	Moderate Support

FIGURE 5.7   Team 6 GEM city market stakeholder analysis.

Stakeholder Analysis					
Identify the internal and external customer roles and concerns to ensure their expectations are addressed. Define roles, impacts, concerns, and receptivity.					
Stakeholder	Type	Primary Role	Potential Impacts/ Concerns	Initial Receptivity	Future Receptivity
Local Farmers	Primary	Supply the restaurant with Ingredients	Supply the restaurant with ingredients grown organic	Strongly Support	Strongly Support
Customers	Secondary	Recieve food in the restaurant	Buy food from Table 33	Neutral	Neutral
Waitresses	Primary	Deliver food	Customer service, quality and efficiency	Moderate Support	Moderate Support
Chiefs	Primary	Make food	Customer satisfaction, provide good quality food	Strongly Support	Strongly Support
Managers	Primary	Manage processes	Manage procedures in the restaurant and deal with local farmers	Strongly Support	Strongly Support

FIGURE 5.8 Team 7 table 33 stakeholder analysis.

Interviews: Interviewing the customers is one of the best ways to collect the VOC. The goals of the interviews should be defined, and then the interview questions should be designed to achieve these goals. The interviews are typically one-to-one or with a customer with the systems engineering team. Personal interviews are an effective way to gain the VOC; however, it can be expensive and training of interviewers is important to avoid interviewer bias. However, additional questioning can occur to eliminate misunderstanding. The objectives of the interview should be clearly defined before the interviews are held.

Focus Groups: Focus groups are facilitated sessions where several customers are included. Several focus groups with multiple customer groups are typically performed. Trained focus group facilitators are required to facilitate the session to ensure that appropriate questions are asked and that everyone gets an opportunity to provide their input. If not well facilitated, strong voices can steer and dominate the discussion. Focus groups are usually held until no new themes are elicited. Focus groups are an effective way to collect VOC data. A small representative group, typically 7 to 10 people, are brought together and asked to respond to predetermined questions. The focus group objective should be developed and the questions should support the objective. The participants should be selected by a common set of characteristics. The goal of a focus group is to gather a common set of themes related to the focus group objective. There is no set sample size for focus groups. Multiple focus groups are typically run until no additional themes are derived.

Advantages of focus groups are:

- They tend to have good face validity, meaning that the responses are in the words of the focus group participants;
- Typically, more comments are derived than in an interview with one person at a time;
- The facilitator can probe for additional information and clarification;
- Information is obtained relatively inexpensively.

Some of the disadvantages of focus groups are:

- The facilitator skills dictate the quality of the responses;
- They can be difficult to schedule;
- It can be difficult to analyze the dialogue due to participant interactions.

Surveys: Customer surveys are a typical way to collect VOC data. The response rate on surveys tends to be low, 20% is a "good" response rate. It can also be extremely difficult to develop a survey that asks the questions that are desired. Customer survey collection can be quite expensive. The steps to create a customer survey are (Malone, 2005):

1. Conceptualization: Identify the survey objective and develop the concept of the survey and what questions you are trying to answer from the survey.
2. Construction: Develop the survey questions. A focus group can be used to develop and/or test the questions to see if they are easily understood.
3. Pilot (try out): Pilot the questions by having a focus group of representative people from your population. You would have them review the questions, identify any unclear or confusing questions, and tell you what they think that the questions are asking. You wouldn't use the data collected during the pilot in the actual results of the surveys.
4. Item Analysis: Item analysis provides a statistical analysis to determine which questions answer the same objectives, as a way to reduce the number of questions. It is important to minimize the number of questions and the total time required to take the survey. Typically, the survey time should be 10 minutes or less.
5. Revision: Revise the survey questions, roll out the customer survey, or pilot again if necessary.

Customer complaint and warranty data: Many organizations collect customer complaint data that can be successfully summarized to provide where problems exist with current products and services. Warranty data also provides insight into problems and defects with current products and services.

Market research is a term that encompasses many different data collection techniques, including searching the internet, viewing social media, reviewing company's websites and annual reports, collecting information on your competitors, and investigating customer buying patterns.

Affinity diagrams organize interview, survey, and focus group data after collection. The affinity diagram organizes the data into themes or categories. The themes can first be generated, and then the data can be organized into the themes or the detailed data can be grouped into the themes. An example of a simple affinity diagram developed themes from initial interviews with the stakeholders, is shown in Figure 5.9.

The themes provided by the Food System stakeholders and the questions that the students asked the stakeholders are included in Figures 5.10 through 5.13. Following is a list of the themes provided:

1) System production and distribution: Enabling the farm to a table system, connecting farmers to the market, finding ways to adjust the price of fresh food.
2) Grow food: Getting the community involved in local gardens, helping people to grow their own food.
3) Community shops to sell food: Finding a way to keep grocery stores open and providing more stores for people to access healthy food options.
4) School meal programs: Providing healthier school meal programs.
5) People over profit: Needing to find a way to not just focus on profit, when distributing and selling food.
6) Relationship to food: Creating knowledge for people to understand their relationship between themselves and food and teaching children where food comes from.
7) Food options at all stores: More access to local, fresh, and healthy food.
8) Access to food: Improving food access to everyone, especially in food insecure areas that were traditionally redlined.

Food System Affinity Diagram

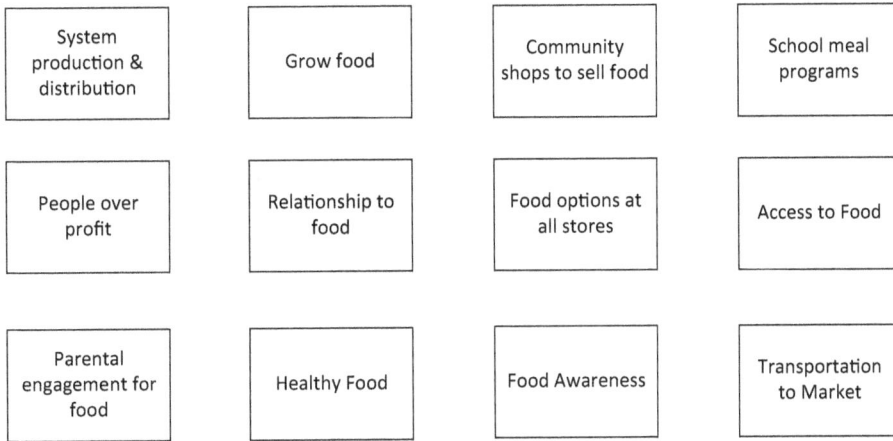

System production & distribution	Grow food	Community shops to sell food	School meal programs
People over profit	Relationship to food	Food options at all stores	Access to Food
Parental engagement for food	Healthy Food	Food Awareness	Transportation to Market

FIGURE 5.9   Food system affinity diagram.

9) Parental engagement to food: Enhancing parental engagement with children and nutrition.
10) Healthy food: Helping people understand healthy food options and the importance of nutrition to life and livelihood.
11) Food awareness: Ensuring that people are aware of Food Justice struggles as well as getting support from the community to tackle the food insecurity issues.
12) Transportation to market: Transporting food safely from farms to markets and helping people get to a grocery store.

Some references refer to identifying the Critical to Quality (CTQ), but the CTS broadens the elements of the CTS by including CTQ, Critical to Delivery (CTD), and Critical to Cost (CTC). There may also be critical elements in the process to measure related not only to quality, delivery, and cost but to time, as well. Not everything should be a CTS. The CTS should be specific to the scope of the project and the system. If there are more than a few CTS identified for the project, the scope is probably too large for a reasonable system project. The CTS should describe the customer need or requirement, not how to solve the problem.

Definitions:
CTS – Critical to Satisfaction: A characteristic of a product or service which fulfills a critical customer requirement or a customer process requirement. CTSs are the basic elements to be used in driving process measurement, improvement, and control
The CTSs can include the following categories:

CTQ – Critical to Quality: These characteristics are product, service, and/or transactional characteristics that significantly influence one or more CTSs in terms of quality.
CTD – Critical to Delivery: These characteristics are product, service, and/or transactional characteristics that significantly influence one or more CTSs in terms of delivery (or cycle).
CTC – Critical to Cost: These characteristics are product, service and/or transactional characteristics that significantly influence one or more CTSs in terms of cost.
CTP – Critical to Process: This is a process parameter that has significant influence or impact on a CTQ, CTD, and/or CTC.

	(1) System production and distribution	(2) Grow	(3) Community shops to sell food
Themes	Farm to table system	They want to make more food	A way to keep stores open
	Locally sourced sustainable syste	Sustainable and community makes money off of it	More stores for people at access
	Connect farmers to market	Get community involved in local gardens	
	Where the food comes from and distributed	People grow their own crops	
	System that creates food in town	More stores to satisfy needs and demands	
	System with more transportation support		
	Process of food getting to dayton more efficient		
	Find a way to adjust the price of fresh food		
Questions	**Questions**	**Questions**	**Questions**
	Where do major stores get their food and produce?	What is the current food availability situation?	What is the number of stores and distribution around the area?
	Are we gonna use our own distribution network?	How much should be grown to satisfy the daily needs?	Are there any cheap/affordable locations available for a store now?
	How can we make these systems economical?	How can we consider food demands?	How can we create a food chain?
	How is food transported now?	How many community gradens exist in Dayton	
	What business/farmers are currently involved?		

FIGURE 5.10 Themes 1 through 3 with questions.

Steps for developing CTSs:

1) Gather VOC data relevant to the product, service, or output.
2) Identify relevant statements in transcripts (verbatims) of customer comments and copy them onto self-stick notes. Focus on statements that relate to why a customer would or would not buy your product/service. This information can be developed from information that was collected during the Process Architecture Mapping process scenarios generation and requirements elicitation activities.

	(4) School Meal Programs	(5) People over Profit	(6) Relationship to food
Themes	Healthier school meal programs	People over profit whenever possible	Create knowledge for people to understand relationship between themselves and food
		Table 33 needs aid to help provide meals and opportunity at a low cost	Teach children where food comes from
		understand the problem	
		We need to find a way to get other	
	Questions	**Questions**	**Questions**
Questions	How can we get teachers to stress the importance of a healthy diet at a young age?	How does a company that is always losing money stay in business?	What is the best way to explain to children where food comes from?
		Where is the money in these organizations coming from?	
		How to motivate people and business to get involved?	
		How does the peoples desires/love for what they do affect profits?	

FIGURE 5.11   Themes 4 through 6 with questions.

3) Use Affinity Diagrams or Tree Diagrams to sort ideas and find themes.
4) Start with the themes or representative comments and probe for why the customer feels that way. Do follow-up with customers to clarify their statements. Be specific to identify why.
5) Conduct further customer contact as needed to establish quantifiable targets and tolerances (specification limits) associated with the need.
6) When you've completed the work step back and examine all the requirements as a set. Fill in gaps as needed.

The CTS criteria for the Women's Healthcare Center is shown in Table 5.10.

The CTSs were extracted from the interviews with the customers and are shown in Figures 5.14 through 5.19 (Table 5.11).

Themes	(7) Food options at all stores		(9) Parental Engagement for food
	More access to local, fresh, healthy food options across Dayton	People in Redline areas can find food	More parent engagement for nutrition especially with kids
		Find ways to clean and store food	
		Left over food at UD is distributed to people who need it	
		Fundraise food for children and give to them	
		Improve food access for everyone	
		Better and healthier food options for all	
Questions	Questions	Questions	Questions
	How much does it cost to distribute more food to more stores?	Are the community gardens open to everyone?	
	Why can't community stores get groceries at the same price as Kroger?		

FIGURE 5.12   Themes 7 through 9 with questions.

	(10) Healthy food	(11) Food Awareness	(12) Transportation to Market
Themes	Understand the healthy options of food	Reduce wasted food (we can start that now)	Provide transportation to help people get to the grocery store
	Nutrition information and lessons for kids	Get information out about the food justice struggles in dayton	Arrange food trucks or vans on the weekends to deliver food
		Get more support by spreading the word	
Questions	**Questions**	**Questions**	**Questions**
	How to make sure healthy food is available?	How can we spread awareness?	How to get more modes of transportation to stores?
	How to know if food is healthy or not?	Education in school about how to plant and grow crops?	How to save the fresh food while it is transported?
		How can we create a survey on where to start? What is most needed?	How to get produce from point A to point B?
			Can we find a way to transport more food in a better way?

FIGURE 5.13   Themes 10 through 12 with questions.

TABLE 5.10
Critical to Satisfaction Criteria for the Women's Healthcare Center

Critical to Satisfaction (CTS)

CTS' are basic elements that can be used in driving process measurement, improvement, and control. They are elements of a process that significantly affect the output of the process. What are the characteristics of the process that are critical as perceived by the customer?

Title	Description
Patient satisfaction	Patients satisfied with service and experience in Women's Center
Customer service	Ability to provide appointments for service in one week or less and high customer service
Timeliness of service	Ability to provide services without delays
Quality of service and results	Provide accurate service and results

Critical to Satisfaction (CTS)	
CTS' are basic elements that can be used in driving process measurement, improvement, and control. They are elements of a process that significantly affect the output of the process. What are the characteristics of the process are critical as perceived by the customer?	
Title	**Description**
Connect farmers and local people	Build a website that easily connects local people with different farmers offering different foods.
Engaging Youth in Food Justice System	Getting more youth involved in the food justice system as a whole.
Access to Food	People will have easy access to healthy food at affordable prices with food of people's choice.
Get More Farmers	Get more and more farmers recruited in the website who offer different variety of foods.

FIGURE 5.14 Team 1 connecting organizations voice of customer.

Critical to Satisfaction (CTS)	
CTSs are basic elements that can be used in driving process measurement, improvement, and control. They are elements of a process that significantly affect the output of the process. What are the characteristics of the process are critical as perceived by the customer?	
Title	Description
Awareness	Provide awareness on the importance of consuming healthy food.
Opportunites	Strives to provide equitable opportunities irrespective of race, class, gender, citizenship
Transportation	Increase the number of stores and buses in order to reduce the time for commutation.
Healthy Food	To provide healthy and nutritious food.

FIGURE 5.15 Team 2 creating food policy voice of customer.

Critical to Satisfaction (CTS)	
CTSs are basic elements that can be used in driving process measurement, improvement, and control. They are elements of a process that significantly affect the output of the process. What are the characteristics of the process are critical as perceived by the	
Title	Description
Quality of product	Microgreens are fully-grown, clean, and visibly appealing
Customer satisfaction	Customer satisfied with product and experience working with program
Timeliness of service to customer	Ability to deliver product to customer same day
Timeliness of processing	Ability to harvest product at correct time and return to distribution center

FIGURE 5.16 Team 3 kids growing food voice of customer.

5.2.2.4 Perform Conceptual Selection

The Pugh Concept Selection Technique is a technique for evaluating and selecting concepts. If you have several different process elements or product concepts to choose from, you could use this technique. You would first brainstorm potential solutions or concepts and generate criteria upon which to compare the concepts. Then you would select one of the concepts as the "candidate" concept. It does not matter which concept you select as the candidate concept, however, for a redesign of a process or product you could select the current process or product. You then compare each of the other (new) concepts to the candidate for each comparison criteria. If the new concept is better than the candidate for those criteria, you would place a plus sign (+) in the cell where the new concept intersects the criteria. If the new concept is worse than the candidate concept for the criteria, a minus sign (−) is placed in the cell. If the new concept is the same as the candidate on those criteria, a zero (0) or S for same is placed in the cell.

Critical to Satisfaction (CTS)	
CTSs are basic elements that can be used in driving process measurement, improvement, and control. They are elements of a process that significantly affect the output of the process. What are the characteristics of the process are critical as perceived by	
Title	**Description**
Temporary storage	As the crop remains in very good condition until its been shifted to the main storage location
Temporary storage packing	Proper bags or wraps should beusedin fields so that food does not contaminate
Food pickup	Food should be picked up within timeline and moved to main storage locations with portable storage conditiosn fulfilling transporation vehhicle
Main storage	Storage location requirements must be met so the food that is stored does not ripen and get spoiled. Instead it lasts long under proper temperature storage of the respective vegetable or food product
Staff	Sensible staff following hygiene of washing hands and surfaces often to keep hygience.
Farm to business / institution model	Producers selling to supermarket chains, independent grocery stores, schools, hospitals, institutions, restaurants and other busineses, directly from farms should maintain healthy and hygenic food.

FIGURE 5.17   Team 4 food storage voice of customer.

Critical to Satisfaction (CTS)	
CTSs are basic elements that can be used in driving process measurement, improvement, and control. They are elements of a process that significantly affect the output of the process. What are the characteristics of the process are critical as perceived by the customer?	
Title	**Description**
Farmer-Restaurant Economy	The relationship that exists now between restaurants downtown and farmers nearby is critical to the modeling of the economy.
Safety of Downtown Occupants	The safety of those downtown is of utmost important, especially in preserving the integrity and culture of the downtown community.
Neighborhood Investment	Investing in neighborhoods to bring about social involvement and equalizing opportunity for food justice
entrepreneur	Food entrepreneur to consistently innovate and invest in the growing economy of downtown.

FIGURE 5.18   Team 5 food hub website voice of customer.

The Pugh matrix is shown in Figure 5.20 for the Women's Healthcare Center. For the Women's Center, five different concepts for providing exceptional care were identified and compared, including: providing a patient navigator that can navigate the patient through the different services and stages of care; providing VIP services to hospital donors; connecting the patients seamlessly to the Cancer Center, when needed; providing imaging same day results; and providing same day screening plus diagnostic services, when needed. A concept is selected as the candidate concept, in this case providing same day results. Each of the other concepts is compared to the candidate concept for each of the criteria, including cost, efficiency, customer satisfaction, and timeliness. These criteria are developed from the stakeholders concerns and impacts derived from the stakeholder analysis. If the alternative concept is better than the candidate concept for the particular criteria, you place a plus sign in the column. If it is worse for the criteria, place a minus sign, and if it is the same, place a zero. Add the number of pluses, minuses, and zeros in each column. The systems engineer can select the

Critical to Satisfaction (CTS)	
CTSs are basic elements that can be used in driving process measurement, improvement, and control. They are elements of a process that significantly affect the output of the process. What are the characteristics of the process are critical as perceived by	
Title	**Description**
Food quality	Local ingredients must have a good quality
Food taste	Local ingredients must be well prepared
Cost	Cost of local ingredients must reduce the prices in the menu to meet customers budget
Local farmers relation	Table 33 must have a good connection with local farmers, so they can supply them with ingredients continuously

FIGURE 5.19   Team 7 table 33 voice of customer.

TABLE 5.11
Team 6 GEM City Market Voice of Customer

Critical to Satisfaction (CTS)

CTs' are basic elements that can be used in driving process measurement, improvement, and control. They are elements of a process that significantly affect the output of the process. What are the characteristics of the process that are critical as perceived by the customer?

Title	Description
Affordable products	Products are affordable to the average community shopper.
Healthy food	The food products provided are healthy and nutritious.
Proximity to residence	The location of the supermarket is accessible to community shoppers.
Quality of products	The products are assured to be of good quality.
Human interaction	Customers deal with human workers providing a "community feel."

Criteria	Concepts for Women's Center				
	Patient Navigator	VIP Services	Connect to Cancer Center	Same Day Results	Same Day Screening + Diagnostic
Cost	+	0	-		+
Efficiency	-	-	+	Candidate	-
Customer Satisfaction	+	+	-	Concept	+
Timeliness	-	-	+		+
Pluses	2	1	2		3
Minuses	2	2	2		1
Zeros	0	1	0		0
Selection					**

FIGURE 5.20   Pugh concept selection matrix for the women's healthcare center.

concept with the highest number of pluses, or they could also combine concepts from elements that contribute to the pluses in a particular concept, in effect, taking the best from each concept. In this example, the same day screening plus diagnostic concept is selected as the most important for the decision criteria used. If there is not enough money or resources to implement all of the concepts, this technique can help to prioritize concepts.

Figures 5.21 to 5.27 show a Pugh's Concept Selection matrix for the Food System. You would select the few concepts with the most pluses and the fewest minuses. You could also attack the weaknesses of the few concepts and enhance them with the strengths of the surviving alternatives. The elements with the pluses can be used to form a hybrid product or process that takes the best aspects of each concept and combines them to form a superior concept. This may result in a shortened list of concepts. The team can then use the Pugh's Concept Selection matrix to further analyze the elements and select the best concept or further incorporate the best elements. In Figure 5.23, Team 3, who was focusing on children growing their own food, developed five different concepts based on interviews with the stakeholders: grow microgreens, grow mature vegetables, raise chickens, grow herbs and spices, manage beehives. The microgreens were selected as the candidate concept. The other concepts were compared to the candidate concepts based on the following criteria: cost, growth cycle, size, time commitment, lucrativeness, simplicity, and the ability to grow the food indoors. Herbs and spices performed better on lucrativeness and the ability to grow food indoors, scored the same on size of the operation, and performed worse on cost, growth cycle, time commitment, and simplicity. The herbs and spices was the highest prioritized concept.

Based upon the concepts selected, we developed a Food Justice System Model that represented the value-added value chain activities, the concepts selected, the student team names, and the community partners who provided stakeholder information and VOC. It is shown in Figure 5.28.

Pugh Concept Selection Technique					
Criteria	Concepts:				
	Community Meetings	Website/Form	Corner Stores	Hartman-Hindsite Farms	Delivery Service
Collect Data about what people want to eat	+	+	0	**Baseline**	+
Develop Self sustaining Market	−	+	0		−
Food Delivery/Ordering Online	−	+	−		+
Create more partners	0	+	0		−
Get younger generations involved	−	+	−		0
Cost	+	0	−		−
Pluses	2	5	0		2
Minuses	3	0	3		3
Zeros	1	1	3		1
Selection	FALSE	TRUE	FALSE	FALSE	FALSE

FIGURE 5.21 Team 1 connecting organizations pugh selection technique.

Pugh Concept Selection Technique-Food Supply						
Criteria	**Concepts:**					
	From Sustainable Urban Agriculture	Import Food	Order fast food Online			
Cost	+	Candidate	+			
Freshness	+		-			
Easy to get	+		+			
Customer Satisfaction	+		-			
Pluses	4		2			
Minuses	0		2			
Zeros	0		0			
Selection	xxx					

FIGURE 5.22   Team 2 creating food policy pugh selection technique.

5.2.2.5   Perform Risk Analysis

The purpose of risk management is to reduce potential risks to an acceptable level before they occur, throughout the life of the system.

Risk management is defined as:

Organized, analytic process to identify what might cause harm or loss (identify risks); to assess and quantify the identified risks; and to develop and, if needed, implement an appropriate approach to prevent or handle causes of risk that could result in significant harm or loss.

(ISO/IEC/IEEE 2010)

Risk is defined as:

Risk is a measure of the potential inability to achieve overall program objectives within defined cost, schedule, and technical constraints and has two components:

1. *The probability (or likelihood) of failing to achieve a particular outcome and*
2. *The consequences (or impact) of failing to achieve that outcome (DAU, 2003).*

The risk assessment and analysis are updated and refreshed throughout the Vee Model Life Cycle, based on the more in-depth knowledge that is collected as the systems engineering moves through the life cycle.

Pugh Concept Selection Technique					
	Concepts:				
Criteria	Microgreens	Mature Vegetables (Carrot, tomato, etc.)	Chickens	Herbs and Spices	Beehives
Cost	Candidate	-	-	-	-
Growth Cycle		-	+	-	+
Size		-	-	0	-
Time Commitment		-	-	-	-
Lucrativeness		+	-	+	+
Simplicity		-	-	-	-
Indoor Production		-	-	+	-
Pluses		1	1	2	2
Minuses		6	6	4	5
Zeros		0	0	1	0
Selection				***	

FIGURE 5.23　Team 3 kids growing food pugh selection technique.

The Systems Engineering risk management process includes the following activities:

- risk planning: The risk planning step develops a strategy for identifying, analyzing, handling, and monitoring risks. The strategy includes both a process and the way that it will be implemented and the risks will be documented and traced throughout the phases of the life cycle.
- risk identification: The risk identification step entails brainstorming and identifying potential risks that could impact the successful completion of the mission. If an existing similar system exists, collecting data on warranty claims and customer complaints can be an excellent way to identify potential risks for the new or revised system.
- risk analysis: The risk analysis assesses the contributing causes, the outcomes, and impacts of the risk events.
- risk handling: The risk handling develops risk handling approaches for each risk to investigate whether the risk likelihood or consequence can be reduced.
- risk monitoring: The risk monitoring is the active monitoring throughout the system development to monitor the risks and reassess the risks, as more knowledge is acquired.

Components of the risk management tools: Risk assessment, risk handling approach, risk prioritization, and the risk cube. Each tool is discussed next.

Pugh Concept Selection Technique

Criteria	Concepts:			
	Storage at farm	Dispersed storage location	Immediately selling food	Centralized storage location
Ease of food distribution	-	-	0	+
Likelihood of working out	+	+	-	+
Time consuming	+	-	+	+
Transportation	+	-	+	+
Cost	-	-	+	-
Community satisfaction	-	+	-	+
Preserving food (practical)	-	+	-	+
Pluses	3	3	4	6
Minuses	4	4	3	1
Zeros	0	0	0	0
Selection				****

FIGURE 5.24   Team 4 food storage pugh selection technique.

5.2.2.6 Risk assessment

The risk assessment tool includes the generation of the potential risk events that could occur related to the system, at this point in time.

- The *risk event* is the potential risk that can occur related to the system.
- The *contributing causes* are those causes that contribute to the risk event possibly occurring.
- The outcomes are the potential results if the risk event was to occur.
- The impact is the effect on the system if the risk event should occur.

An easy way to generate the risk assessment is to write a sentence including the terms:

"As a result of <contributing causes>, <risk event> may occur, causing <outcomes>, which can be expressed as <impact>."

(Sugarman, 2015)

For example: As a result of <community apathy>, <the community may not participate in helping reduce food insecurity>, causing <farmers to make few sales of their locally sourced products>, which can be expressed as <farmers losing interest in participating in local food sourcing>.

Pugh Concept Selection Technique				
Criteria	**Concepts:**			
	Website	Marketing	Invest	Pricing
Brings in Entrepenuers	0	+	+	Candidate
Expands Current DDP System	+	0	+	
Uses Commercial Space	0	+	-	
Provides Fresh Food	+	0	-	
Creates Economic Growth	0	+	0	
Pluses	2	3	2	
Minuses	0	0	2	
Zeros	3	2	1	
Selection	2	3	0	
A: Create a Website connecting farmers, groceries, and restaurants				
B: Marketing to food entrepenuers				
C: Local Restaurant owners invest in downtown markets to share with public				
D: Pricing Strategy to incentivize real estate development of groceries				

FIGURE 5.25   Team 5 food hub website pugh selection technique.

5.2.2.7   Risk Handling Approach

Risk handling is the process that identifies and selects options to manage the identified risks to an acceptable level during the systems project. Risk handling options include assumption, avoidance, control (mitigation), transfer, research, and monitor (SEBoK, 2020), (Sugarman, 2015).

5.2.2.7.1   Risk Handling Options

1) Assumption: Assumption accepts the risks and takes no actions at this time. It is typically assumed that the risk is rated low or it would take an inordinate amount of resources to reduce the risk to an acceptable level.

Pugh Concept Selection Technique					
Criteria	Concepts:				
	Opening grocery branch	Transportation to supermarkets	Food trucks	Mobile shopping of food	GEM City market
Cost	+	-	-	-	**Candidate**
Efficiency	+	-	-	+	
Customer satisfaction	+	0	+	+	
Saving Time	0	-	0	+	
Stock	0	0	-	0	
Pluses	3	0	1	3	
Minuses	0	3	3	1	
Zeros	2	2	1	1	
Selection	*****				

FIGURE 5.26   Team 6 GEM city market pugh selection technique.

An example of assumption risk handling approach for our food system is:

The risk is that a student decides not to participate in the school food growing program. It would be assumed that the child has the prerogative not to participate in the program.

2) Avoidance: In the avoidance handling approach, action is taken to eliminate the risk to reduce the likelihood of the risk to 5% and/or eliminate the negative consequence to the system.
   An example of avoidance risk handling approach for our food system is:

The risk is that they are overusing commercial space and there is little space for urban farms. To avoid the risk, take preventive measures to assess and ensure that commercial zoning is appropriate and adhered to by commercial users.

3) Control (mitigate): In the control or mitigation handling approach, activities are performed to reduce the likelihood and/or the consequence to an acceptable level. It does not completely eliminate the risk, as avoidance does.

Pugh Concept Selection Technique

Criteria	Concepts:				
	cooking equipment	green houses	growing classes	farmers market	Purchase local ingredients
economic	(-)	(-)	0	(+)	(+)
quality	(+)	(+)	0	0	(+)
customer satisfaction	(+)	(+)	(+)	(+)	(+)
customer knowledge	0	0	(+)	0	0
Pluses	2	2	2	2	3
Minuses	1	1	0	0	0
Zeros	1	1	2	2	1
Selection					***

FIGURE 5.27 Team 7 table 33 pugh selection technique.

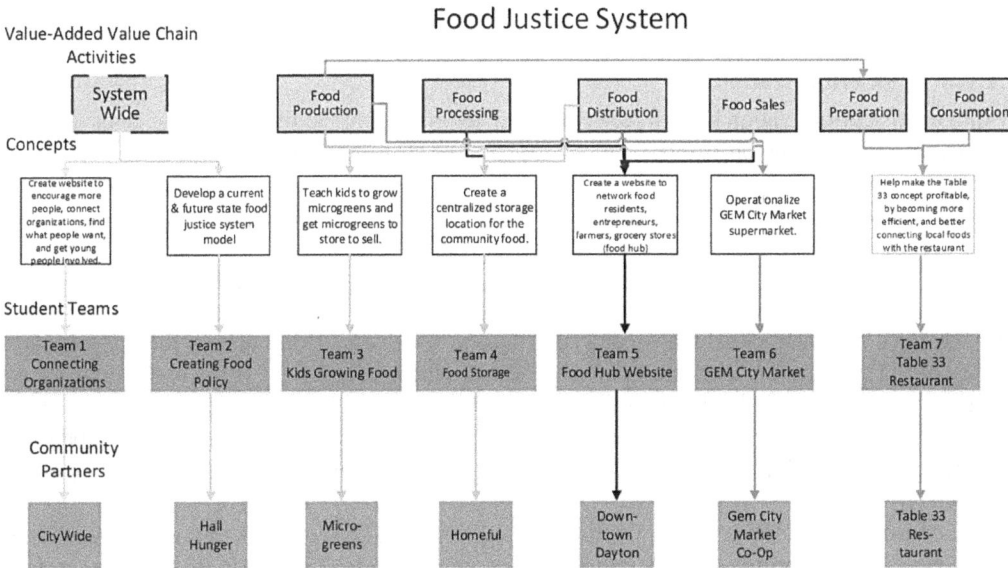

FIGURE 5.28 Food justice system model.

An example of control (mitigate) risk handling approach for our food system is:

The risk is that the website has a poor user interface and is not easy to use. The team would run tests and trials with volunteers to test the user interface throughout development to control or mitigate this risk.

4) Transfer: In the transfer handling approach, the risk event or impact is shifted away from the program. The risk could be transferred between stakeholder or transfer risk within a system itself. This approach does not change the likelihood or the impact of the risk.
   An example of the transfer risk handling approach for our food system is:

   The risk event is that the website is hacked. To transfer the risk, hire a cyber security firm to manage the website.

5) Research: The research handling approach includes collecting additional information before a decision is made to reduce the risk.
   An example of research risk handling approach for our food system is:

   The risk is that zoning doesn't permit farm stands. Additional research is needed to investigate these zoning ordinances further and determine what we can do to change them before a decision can be made related to this risk.

6) Monitor: The monitor handling approach is to take no immediate action but monitor changes related to the risk.
   An example of monitor risk handling approach for our food system is:

   The risk is that the weather doesn't cooperate with allowing food on farms to be grown for local sourcing. The weather is monitored, as there may not be much else that can be done to mitigate weather conditions.

5.2.2.8   Risk Prioritization

Within the risk prioritization tool, the team assesses the risk event likelihood and the consequences. The risk likelihood is the probability that the risk will occur assessed typically on a scale of 1 to 5, as shown in Table 5.12. Level 1 is not likely that the risk will occur. Level 2 represents a low likelihood that the risk will occur. Level 3 is that the risk is likely to occur. Level 4 is that the risk is highly likely to occur, and the level 5 is a near certainty that the risk will occur.

5.2.2.9   Risk Cube

The risk cube is a matrix that maps each risk aligned to its risk likelihood and the consequences of the risks. This provides a simple graphical view of where to focus the risks, based on the color coding of the risk cube: red is for high risk, these risks should be closely monitored and continually reassessed; yellow is for medium risk, these risks should be monitored and reassessed periodically; and green is for low risk, occasional monitoring and reassessment should be applied for these risk events.

The consequences represent the magnitude of the impact of the risk. It is also typically assessed on a scale of 1 to 5. The impact or consequences is typically assessed related to the technical aspects, the system project schedule, and/or the cost of the program. The impact matrix can be found in Table 5.13.

Once the risk likelihood and consequences are assessed, the risk cube is used to display the risks (by risk event number) on the space in the cube based on the intersection of the risk likelihood level and the impact number. The risk cube does a nice job of visually showing which risks are most important to incorporate a satisfactory risk handling approach if needed to reduce the risk likelihood or impact.

The risk assessment for the Women's Healthcare Center is shown in Figure 5.29.

The risk assessments, risk handling approaches, risk prioritization, and the risk cubes for the food system projects are shown in Figures 5.30 to 5.36.

TABLE 5.12
Risk Likelihood

What Is The Likelihood The Risk Will Happen?

Level		Your approach and processes...
1	Not likely	... Will effectively avoid or mitigate this risk based on standard practice
2	Low likelihood	... Have usually mitigated this type of risk with minimal oversight in similar cases
3	Likely	... May mitigate this risk, but workarounds will be required
4	Highly likely	... Cannot mitigate this risk but a different approach might
5	Near certainty	... Cannot mitigate this type of risk: no known processes or workarounds are available

TABLE 5.13
Magnitude of Impact or Risk Consequences

Given the Risk is Realized, What Would be the Magnitude of the Impact?

Level	Technical	Schedule	Cost
1	Minimal or no impact	Minimal or no impact	Minimal or no impact
2	Minor performance shortfall, same approach retained	Additional activities required, able to meet key dates	Budget increase or unit production cost increase < 1%
3	Moderate performance shortfall, but workarounds available	Minor schedule slip, will miss needed dates	Budget increase or unit production cost increase < 5%
4	Unacceptable, but workarounds available	Project critical path affected	Budget increase or unit production cost increase < 10%
5	Unacceptable, no alternatives exist	Cannot achieve key project milestones	Budget increase or unit production cost increase > 10%

Risk Assessment

Risk #	Risk Event	Contributing causes	Outcomes	Impact
1	May not get donations that cover cost of construction of women's center	Recession economy	May not be able to make construction changes and services per desired schedule.	May have to reduce desired services
2	Regulatory requirements new or changing regulations require changes to policies, processes or procedures.	New or changing regulatory requirements	May add to project schedule, to make appropriate changes.	May require additional resources, technology and changes to policies, processes and procedures.
3	Software technology may not be available within schedule	Technology availability	May not be able to make technology changes for Business Process Management Systems, and VIP alerts	May not meet customer service goals.

Risk Handling Approach

Risk #	Risk Event	Risk Handling Type	Risk Handling Description
1	May not get donations that cover cost of construction of women's center	Control	Enhance fund raising
2	Regulatory requirements new or changing regulations require changes to policies, processes or procedures.	Monitor	Regulatory compliance monitors regulations
3	Software technology may not be available within schedule	Monitor	Monitor projects

Risk Prioritization

Risk #	Risk Event	Risk Likelihood	Risk Impact
1	May not get donations that cover cost of construction of women's center	3	4
2	Regulatory requirements new or changing regulations require changes to policies, processes or procedures.	4	4

FIGURE 5.29 Risk assessment for the women's healthcare center.   (*Continued*)

3	Software technology may not be available within schedule	3	3

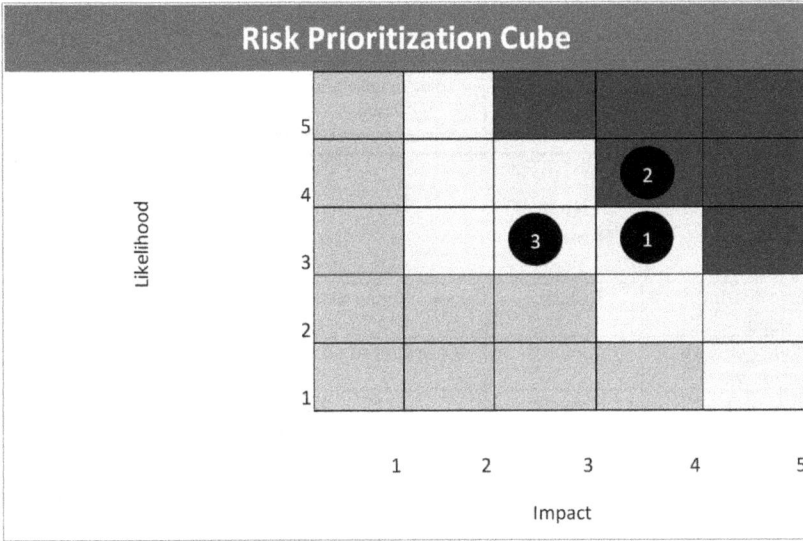

Risk Prioritization Cube

(Likelihood vs. Impact scatter: point 2 at Impact 4, Likelihood 4; point 3 at Impact 2, Likelihood 3; point 1 at Impact 3, Likelihood 3)

FIGURE 5.29 **(Continued)** Risk Assessment for the Women's Healthcare Center.

Risk Assessment				
Risk #	Risk Event	Contributing causes	Outcomes	Impact
1	Zoning doesn't permit farmstand	City ordinances	Farmer can't set up in the local community	Not able to provide food to the needy community
2	Community doesn't participate	Community apathy	Farmers make little sales	Farmers lose interest in participating
3	Website hacked	Hackers	Website is down	Data is lost and inability to fill orders
4	Unsuccessful advertising	Poor marketing/ lack of funding/ market competition	Less people are aware of the program	Farmers lose interest in participating
5	Weather doesn't cooperate	Inclement weather	Poor quality food, people don't come to the market	Sales reduced and unavailability of food
6	Food Poisoning	Food quality	People get sick	People lose trust in program, could possibly sue
7	Poor data analysis	Incorrect data entry, lack of appropriate measures	Bad information distributed to farmers	Miss recruitment opportunities or farmers lose trust in the program
8	Lack of internet access	Expensive, internet may be down	People can't access the website	Less engagement form the community
9	Website server runs out of available memory	Didn't buy large enough package from web host	Website bogs down/can't process new data/orders	Data is lost and inability to fill orders

FIGURE 5.30 Team 1 connecting organizations risk analysis & prioritization.   (*Continued*)

10	Website has a poor user interface (UI)	Not enough time was spent making the UI user friendly	People get frustrated with the website and don't use it	Website is unsuccessful and can't be used to expand the food program
11	Data output isn't presented in a useful manner	Lack of communication in software development	Output is not in useful terms for Farmers	Miss recruitment opportunities or farmers lose trust in the program
12	Malicious use of website forum	Lack of forum moderation	Toxic or unhelpful forum	Good relationships between farmers and community members won't be as easily started or maintained

Risk Handling Approach			
Risk #	Risk Event	Risk Handling Type	Risk Handling Description
1	Zoning doesn't permit farm stand	Research	look into the local ordinances and see what you can do to change them.
2	Community doesn't participate	Control	More door to door solicitations, and continue to advertise
3	Website hacked	Transfer	Hire a cyber security firm
4	Unsuccessful advertising	Research	More effective means to reach the public
5	Weather doesn't cooperate	Monitor	Always be aware of the weather and plan accordingly
6	Food Poisoning	Transfer	Establish a contract with the farmer where they guarantee quality assurance
7	Poor data analysis	Control	Training of the volunteers that with run the website
8	Lack of internet access	Avoidance	Use paper forms
9	Website server runs out of available memory	Research	The size other data collection websites handle large quantities of data and how much server space they operate with
10	Website has a poor user interface (UI)	Control	Run trials with volunteers to test UI along the way through development
11	Data output isn't presented in a useful manner	Research	Run a study to find out what information farmers want to see and how they would like it presented
12	Malicious use of website forum	Control	Hire or have a volunteer actively moderate the forum

Risk Prioritization			
Risk #	Risk Event	Risk Likelihood	Risk Impact
1	Zoning doesn't permit farmstand	2	2
2	Community doesn't participate	3	3
3	Website hacked	2	4
4	Unsuccessful advertising	3	2
5	Weather doesn't cooperate	3	2
6	Food Poisoning	2	4
7	Poor data analysis	2	3
8	Lack of internet access	3	3
9	Website server runs out of available memory	2	4
10	Website has a poor user interface (UI)	2	4
11	Data output isn't presented in a useful manner	3	3
12	Malicious use of website forum	4	2

FIGURE 5.30 (**Continued**) Team 1 connecting organizations risk analysis & prioritization.

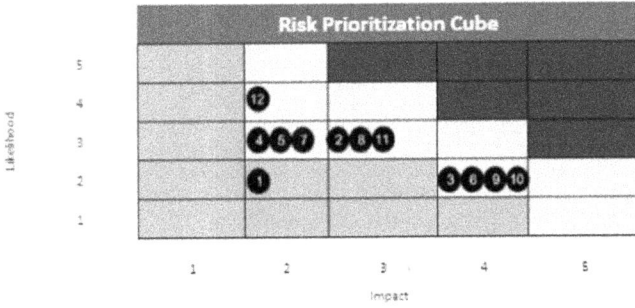

FIGURE 5.30   (**Continued**) Team 1 connecting organizations risk analysis & prioritization.

Risk Assessment				
Risk #	Risk Event	Contributing causes	Outcomes	Impact
1	New policy idea may get rejected by commission	Not enough discussion	May not be able to change the policy	Problem will not be solved
2	May not get enough support	Less communication	May not be able to come up with the new policy	Problem is not identified properly
3	Software technology may not be available on website	Technology availability	May not be able to show calendar and schedule on official website	Information can't be shared
4	Champion is not elected	No proper qualification	Policy change idea cannot be presented to the Commission	Problem is not solved
5	Champion not being able to take the discussion forward effectively	Champions inability	Champion might not be able to drive the request forward	Fail to get the desired changes in the policy
6	Data presented might not be enough	Insufficient data gathered	Not be able to present the case commendably	Fail to get the desired changes in the policy

Risk Handling Approach			
Risk #	Risk Event	Risk Handling Type	Risk Handling Description
1	New policy idea may get rejected by commission	Control	Have more discussion and communication with customers and stakeholders
2	May not get enough support	Control	Be patient and kind to customers and stakeholders and listen to their needs

FIGURE 5.31   Team 2 creating food policy risk analysis & prioritization.   (*Continued*)

3	Software technology may not be available on website	Monitor	Monitor system
4	Champion is not elected	Research	Work on the required qualifications to elect a champion
5	Champion not being able to take the discussion forward effectively	Control	Make sure Champion is apprised with enough information and he has the control over the request/issue
6	Data presented might not be enough	Avoidance	Ensure to explore every option to collect the data

Risk Prioritization			
Risk #	Risk Event	Risk Likelihood	Risk Impact
1	New policy idea may get rejected by commission	2	4
2	May not get enough support	3	4
3	Software technology may not be available on website	1	3
4	Champion is not elected	2	2
5	Champion not being able to take the discussion forward effectively	2	4
6	Data presented might not be enough	3	4

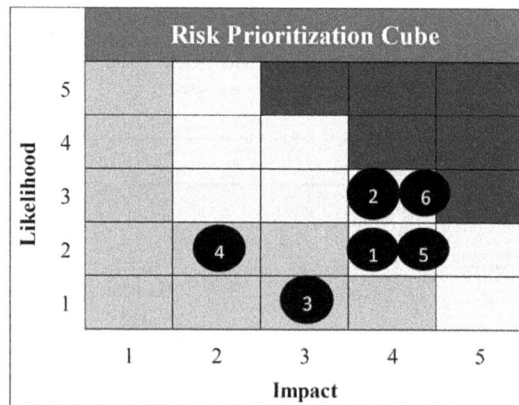

FIGURE 5.31   (Continued) Team 2 creating food policy risk analysis & prioritization.

Risk Assessment				
Risk #	Risk Event	Contributing causes	Outcomes	Impact
1	Child kills microgreens	Not soaking the seeds for long enough, not watering the microgreens enough, overwatering the microgreens, not exposing to light at proper time, exposure to hazardous material	Microgreens are dead, child does not receive payment, donated supplies are wasted	Repeated killing of microgreens would end the project, child drops out of project
2	Microgreens are exposed to unsanitary/hazardous conditions and are still accepted	Watering with contaminated water source, exposure to household pets, exposure to smoke, exposure to mold or fungus	People get sick, microgreens exhibit poor taste	Detrimental reputation to the program, Table 33 ceases cooperation with program
3	Child chooses not to participate in program	Guardian will not sign permission form, lack of guardian support, children unmotivated, children have other priorities	Fewer microgreens are grown to be sold to Table 33	Program is not sustainable
4	Child expresses no interest in program	Children have other priorities, lack of guardian support, "uncool" reputation	Fewer microgreens are grown to be sold to Table 34	Program is not sustainable
5	Table 33 ceases support	Table 33 closes, Table 33 partners with different organization, Table 33 no longer has ability to fund, microgreens from program are consistently of poor quality	No customer for microgreens, no compensation for children or teacher	Program is not sustainable
6	Child does not follow directions	Children do not attend demonstration, children do not pay attention during demonstration, children are rebellious	Potentially damaged or unsanitary microgreens, unsafe conditions for children	Children removed from program, program production rate decreases

Risk Handling Approach			
Risk #	Risk Event	Risk Handling Type	Risk Handling Description
1	Child kills microgreens	Transfer	If children kill microgreens, they must stay out of program for the remainder of the quarter
2	Microgreens are exposed to unsanitary/hazardous conditions and are still	Control	Demonstrate and emphasize the use of the lid during microgreen growth process so that unsanitary conditions cannot reach microgreens. Also, microgreens are cleaned using hydrogren peroxide and following a pre-established process
3	Child chooses not to participate in program	Assumption	Program established to help children so they are only hurting themselves
4	Child expresses no interest in program	Control	Strongly emphasize program benefits and share success stories
5	Table 33 ceases support	Research	Identify other possible microgreen customers in the Dayton area and begin to establish relationships with them. Could potentially lead to multiple partnerships at the same time.
6	Child does not follow directions	Control	Supply assessment at the end of microgreen growing demonstration to ensure student competency

Risk Prioritization			
Risk #	Risk Event	Risk Likelihood	Risk Impact
1	May not obtain required funds to cover construction cost of store	3	4
2	Supplier might not be reliable and efficient	2	3
3	Customers may not like the food products in the store	2	3
4	Farmers don't provide good quality products, unable to meet demand	3	3
5	Theft from store	2	2
6	Fire	2	4

FIGURE 5.32 Team 3 kids growing food risk analysis & prioritization. *(Continued)*

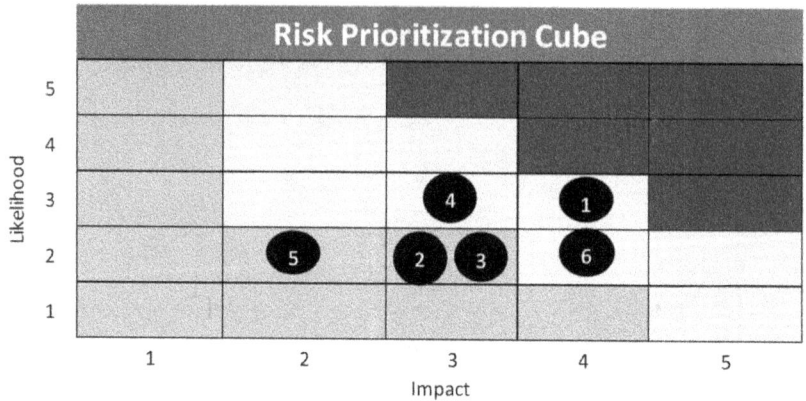

FIGURE 5.32   (**Continued**) Team 3 kids growing food risk analysis & prioritization.

Risk Assessment				
Risk #	Risk Event	Contributing causes	Outcomes	Impact
1	The funds may not be released for the community development	Recession economy	May not be able to reach the goals of the community	May have to keep the goals limited
2	Changing weather condtions is a threat to the crop yield and quality	Changing weather condition	May not be able to get the desired crop yield	Economic loss
3	Changing minds of consumer	Food loss	May lead to food loss and impact on the community goal	Will have to preserve food for long or food wasteage
4	Improper storage facility or lack of place for food storage	No storage space	May lead to food loss and cause economic loss	Reduction in the food production
5	The staff not able to organize storage of food	Unskilled staff	May lead to delays for distribution to markets	Not able to reach goals of community improving sufficient food to markets

Risk Handling Approach			
Risk #	Risk Event	Risk Handling Type	Risk Handling Description
1	The funds may not be released for the community development	Control	Work on gaining more sponsors
2	Changing weather condtions is a threat to the crop yield and quality	Monitor	Monitor the weather conditions to overcome issue
3	Changing minds of consumer	Research	Research the people's interest in that region to understand their interest in food preferences
4	Improper storage facility or lack of place for food storage	Monitor	Create a storage unit or monitor the production with available storage
5	The staff not able to organize storage of food	Control	Train the staff prior to appointing and make them skillful at their job.

Risk #	Risk Event	Risk Likelihood	Risk Impact
1	Not able to find all ingredients locally	3	4
2	Ingredients transportation	2	4

FIGURE 5.33   Team 4 food storage risk analysis & prioritization.   (*Continued*)

3	Insufficient Cash flow	1	5
4	Unskilled employees	3	3
5	Bad ingredients quality	1	3
6	New regulations	2	1

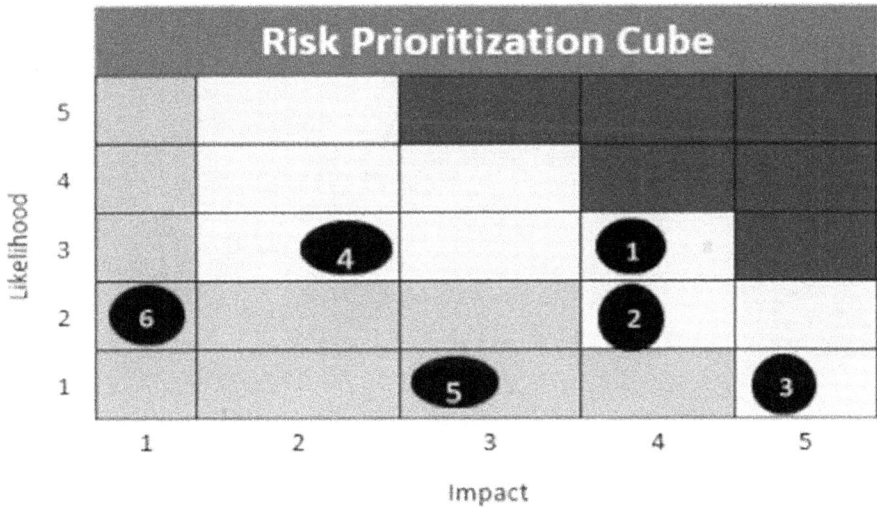

FIGURE 5.33   (Continued) Team 4 food storage risk analysis & prioritization.

Risk Assessment				
Risk #	Risk Event	Contributing causes	Outcomes	Impact
1	Overwhelm Farmers	Too High Demand	Farmland unusable due to high yield, underproduce to meet demand	Farmers may not want to sell within network anymore
2	Overusing Commercial Space	Poor City Planning, too many investors	Overcrowding downtown, hike in cost of living	Gentrifying downtown, creating another issue
3	Website Failure	Too much traffic	Temporary shutdown, possible delays on deliveries	People flee from utilizing website as a tool, chaotic
4	Distributor Failure	Lack of communication and logistics	Possible delay on deliveries	People flee from utilizing website as a tool, chaotic
5	Corruption	Lack of oversight	Embezzlement, Price Gouging	Businesses fail, produce not available
6	Drought	Poor Weather	Death, Famine, Barren Land	No longer a viable food system

FIGURE 5.34   Team 5 food hub website risk analysis & prioritization.   (Continued)

Risk Handling Approach			
Risk #	Risk Event	Risk Handling Type	Risk Handling Description
1	Overwhelm Farmers	Avoidance	Take preventative measures to ensure yield matches expectations and clear communication is key
2	Overusing Commercial Space	Avoidance	Take preventative measures to ensure that commercial space is used as intended
3	Website Failure	Control	Impossible to guarantee it wont happen, but with enough attention can be diligently prevented.
4	Distributor Failure	Control	Impossible to guarantee it wont happen, but with enough attention can be diligently prevented.
5	Corruption	Control	Consistently watch to ensure that all parties are in check and not attempting out of malice to short each other.
6	Drought	Assumption	We are going to assume this wont happen, because the odds are very rare that it would occur. If it does, there is not much in our control.

FIGURE 5.34   (Continued) Team 5 food hub website risk analysis & prioritization.

Risk Assessment				
Risk #	Risk Event	Contributing causes	Outcomes	Impact
1	May not obtain required funds to cover construction cost of store	Economic depression contributing to scarcity of funds	May not be able to construct the store	May have to reduce construction cost
2	Supplier might not be reliable and efficient	Inefficiency of supply management	Not enough variety and proper quality of food products	Will need to pick a better supplier
3	Customers may not like the food products in the store	Not meeting customer expectations by not providing a wide range of	May not meet financial targets	Will need to improve food products (quality and quantity) from customer
4	Farmers don't provide good quality products, unable to meet demand	Not enough knowledge and funds to grow required crops	Shortage of fresh vegetables	To provide more knowledge and funds to farmers
5	Theft from store	People who don't want to pay and weak security system	Bad reputation for the store	People might be afraid to get robbed in the store
6	Fire	Not having a good sprinkler system	Losing a part of the store	Money will go to rebuilding instead of improving the store

Risk Handling Approach			
Risk #	Risk Event	Risk Handling Type	Risk Handling Description
1	May not obtain required funds to cover construction cost of store	Control	Improve fundraising effort
2	Supplier might not be reliable and efficient	Control	Select a more reliability supplier
3	Customers may not like the food products in the store	Research	Improve according to customer reviews
4	Farmers don't provide good quality products, unable to meet demand	Control	Providing training to farmers and needed funds
5	Theft from store	Monitor	Good surveillance and security systems
6	Fire	Avoidance	Good sprinker system

FIGURE 5.35   Team 6 GEM city market risk analysis & prioritization.

Risk Assessment				
Risk #	Risk Event	Contributing causes	Outcomes	Impact
1	Not able to find all ingredients locally	Lack of local farms	Lack of local ingredients	Increasement in food price/less profit, look for non-local ingredients
2	Ingredients transportation	Lack of transportation vehicles	Delay in ingredients delivery	decrease sales and profit
3	Insufficient Cash flow	Investors pull out	Not able to meet the restaurant's ingredient needs	Less meals, less profit, not able to help the community
4	Unskilled employees	Hiring people without looking at their resumes	Hiring non-experienced workers	Decrease efficiency of the restaurant and customer satisfaction
5	Bad ingredients quality	non-organic growing practices	unhealthy ingredients	bad food quality and decreasement in sales
6	New regulations	Food and Drug Administiration (FDA) rules	Limited ingredients resources	Limited food choices, which can decrease sales

Risk Handling Approach			
Risk #	Risk Event	Risk Handling Type	Risk Handling Description
1	Not able to find all ingredients locally	Research	Make researches for all type of ingredients needed locally. In case of missing ingredients, try to communicate with farmers and ask for possibilities to grow those ingredients
2	Ingredients transportation	Monitor	Provide transportation/delivery vehicles that trasport whatever ingredients needed to the restaurant. Or make arrangements with farmers for delivery options
3	Insufficient Cash flow	Control	Communicate with investors to avoid pull outs.
4	Unskilled employees	Research	Make researches and put advertisments to hook experienced and interested people.
5	Bad ingredients quality	Control	Send experienced emplyees who can identify the quality of ingredients before making the purchases
6	New regulations	Monitor	Make sure to follow all regulations and rules that are related to food in the area.

FIGURE 5.36   Team 7 table 33 risk analysis.

5.3   SUMMARY

In this chapter we covered the mission analysis within the Concept of Operations Phase, in addition to describing the risk analysis.

5.4   ACTIVE LEARNING EXERCISES

Develop a project charter for the problem or opportunity, proposed system, and concepts. Include a project overview, problem statement, and project scope.

2) Generate interview questions to ask the stakeholders for performing a stakeholder analysis.

3) interview the proposed stakeholders and create a stakeholder analysis, identifying the stakeholder groups, their roles related to the system, how they are impacted and concerns they have related to the system, and their initial and future needed receptivity.

4) Generate proposed Critical to Satisfaction criteria from the stakeholder interviews.

5) Plan and run a focus group to generate VOC and the Critical to Satisfaction criteria.

6) Plan and develop a customer survey to generate VOC and the Critical to Satisfaction criteria.

7) Use the Pugh Concept Technique to generate and select alternative system concepts. Brainstorm potential concepts with the stakeholders. Which concept did you prioritize the highest based on the Pugh Concept Technique?

8) Perform a risk analysis for your system.
   (a) Develop a list of risk events for your system in the following format:

   "As a result of <contributing causes>, <risk event> may occur, causing <outcomes>, which can be expressed as <impact>."

   (b) Develop a risk handling approach, using the risk handling approach tool.
   (c) Prioritize the risks based on risk likelihood and risk impact.
   (d) Display the risks on the risk cube. Identify which was the highest risk based on likelihood and impact? Does the risk prioritization and cube change your risk handling approach for any of the risks identified?

BIBLIOGRAPHY

DAU. 2003. *Risk Management Guide for DoD Acquisition*: Fifth Edition. Ft. Belvoir, VA, USA: Defense Acquisition University (DAU)/U.S. Department of Defense, Version 2.

Duffy G.L. and Furterer, S. L. 2020. *The ASQ Certified Quality Improvement Associate Handbook*, 4th. Milwaukee: ASQ Quality Press.

Malone, L. 2005. Class Notes Guest Lecture, ESI 5227, University of Central Florida, Department of Industrial Engineering and Management System, Orlando, FL.

George, M., Rowlands, D., Price, M., and Maxey, J. 2005. *Lean Six Sigma Pocket Toolbook*. New York: McGraw-Hill, Print.

ISO/IEC/IEEE. 2010. Systems and Software Engineering - System and Software Engineering Vocabulary (SEVocab). Geneva, Switzerland: International Organization for Standardization (ISO)/International Electrotechnical Commission (IEC)/ Institute of Electrical and Electronics Engineers (IEEE). ISO/IEC/IEEE 24765:2010.

SEBoK Editorial Board. 2020. The Guide to the Systems Engineering Body of Knowledge (SEBoK), v. 2.3, R.J. Cloutier (Editor in Chief). Hoboken, NJ: The Trustees of the Stevens Institute of Technology. Accessed [DATE]. www.sebokwiki.org. BKCASE is managed and maintained by the Stevens Institute of Technology Systems Engineering Research Center, the International Council on Systems Engineering, and the Institute of Electrical and Electronics Engineers Systems Council.

Sugarman, R. 2015. University of Dayton ENM 505 Course Materials, University of Dayton.

6 Requirements and Architecture Phase 2
Develop Logical Architecture

In this chapter, we develop the logical architecture that is part of Phase 2 – Requirements and Architecture.

6.1 PURPOSE

The Requirements and Architecture Phase is the second phase in the Systems Engineering Vee Life Cycle Model. The purpose of the phase is to develop the logical architecture for the system and then elicit the customer, system, and specialty engineering requirements. In this chapter, we apply tools that develops the logical architecture for the system. These models and tools help to illustrate a view of the high-level functionality to be provided by the system. The tools provide a boundary and scope for what is part of the system and what is not. It is not yet a physical manifestation but a conceptual one.

6.2 ACTIVITIES

The activities performed and the tools applied in the requirements and architecture phase are shown in Table 6.1. We cover "Develop Logical Architecture" activity and the Logical Architecture tools including the SIPOC, Value Chain, Functional Decomposition Model (FDM), and use case diagrams. In the next chapter, we discuss the "Elicit Requirements" activities and the associated requirements definition tools. In Chapter 8, we discuss the derivation of the requirements, the measurement plan, information models, and quality management plan.

The systems architecture model for the Requirements and Architecture phase, discussed in Chapter 3, is shown in Figure 6.1.

6.3 DEVELOP LOGICAL ARCHITECTURE

The logical architecture represents the functions that the system will perform. Describing the functions help us to better elicit the customers' and systems requirements within this phase.

The definition of an architecture used in ANSI/IEEE Std 1471-2000 is (TOGAF 8.0, 2006).

> The fundamental organization of a system, embodied in its components, their relationships to each other and the environment, and the principles governing its design and evolution.

The logical architecture tools that are developed in this phase are:

- SIPOC – defines the boundaries of the functions or processes included in our system as well as the system stakeholders (Elements, behavior)
- Value Chain – describes the value-added and support activities to be performed by the system
- Functional Decomposition– decomposes the system functions (behavior)
- Use Case Diagrams – identifies how the system will be used

TABLE 6.1

Activities and Tools in the Requirements and Architecture Phase

Vee Phase	Activities	Tools	Principles
Phase 2: Requirements and Architecture	Develop Logical Architecture 1. Develop Logical Architecture Elicit Requirements 2. Elicit customer requirements 3. Derive requirements • Develop system requirements • Develop specialty engineering requirements 4. Develop a measurement plan 5. Develop an information model 6. Develop a quality management plan	Logical Architecture: • SIPOC • Value Chain and Functional Decomposition • Use Case Diagrams Requirements: • Process Scenarios via Process Architecture map • Use Cases • Customer requirements • System requirements • Specialty engineering requirements: software engineering; environmental engineering; safety and security engineering; Human Systems Integration • Data Collection Plan, with operational definitions • Class Diagram with information model and hierarchy model • Quality management plan	• System dynamics (behavior, system elements) • Cybernetics (information flow) • Systems thinking • Abstraction • Views

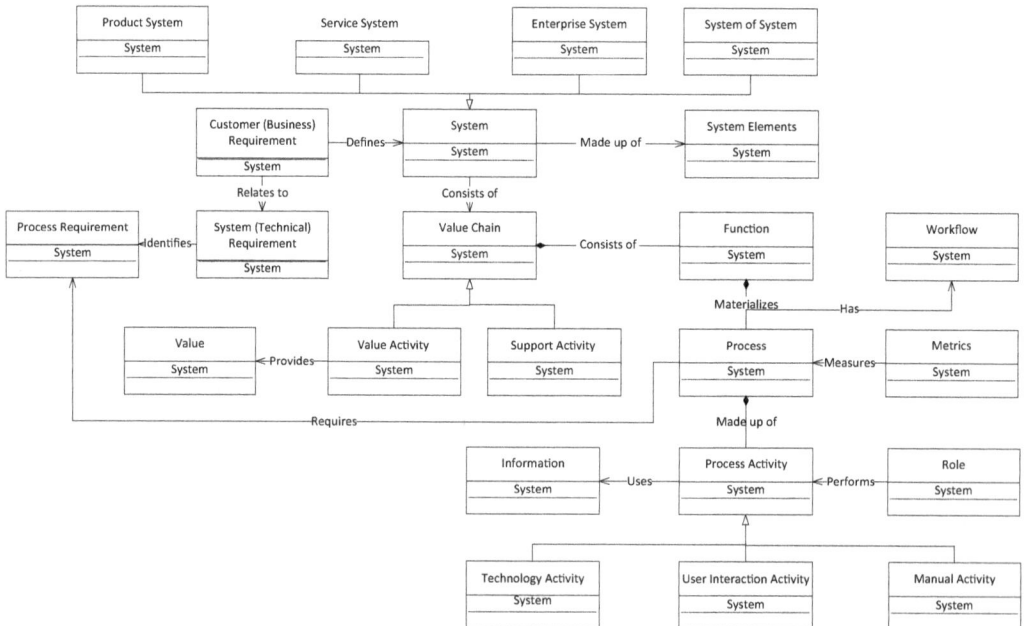

FIGURE 6.1   System architecture and requirements model.

Figure 6.2 shows the relationships between the tools of the system architecture including the SIPOC process steps, the value chain activities, the functions in the Functional Decomposition, and the uses or functions in the use case diagrams.

Figure 6.3 illustrates the relationship between the logical architecture, the requirements, and the physical architecture. The logical architecture consists of the functions performed by the system. The value chains and the FDM provides an inventory of the functions performed by the system. These functions are used to elicit the customers' requirements and derive the systems or technical requirements. In the design phase, the physical architecture is designed. The functions are actions or verbs, while the physical architecture consists of system elements that are nouns (parts, components, information, roles, etc.). There should be traceability between the logical architecture, through the requirements to the physical architecture, and the verification and validation of test cases, to the implemented system.

Figure 6.4 shows the decomposition from requirements through logical architecture to the system architecture. We first develop the logical architecture, to provide boundaries of the functions that the system will possess. This progresses from "what" the customer wants, to "how" the system will deliver what the customer wants. Then we describe the requirements elicitation process in Chapter 7 and the derivation of the requirements in Chapter 8.

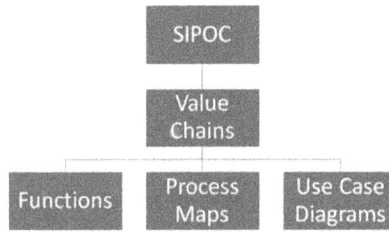

FIGURE 6.2   Relationship between system architecture tools.

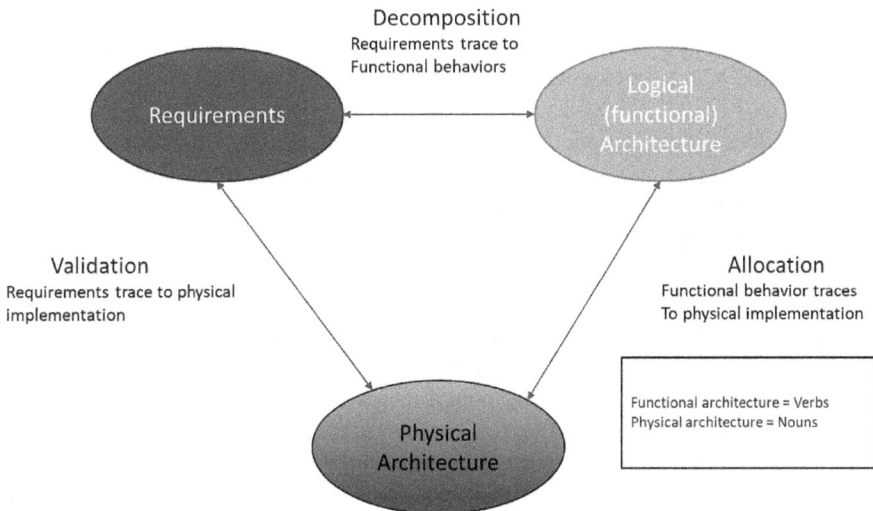

FIGURE 6.3   Relationship between the logical architecture, requirements, and the physical architecture. (Derived from SEBoK, 2020.)

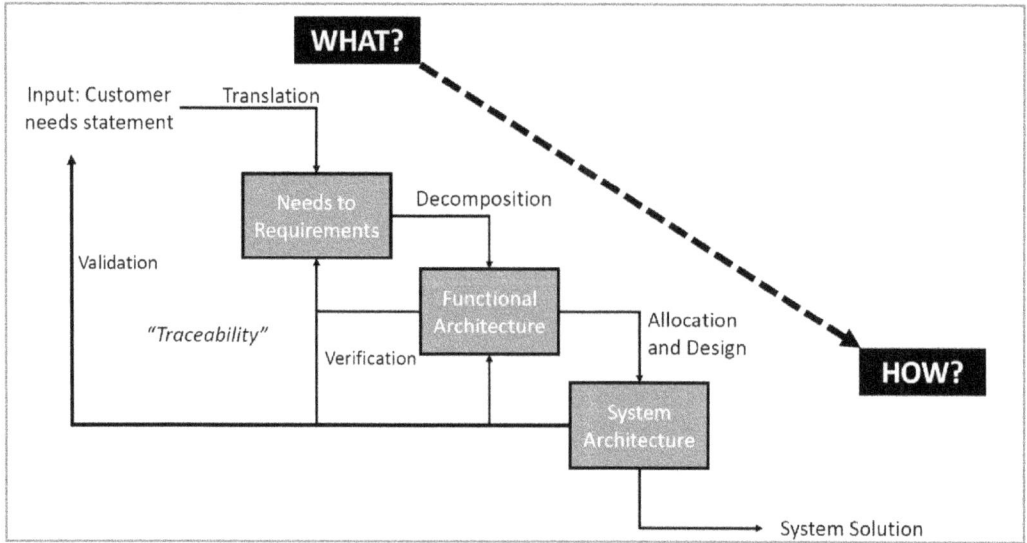

FIGURE 6.4 Decomposition of requirements to architecture. (Derived from the SEBok, 2018.)

6.3.1 SIPOC Tool

The SIPOC is a useful tool in this phase to help scope the boundaries of the system from a functionality perspective. SIPOC stands for suppliers, inputs, process, outputs, and customers. The SIPOC shows the interrelationships between the customers and suppliers of the system and how they interact with the functions. The SIPOC tool is typically applied within process improvement and/or Lean Six Sigma projects to identify the high-level processes that are part of the process to be improved. In the systems engineering life cycle, it is effective to identify the proposed functions of the system, instead of the process steps of the process to be improved. It also identifies the inputs used in the functions and the outputs of the functional steps. The functional steps transform the inputs into the outputs. The best way to construct the SIPOC is to identify the five to seven high-level functions that the system will perform. For each function, they identify the inputs to the functions and who supplies the inputs. Next, they identify the outputs of each function and the customer of the output. An example of a SIPOC for developing a women's center healthcare service processes is shown in Figure 6.5.

Figure 6.6 provides a list of suppliers and explains their interaction with the customers including the process, inputs, and output during this process for Team 1. The initial supplier in the system is the shopper. Shoppers participate in the system by placing orders and suggestions online or in paper form. After Citywide analyzes and prioritizes the data and orders, the analyzed information will be transferred to the farmer. Additionally, the information will be posted on a website to allow farmers to evaluate what the desired products are. Farmers can use the information and complete the orders. Under this condition, farmers become the suppliers and the shoppers become the customers.

For Team 2, Table 6.2 shows the SIPOC. In the SIPOC, the community members face different challenges for various reasons like existing policies, lack of facilities, and so on. The community members gather and discuss the challenges faced by them. They identify the root causes of the challenges. To take the problem to a higher level, community members decide to elect a champion. The champion takes the idea of a policy change to the Commission. The Commission discusses the Policy Change idea, approves the idea, and prepares a draft policy. The draft is presented in the Legislative body and members of the body discuss and approve it. Once the Policy change idea is

SIPOC				
Explain the Customer / Supplier relationship in the process. Identify where the process begins and where it ends, and the activities included within the				
Suppliers	**Inputs**	**Process**	**Outputs**	**Customers**
Patient	* Patient information * Medical need * Order	Schedule Screening	* Screening appointment	Patient
Patient	* Insurance information * Patient information	Authorize Screening	* Insurance authorization for service	Patient Center
Patient	* Patient information * Order *Authorization * Insurance information * Payment	Perform Screening	* Screening films	Radiologist
Radiologist	* Screening films	Read/Send Results	* Results report * Screening films	Physician
Physician	* Results report * Screening films	Review Screening Results	* Results report * Diagnosis	Patient
Physician Center	* Results report * Diagnosis * Order(s)	Provide Results	* Results report	Patient

FIGURE 6.5 SIPOC example of a women's healthcare center.

SIPOC				
Explain the Customer / Supplier relationship in the process. Identify where the process begins and where it ends, and the activities included within the scope of the				
Suppliers	**Inputs**	**Process**	**Outputs**	**Customers**
Shopper	- Online food order form	Input info onto website	Order form to Farmer	Shopper
Shopper	- Online suggestions/survey	Input info onto website	List of Desired Produce	Farmer
Shopper	- Paper food order form	Input info onto form	Order form to Farmer	Shopper
Shopper	- Paper suggestions/survey	Input info onto form	List of Desired Produce	Farmer
Farmer	- Harvested food - Transportation	Delivery of the food	Food	Shopper
Shopper	- Feedback form	Input info onto form	Feedback to farmer	Farmer

FIGURE 6.6 Team 1 SIPOC – connecting organizations.

approved, the government makes it a law. Although this is somewhat simplified, the critical elements are provided in the SIPOC.

Team 3s SIPOC describes the processes and the related customers, suppliers, inputs, and outputs related to kids growing their own food. The SIPOC is shown in Table 6.3.

Team 4 used the SIPOC system to recognize various tasks to understand the internal process between customers and suppliers. Their SIPOC is shown in Table 6.4.

The SIPOC's purpose was to define the boundaries of the functions in Team 5's system. Given the functions, the boundaries begin with the food entrepreneurs investing in the system. The boundaries

TABLE 6.2

Team 2 SIPOC – Creating Food Policy

SIPOC

Explain the Customer/Supplier relationship in the process. Identify where the process begins and where it ends and the activities included within the scope of the process to be improved.

Suppliers	Inputs	Process	Outputs	Customers
Community Members	Challenges faced	Identifying the problem	Problem Identification	Community Members
Community Members	Discussions, Committee	Electing a Champion	Champion is elected	Community Members
Champion	Policy change idea	Presenting Policy change idea to Commission	Policy change idea is presented to the Commission	Commission
Commission	Discussions, Policy change idea	Commission discusses on the idea of policy change proposed by the Champion	Commission approves the Policy change idea and prepares a draft	Community Members
Legislation	Approved Policy change draft	Legislative Members discuss on the idea and, based on the majority, the idea is approved	Policy change idea is approved by the legislative members	Community Members
Government	Approved Policy change idea	The approved idea is made as a law by the Government	Law is made	Community Members

TABLE 6.3

Team 3 SIPOC – Kids Growing Food

SIPOC

Explain the Customer/Supplier relationship in the process. Identify where the process begins and where it ends and the activities included within the scope of the process to be improved.

Suppliers	Inputs	Process	Outputs	Customers
Teacher	Prototype	Garner children's interest	Interest, Independence, Fun	Child
Teacher	Container, seeds, dirt	Growing microgreens	Microgreens	Child, teacher
Child	Microgreens	Selling microgreens	Money	Table 33

end with restaurants and stores selling the produce produced within the system. The SIPOC for Team 5 is shown in Table 6.5.

Team 6 developed the SIPOC related to the functions of the GEM City Market, shown in Table 6.6. The SIPOC for the Table 33 Restaurant is shown in Table 6.7.

With the process boundaries defined in the SIPOC, we move into developing the value system and Value Chain Models.

TABLE 6.4
Team 4 SIPOC – Food Storage

SIPOC

Explain the Customer/Supplier relationship in the process. Identify where the process begins and where it ends and the activities included within the scope of the process to be improved.

Suppliers	Inputs	Process	Outputs	Customers
Farmers	• Food Containers Farmers	Packaging the food	Packaged food	Collector Homefull Staff
Farmers	• Temporary Storage Carriers Temperature system	Store food at temporary storage	Stored food	Collector Homefull Staff
Collectors	• CarrierDestination Information Request to collect food	Collect food from temporary storage	Collected food	Collector Homefull Staff
Collectors	• Truck with refrigerator trailer Temperature system	Transport food to Central Storage location	Collected food	Central Storage location staff
Homefull Staff	• Temperature system Inventory Packaged food	Store packaged food at Central Storage location	Stored food	Farmers Market
Homefull Staff	• Truck with refrigerator trailer Travel time Market information Request to distribute	Distribute food to market	Delivered food on time and fresh	Customers

TABLE 6.5
Team 5 SIPOC – Food Hub Website

SIPOC

Explain the Customer/Supplier relationship in the process. Identify where the process begins and where it ends and the activities included within the scope of the process to be improved.

Suppliers	Inputs	Process	Outputs	Customers
Food Entrepreneurs	Invest in Food Markets	Build New Food Markets	New Groceries, Food Stands	Local Residents
Realtors	Commercial Properties	Renting Properties	New Business i.e. Grocery or Food Market	Local Residents
Customer	Money	Purchasing Goods	Economic Growth in City	City of Dayton
Farmer	Produce	Farming Produce	Fresh Food	Restaurants, Supermarket
Supermarket	Fresh Goods	Sell those goods	Fresh Food provided to Locals	Local Residents
Restaurants	Provide Food	Cook/Prepare Meals	Meals	Patrons

TABLE 6.6
Team 6 SIPOC – GEM City Market
SIPOC

Explain the Customer/Supplier relationship in the process. Identify where the process begins and where it ends, and the activities included within the scope of the process to be improved.

Suppliers	Inputs	Process	Outputs	Customers
Community Farmer Distributor	Request for food products	Obtain food products	Food products	Community workers Management
Community workers	Services	Manage store	Good customer service	Customers
Customers	Request for food products	Purchase food products	Food products	Community Customers
Customers	Reports Review Feedback	Review results	Results	Customers Management
Management	ReportsResults	Meet sales goals	Met goals	Management

TABLE 6.7
Team 7 SIPOC – Table 33 Restaurant
SIPOC

Explain the Customer/Supplier relationship in the process. Identify where the process begins and where it ends and the activities included within the scope of the process to be improved.

Suppliers	Inputs	Process	Outputs	Customers
Local farmers	Food Ingredients	Providing fresh ingredients for the restaurant	Organic ingredients	Table 33
Table 33	Meals	Preparing food	Money/Profit	Restaurant customers
Inventory manager	Report missing/ needed ingredients	Check the inventory to identify missing/needed ingredients to fulfill the menu	Provide a list of missing ingredients	Restaurant manager
Restaurant manager	Connect with local farmers	Look for local farmers to supply the restaurant	Local farmers	Table 33

6.3.2 VALUE SYSTEM AND VALUE CHAIN MODELS

The value chain is a chain of activities that provide value to your customer and enable competitive advantage. The concept of a value chain was first described and popularized by Michael Porter (Porter, 1985). Porter terms this larger interconnected system of value chains the "value system." A value system includes the value chains of a firm's supplier (and their suppliers, etc.), the firm itself, the firm's distribution channels, and the firm's buyers (and presumably extended to the buyers of their products and so on) (Porter, 1980). Value systems can consist of value activities, also called primary activities, and support activities. Value activities are the key, critical activities that provide value to the customer and can differentiate our organization from our competitors. The support activities are necessary but are not core to providing the value to our customers.

Porter's value chains include the following primary activities for a manufacturing organization:

- Inbound logistics: includes receiving the materials, storing and managing inventory, and scheduling transportation.
- Operations: all of the operational processes required to make the product.
- Outbound logistics: the distribution activities to ship the finished product including warehousing, order management, transportation, and distribution.
- Marketing and Sales: marketing and sales of the product.
- Service: customer service, repair, installation, and training.

The support activities in Porter's model include:

- Procurement: purchasing of raw materials and components.
- Technology development: technology development, including research and development and product design.
- Human Resource Management: recruiting, development, retention, and employee compensation activities.
- Firm Infrastructure: includes all of the managerial support functions including legal, accounting, finance, quality management, strategic planning, etc.

In our value system we define the concept somewhat differently than Porter's model. We define a value system as an aggregation of the value chains that provide value to the customers. The value chains are the sets of activities that provide a specific service or product line to the customer. In a manufacturing organization, the majority of the product lines would use the same activities and value chains to create the product, so multiple sets of value chains would not be necessary. For a service organization, the activities that are part of the value chain could vary for different service lines more than a manufacturing organization's value chains. By defining our value system differently, we enable our model to be used for a manufacturing and a service organization.

The Value System Model is related to the FDM through the value chains. The value chain consists of business functions. The value system is composed of the value chains. The value chain consists of primary value-creating activities and support activities. The value activities provide value to the organization and to the customers. The value system is leveraged by the business capabilities through the business processes.

Value System Model Element Definition:

- Value System: We define a value system as an aggregation of the value chains that provide value to the customers.
- Value Chain: The value chains are the sets of activities that provide a specific service or product line to the customer.
- Value Activity: An activity within a value chain that provides value to the customer.
- Support Activity: An activity that supports value activities in a value chain. It does not directly add value to the customer.

The Women's Healthcare Center value chain model is shown in Figure 6.7.

The food system's value chain model is shown in Figure 6.8. Each of the teams will be aligned to one or more of these activities from the value chain model perspective.

Table 6.8 illustrates the value chain activities that relate to each team's scope of their system. Their functional decompositions assume that their first level functions relate to these value chain activities, as level 0.

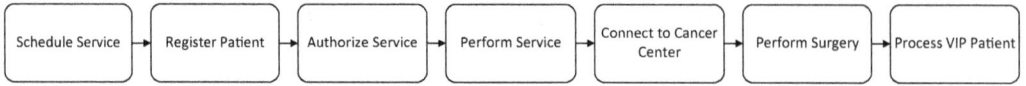

FIGURE 6.7 Women's healthcare center value chain model.

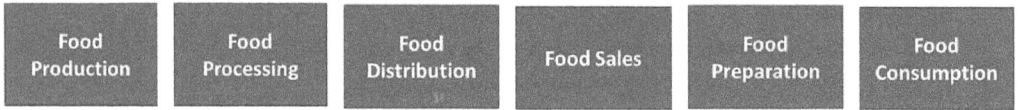

FIGURE 6.8 Food system value chain model.

TABLE 6.8
Teams and Related Value Chain Activities

		Related Value Chain Activities					
#	Team Name	Food Production	Food Processing	Food Distribution	Food Sales	Food Preparation	Food Consumption
1	Connecting organizations	X	X	X	X	X	X
2	Creating food policy	X	X	X	X	X	X
3	Kids growing food	X		X	X		
4	Food storage	X	X	X			
5	Food hub website	X		X	X		
6	GEM City Market	X		X	X		
7	Table 33 Restaurant	X		X		X	X

6.3.3 FUNCTIONAL DECOMPOSITION MODEL

The next model that we describe is the FDM. A FDM performs the following:

- Identifies and decomposes the functions of our system
- Decomposition: breaking down the functions into smaller functions
- Enables modeling of the functions into process maps

The FDM is used to identify system functions and the processes that comprise them. A function consists of those operations performed by the system. The functional decomposition diagram itself does not depict process flows but rather the hierarchical organization of functions and the processes that they include. The FDM then is in essence a taxonomy or inventory of system functions in that each top-level function generalizes its lower-level functions.

Traditionally a FDM has been used to identify the functions which allow us to define our processes and business systems. The intent of a FDM is related to the value chain and to capture and understand the current state or AS – IS structure of system functions and their relationship to processes.

A FDM model has a relationship to the Process Model through the set of processes identified at the lowest levels of the hierarchy. The value chain is made up of the system functions.

The FDM starts with the high-level functions in the value chain.

FDM Element Definition:

- System Function: A system function consists of those operations performed in the organization.

  At some level, system functions break down into processes. The conceptual dividing point is somewhat arbitrary, but you can usually differentiate a process from a function by the amount of activity that it represents. A process represents a tangible activity that occurs within the system. A system function is usually described in a noun-verb pairing, in that order, such as "Patient Registration."

  Steps to develop the FDM:

1. Identify the scope of the system to be modeled.
2. Identify the highest-level functions, using noun-verb names.
3. Decompose these functions into lower-level functions, again using noun-verb names.
4. Stop when there becomes a natural sequence of the activities, usually 2 to 4 levels. Next is the time to use a process map.

Note: "Don't let perfect get in the way of better." This functional decomposition doesn't need to be perfect. It's better to get something on paper for the stakeholders to validate and augment through interviews and discussions.

The FDM for the Women's Healthcare Center is shown in Figures 6.9 and 6.10.

The FDMs for the food system are shown in Figures 6.11 through 6.17. These can be shown in a table or a graphical diagram depending on the team's preference. For the table each row is a function, where the first level functions are listed on the far left and the decomposed second level functions on the same row to the right.

For Team 1's system's functional decomposition (Figure 6.11), the functions are defined at two separate levels. The first level includes food sale and food production. These two aspects reflect the hierarchy definition of our functional processes for the system. The second level lists the functional steps taken that aid in development of the process scenarios created.

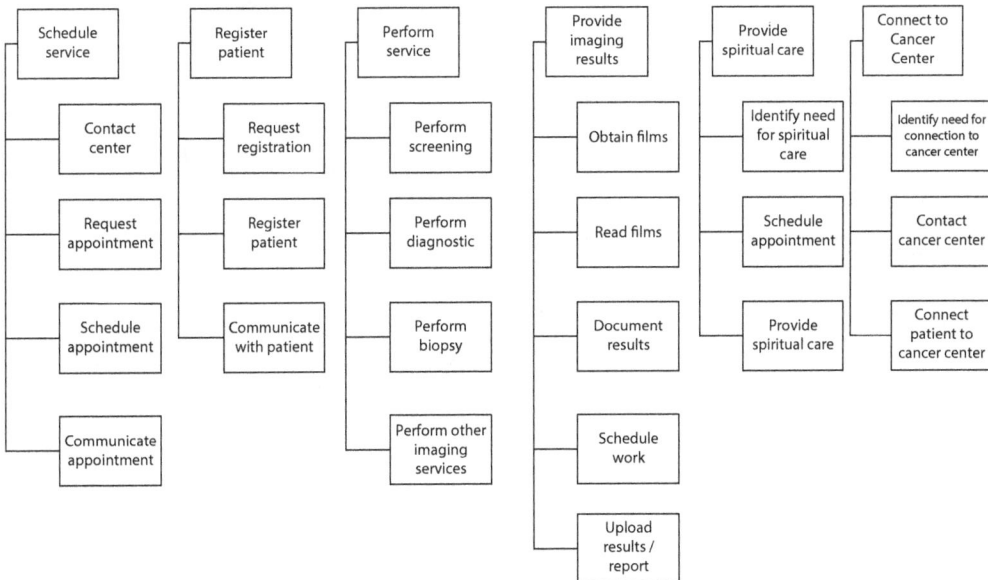

FIGURE 6.9 Functional decomposition diagram women's healthcare center –Part 1.

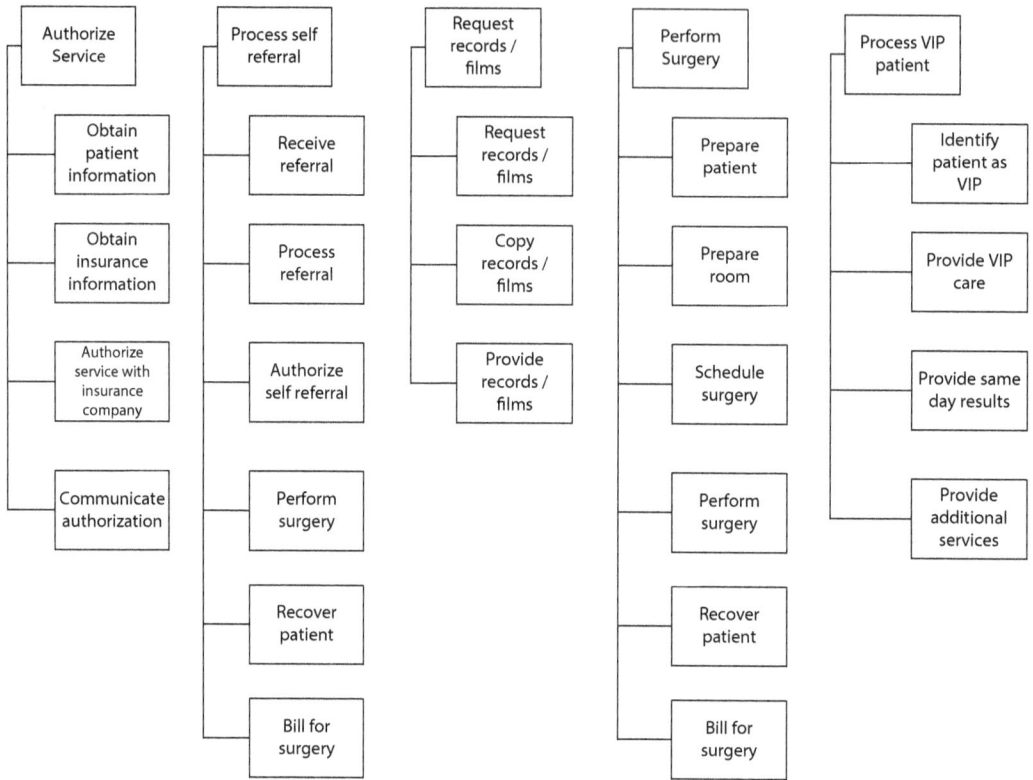

FIGURE 6.10 Functional decomposition diagram women's healthcare center –Part 2.

First Level	Second Level			
Food Sales	Obtain Form	Fill out order	Fill out survey	Return/Submit form
Food Sales	receive data	Analyze Data	Deliver order to farmer	Prioritize suggested items
Food Sales	digital/physical advertising	host event (provide food)	contact community partners	Provide requested goods list to farmers
Food Sales	identify community member/organization	connect people to each other and to organizations	Self-advertise / word of mouth	Forum
Food Production	Analyze Data	Report Data to Farmers	Identify types of food needed	Request farmer that will provide needed food

FIGURE 6.11 Team 1 functional decomposition model – connecting organizations.

The second level of food sales includes completing forms and surveys, addressing data, advertising, and connecting the community members and organization. These functions reflect the inside and outside activity for the whole website. These functions include the following:

1. Completing forms and surveys: The team will obtain forms via both papers and electronic versions. Next, community members can fill out their form and survey. Then, members can send their form to the collecting location or by email to the website administrator.

2. Addressing data: Website administrators will receive the data in the database and the data will be analyzed be Citywide volunteers. Analysts analyze and prioritize the data. Then, the data is transferred to the farmer to fulfill orders and supply requested items.

3. Advertising: Advertisement via digital and physical advertisements are used to promote the program and website. This includes hosting events where foods may be purchased but also spreading the word about the program. Contacting community partners and providing requested goods lists to farmers are also two important activities to promote food sales.

4. Connecting the community members and organization: For this part, the website allows the connection of community members, farmers, and organizers via a forum. Additionally, organizers may promote their companies as well as perform self-advertisement on the website forum and by word of mouth.

The second level of food production is analyzing data received from surveys completed by community members. It is the main function which connects the community members, organizations, and farmers. Additionally, the data prioritizes the most requested foods by the communities.

For Team 2, the FDM helps to identify and decompose the functions of their system which is generally accomplished through thoughtful analysis and team discussions. This enables modeling of the functions into process maps. The FDM, shown in Figure 6.12, describes the various functions of the policy change process and its decomposition. Each of the functions has the smaller individual functions that detail out the main function for a better and deeper understanding of the system. Of the seven functions, the first three, problem identification, champion election, and gather support data are crucial, because the rest of the functions largely depend on the champion and the data collected initially.

The FDM is shown in Figure 6.12 that describes the food growing functions.

Team 4 utilized the functional decomposition to define the high-level functions of the central storage location. Figure 6.14 shows the functional decomposition the team created for the central storage location.

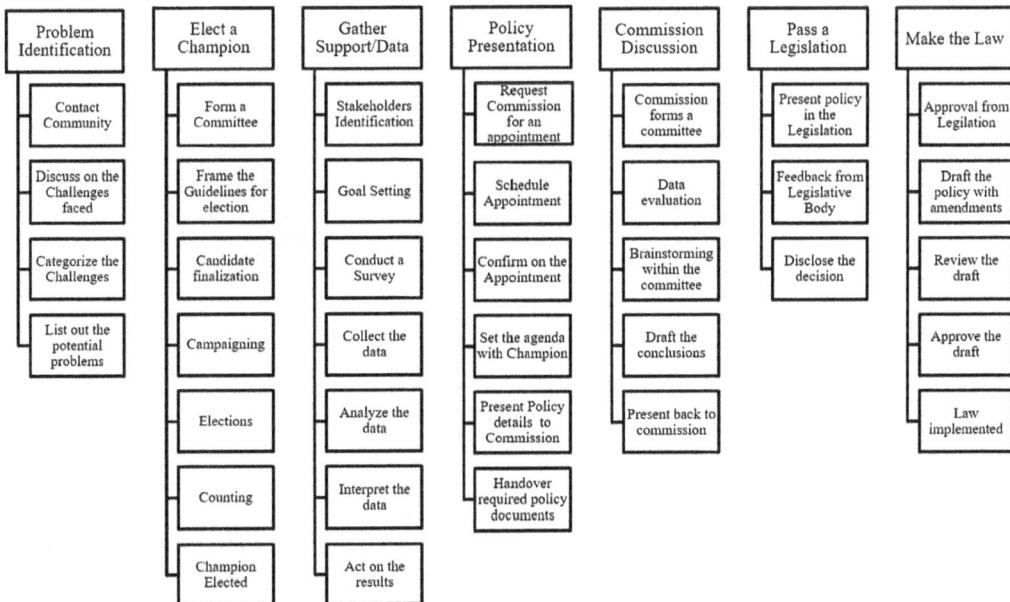

FIGURE 6.12 Team 2 functional decomposition model – creating food policy.

Functional Decomposition					
The project functional decomposition is used to identify and decompose the functions of the system and break the functions down into smaller functions.					
First Level	**Second Level**				
Food and Growing Instruction	Teach children about microgreens	Show how to grow microgreens	Children practice set up for growing		
Food Production	Acquire and distribute supplies	Children grow microgreens at home			
Food Processing	Transfer microgreens back to school	Assess microgreens for quality	Acquire compensation for microgreen crop	Harvest (cut) microgreens	Clean microgreens
Food Distribution	Store in clean refrigerator in damp paper towels for distribution	Monitor length of time in storage	Retrieve from storage in a timely manner	Transport in a sanitary manner (cooler) to Table 33	Acquire compensation for microgreen crop from end user

FIGURE 6.13   Team 3 functional decomposition model – kids growing food.

Functional Decomposition				
First Level	**Second Level**			
Storing the food in temporary storage at the farm	Harvesting the produce and collecting it	Culling the produce	Packaging the food with suitable containers	Storing the food at temporary stoarge until food collection time
Collecting food from temporary storage	Preparing the produce to be collecting by packing in	Collecting the packaged food	Requesting the pick up time	Storing the package food on the trailer with the refrigerator
Storing packaged food at a central storage lcoation	Recording the inventory of the food	Stored the packaged food until distributing	Setting up the temperature system to maintain for the long term	
Distribute food to the market	Using electrical bike to distribute	Distribute the food by mobile grocery truck	Distribute the food through the corner market	

FIGURE 6.14   Team 4 functional decomposition model – food storage.

The main functions for Team 5 are food production, food distribution, and food sales. These three are then broken down into more specific functions that relate to the main functions. The FDM is shown in Figure 6.15.

Team 6's FDM (Figure 6.16) consists of two levels, the First level and the Second level. The First Level consists of the main aspects such as improving according to demand, hybrid pricing, marketing, and delivery and pickup. The Second Level consists of the activities that are associated with each of the First Level functions.

Team 7's FDM is shown in Figure 6.17. The high-level functions are the inventory management systems, local farms, transportation, kitchen, and finally Table 33 Restaurant. These functions have been decomposed as we move to the lower-level function in the second level.

Functional Decomposition	
First Level	**Second Level**
Food Production	Utilize local farmers and their goods to be sold
	Market demand across all levels of those involved in the system
	Will increase quality/efficiency of food delivery
Food Distribution	Allow cross communication between all three groups: distribution , produce, and buyers
	Will increase efficiency of produce delivery
	Create direct communicaton between producers and distributor
	Shows real time need for distributors
Food Sales	Directly relay the costs of goods for markets
	Help customer know what food is available and where

FIGURE 6.15   Team 5 functional decomposition model – food hub website.

Functional Decomposition					
First Level	**Second Level**				
Improving according to demand (Obtaining food products)	Checking products	Instore surveys	Online surveys	Getting feedback	Improving the store
Hybrid Pricing (Promoting the pricing and managing the store)	Price selection	High end products purchased	Basic products purchased	Sales Statistics	Adjusting price if needed
Delivery and Pickup (Purchasing the food products)	Make an Application (Mobile/website)	Shopping list	Ordering online	Packaging of products	Trasport to customer
Marketing (Meeting the requirements and getting feedbacks)	Buying food products	Getting food sales from the sale statistics	Promote the food products	Sales Statistics	Based on the sales made provide offers and discounts to attract customers and improve food sales

FIGURE 6.16   Team 6 functional decomposition model – gem city market.

6.3.4   Use Case Diagrams

Use cases diagrams are a graphical representation of use cases or functions organized from the users' perspective. They are used to generate the system's requirements. The use cases, which are covered in the design phase, are narrative descriptions of the use case diagrams. There are several advantages of use case diagrams and use cases:

- The system requirements are placed in the context of the users work and domain of knowledge. Each requirement is part of a logical sequence of actions or functions to accomplish a goal of a user.
- Use cases are easy to understand. The model expresses requirements from a user's perspective and in the user's own vocabulary. The model shows how users think of the system and what the system should do.
- Use cases show why the system is needed. Use cases show what users can accomplish by using the system.
- Use cases show the interactions between the functionality and the actors in the system.

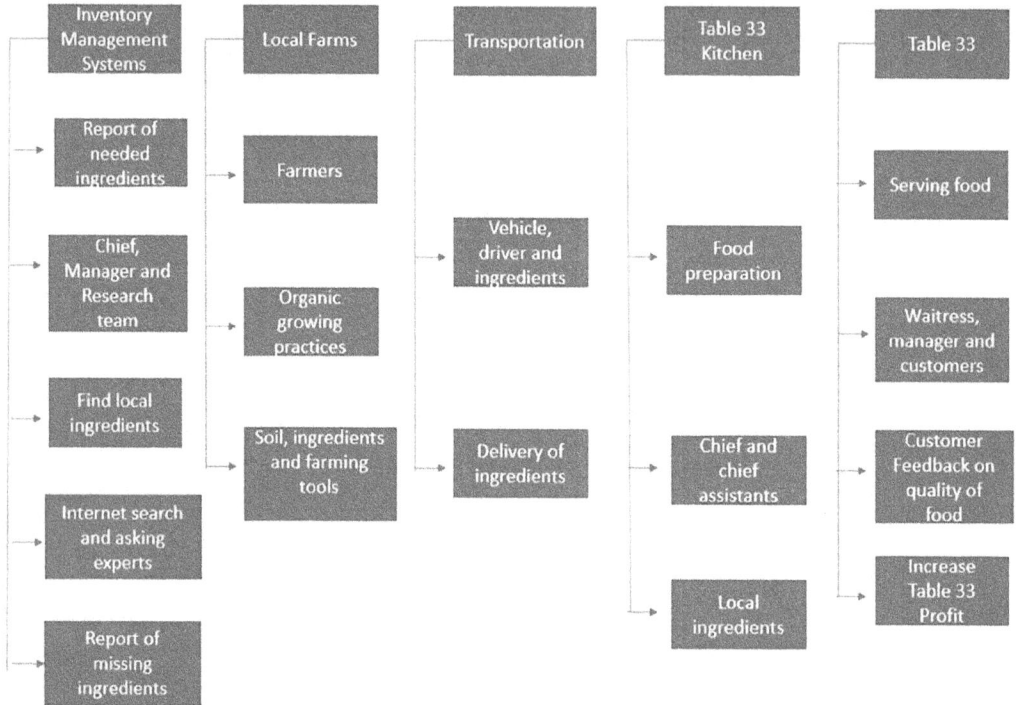

FIGURE 6.17 Team 7 functional decomposition model – table 33 restaurant.

Definitions of use case diagram elements:

Uses or functions: The actions performed by the system.
Use case: Defines and describes a reasonable piece of system functionality.
Actors: The roles or systems that perform functions from outside of the system that interacts
        with the system. Actors can be people-related roles, for functions that are performed
        without an automated system. Actors can be systems that people interact with, to pro-
        vide semi-automated functionality. Actors can also be a system that fully automates the
        functionality.
System: The product or service that is being described by the collection of use cases.
Associations: Shows the actors and use cases that interact.

- Extends: the extends relationship between two use cases indicates that an instance of use case
  B can include the behavior specified by use case A. In other words, B is a special representation of A due to unique circumstances. The arrowhead in the extends relationship points
  toward the base use case, not toward the extension.
- Uses: the uses relationship between two use cases indicates that an instance of use case A will
  also include behavior as specified by use case B. In other words, the process of doing A always
  involves doing B at least once. The arrowhead points toward the use case being used.

An example of a use case diagram for the women's healthcare services is shown in Figure 6.18.
    Sample food system use case diagrams are shown in Figures 6.19 through 6.26. Each sub-system
or team may have one or several use case diagrams to describe their entire sub-systems. These use
case diagrams are shown for each sub-system to give the reader an idea of how to construct the use
case diagrams.

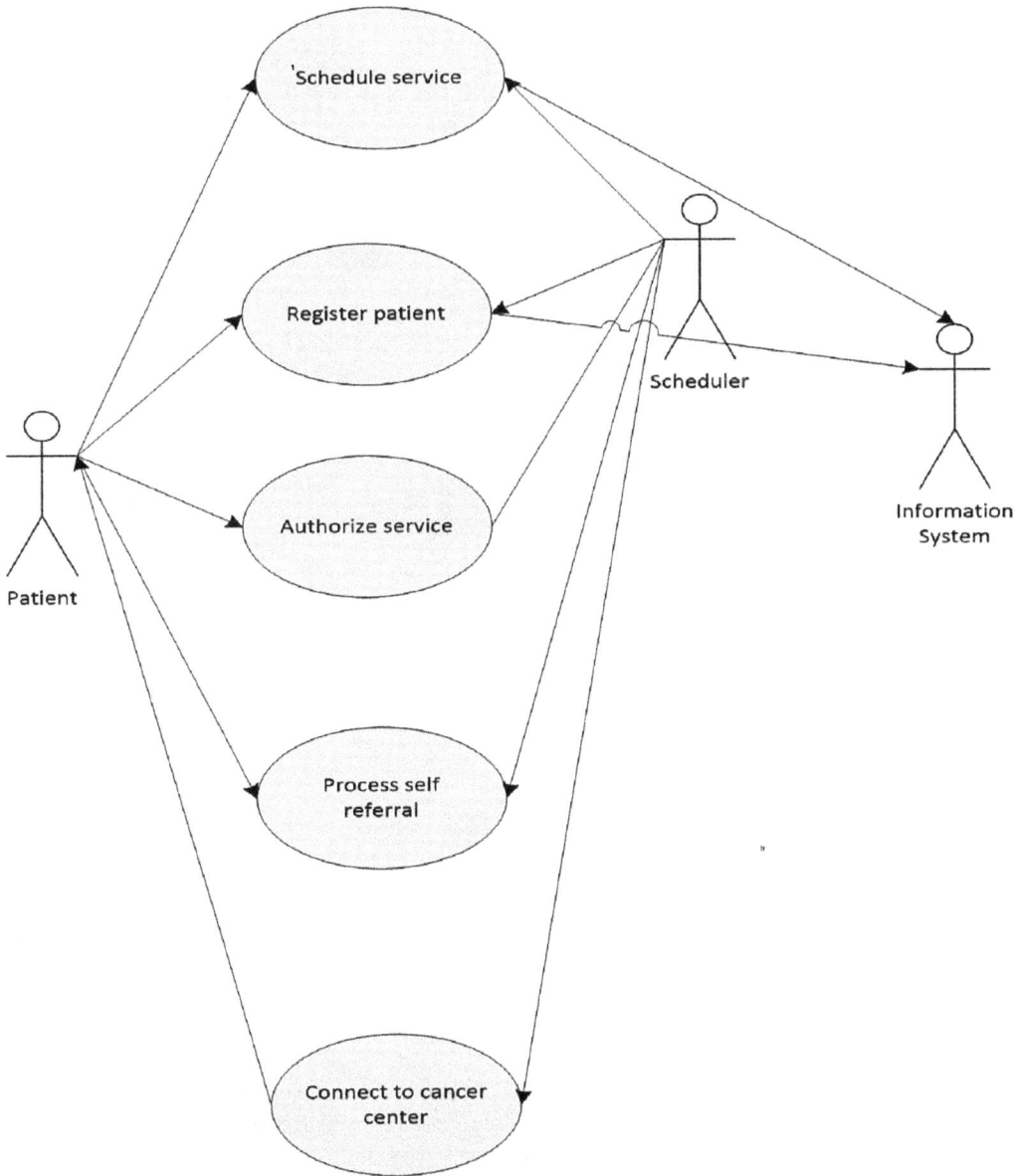

FIGURE 6.18 Women's healthcare center use case diagram.

Team 1's use case diagram, shown in Figure 6.19, identifies three main actors: Farmers; Citywide, and Community Partners. Arrows represent influence of these actors on certain departments or functions. Note that some arrows are coming toward the actors which mean there are some functions which require feedback from these actors.

The class diagram shown in Figure 6.19 and 6.20 is for the policy change system. The class diagram for the policy change shows that the two main actors in the system are the "champion" and the committee commission that votes on the potential legislation.

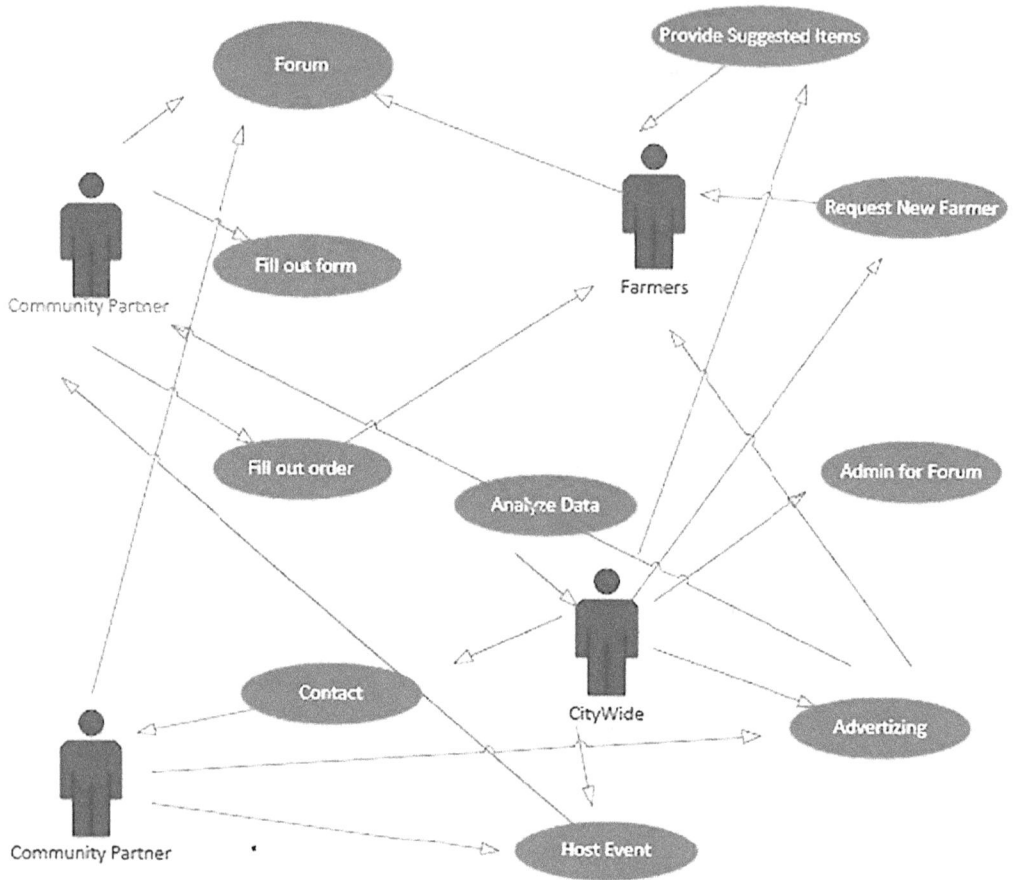

FIGURE 6.19 Team 1 use case diagram for connecting organizations.

Team 3 utilized the use case diagram (Figure 6.20 and 6.21) to determine the various relationships between actors, including the teacher, children, and Table 33 Restaurant, and the functions including producing, processing, and distributing the microgreens.

Figure 6.22 shows the use case diagram for the centralized storage location system.

For Team 5, the use case diagram (Figure 6.23) maps how their system will actively run. In the system, the first action is that an entrepreneur invests in a business owner. From there, the website can help the other users interact and make the system run smoothly. Similarly to the other parts of this report, this diagram helps us keep each user and each function in mind when the system is designed.

For Team 6, the food sales consist of two use case diagrams to illustrate the coordination between the management and the customer. The tool identifies the inputs of the management team, community farmers and workers, customers, distributors, and suppliers. The first use case diagram (Figure 6.24) covers the flow in the food sales system from how food production begins to its delivery in the store and how it reaches the customer. The second use case diagram (Figure 6.25) covers the customer and the in-store management which depicts the flow of how the customer buys the product in the store as well as how the workers would manage and act in the use case.

Team 7's system use case diagram is illustrated in Figure 6.26. The diagram shows how the system is used and defines the relationships and interactions between the actors in the system.

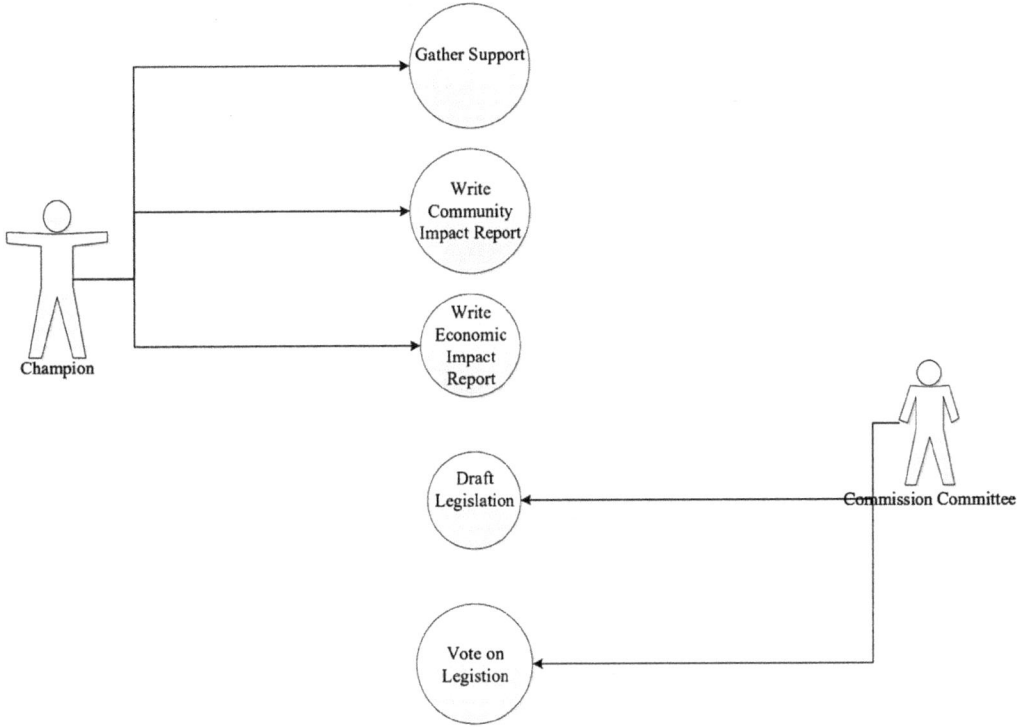

FIGURE 6.20   Team 2 use case diagram creating food policy.

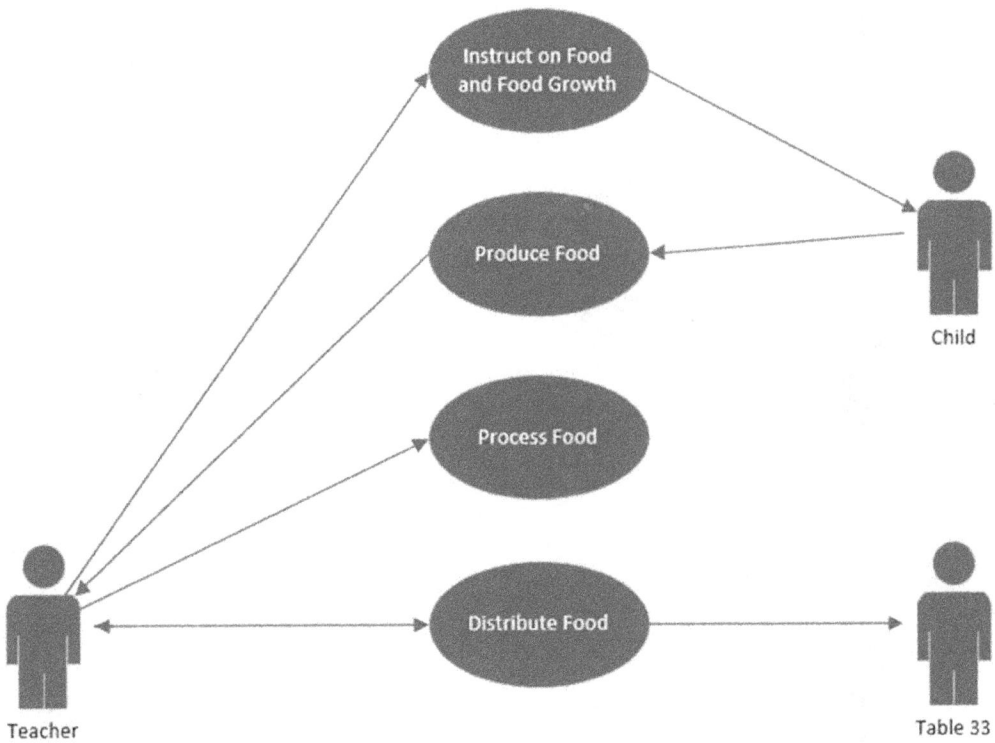

FIGURE 6.21   Team 3 use case diagram for kids growing food.

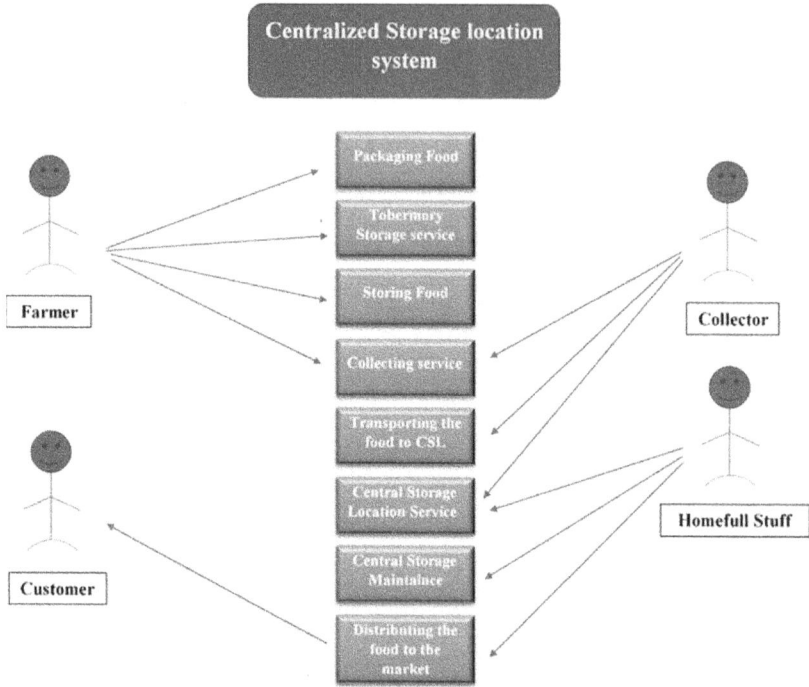

FIGURE 6.22   Team 4 use case diagram for food storage.

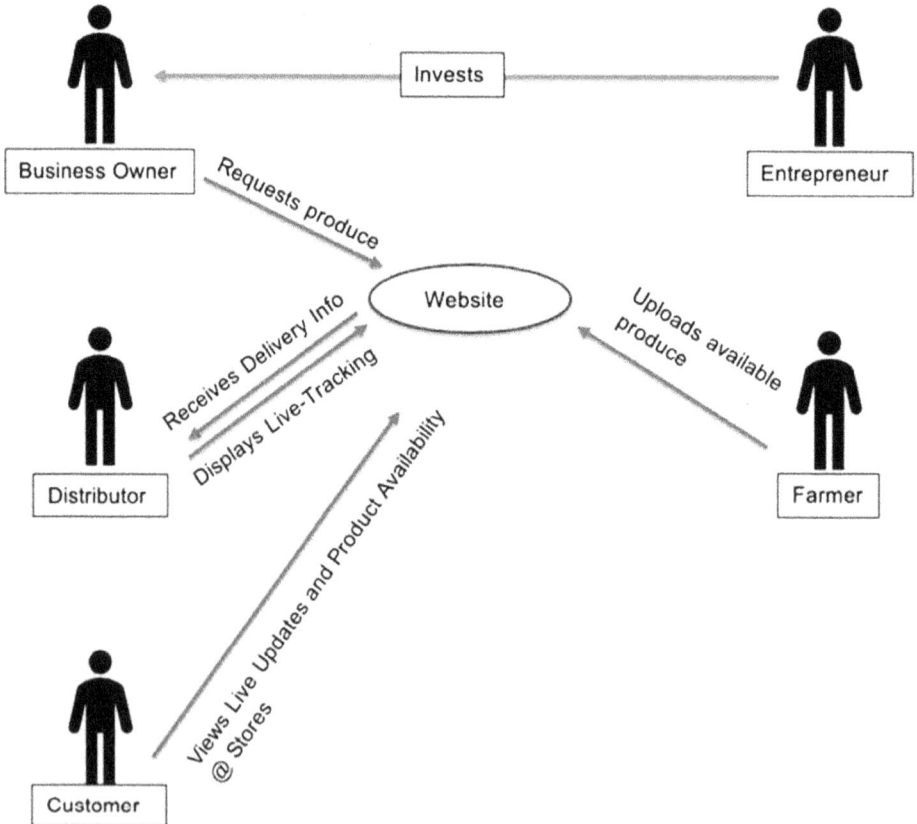

FIGURE 6.23   Team 5 use case diagram for food hub website.

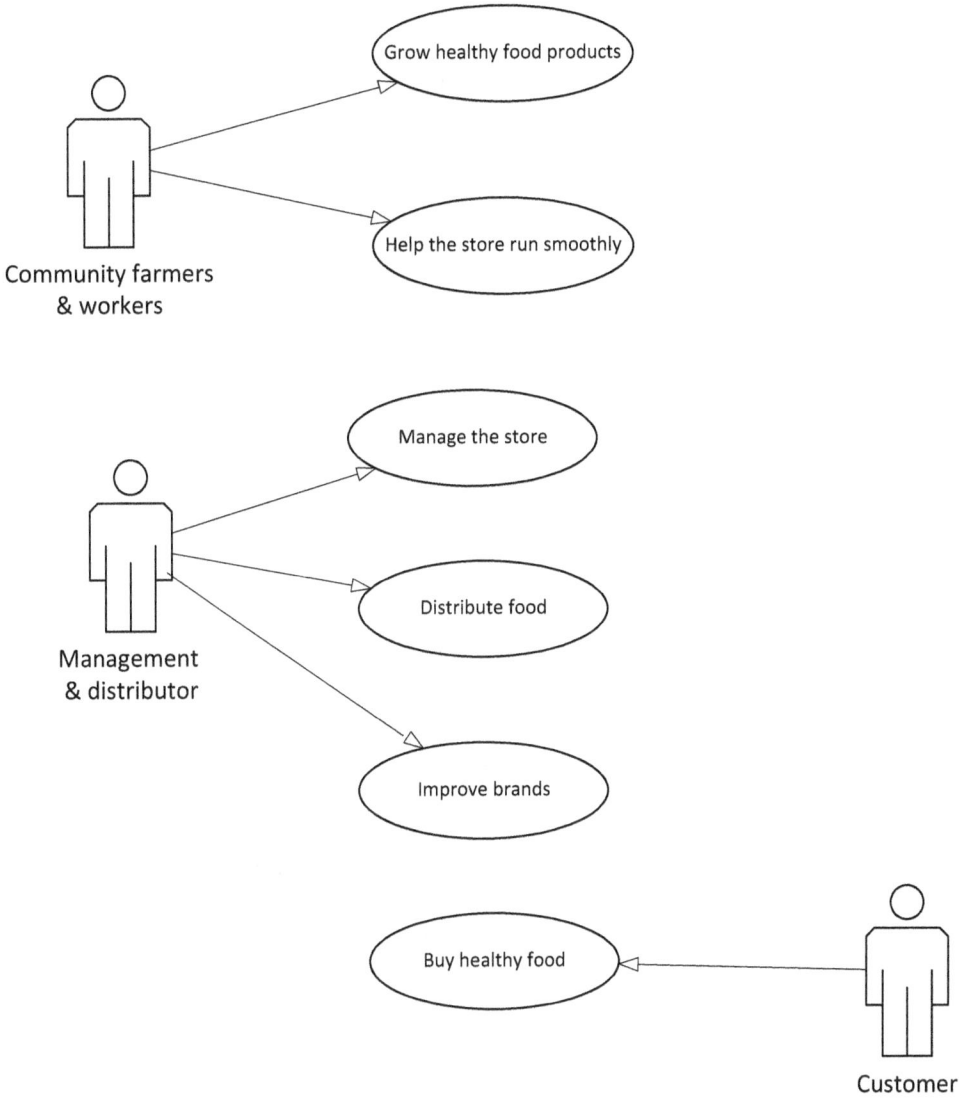

FIGURE 6.24   Team 6 use case diagram for gem city market – food sales.

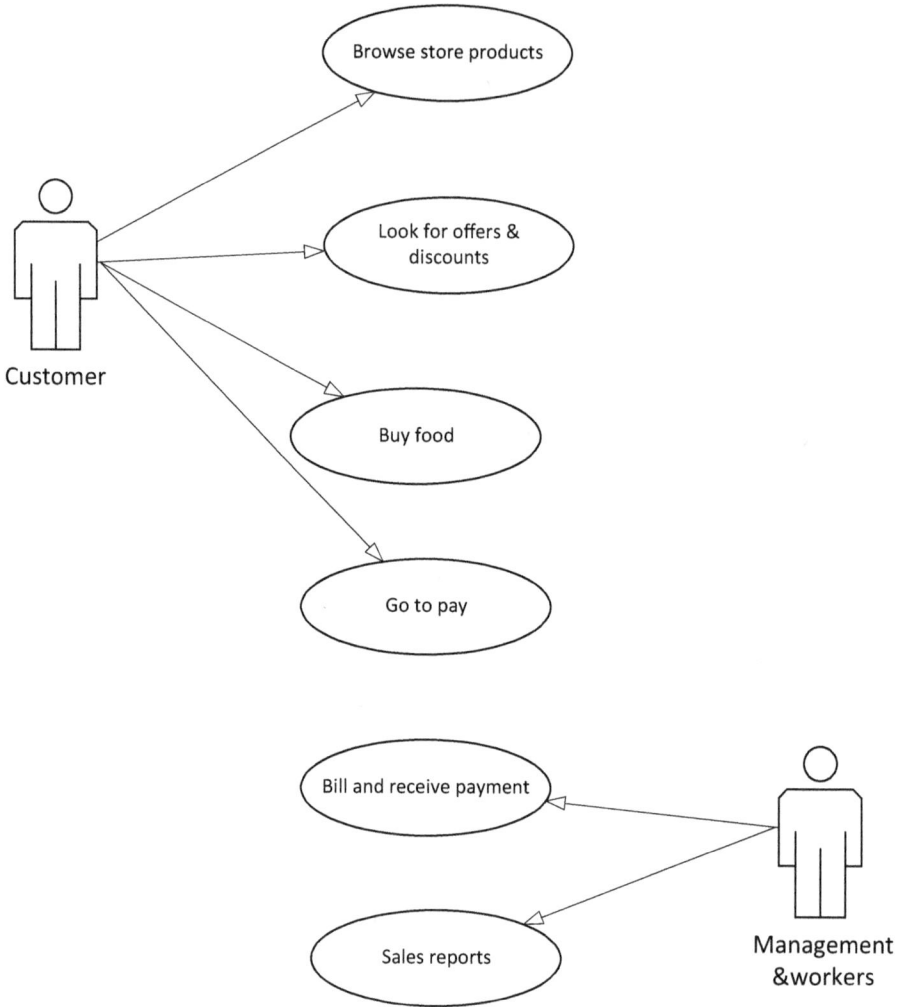

FIGURE 6.25   Team 6 use case diagram for gem city market – in-store management.

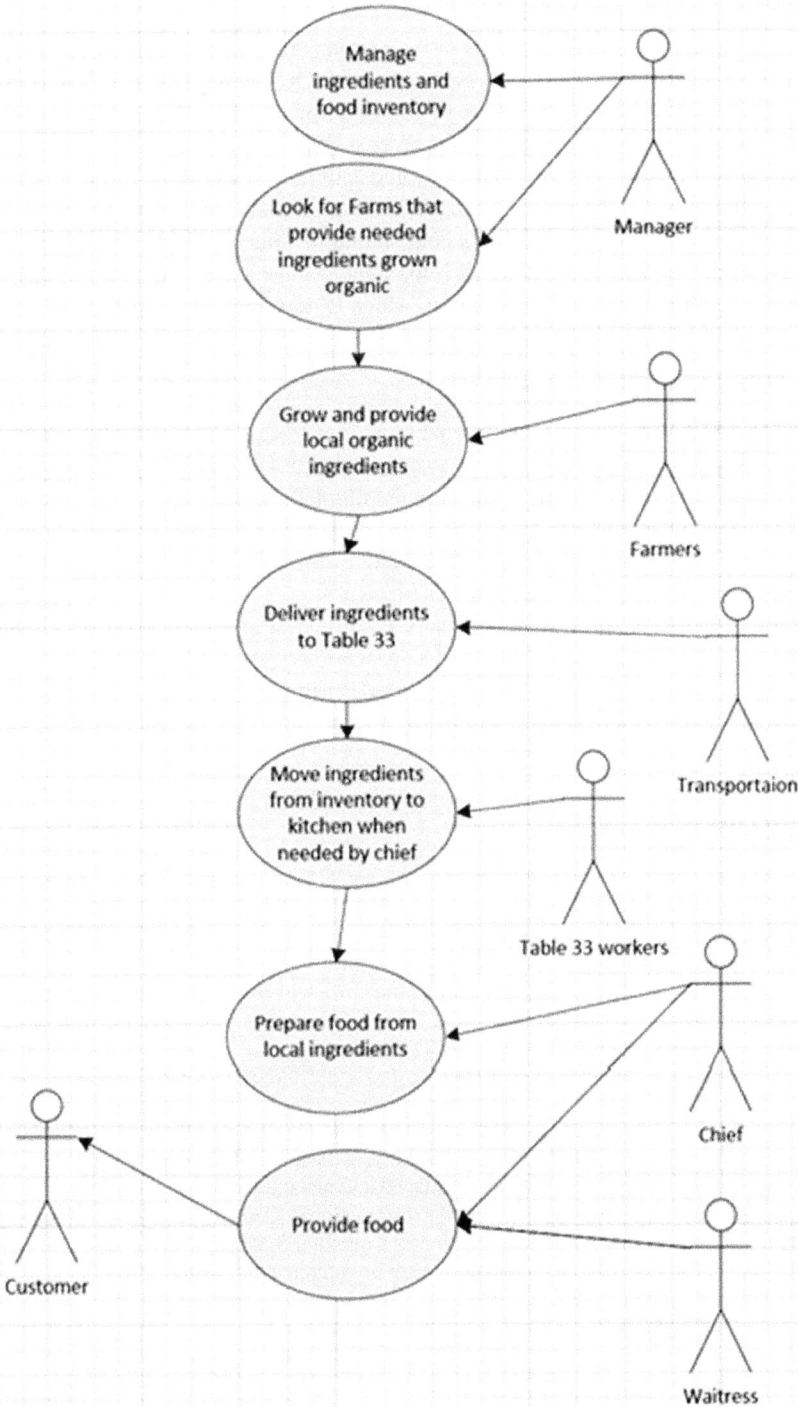

FIGURE 6.26   Team 7 use case diagram for table 33 restaurant – in-store management.

6.4 SUMMARY

The tools in the systems architecture part of the architecture and requirements phase help to define the functions to be performed by the system. The SIPOC, value chains, FDM, and the use case diagrams all interrelate but tell the store of the functionality in different, but interrelated views. The SIPOC provides five to seven high-level process steps or functions that bound the system. The SIPOC functions are typically the high-level value activities in the value chain, which then become the first level functions in the FDM. These functions are then used in the use case diagrams. The stakeholders, customers, or users of the system should be interviewed to understand how they want to use the system, which will enable the development of these tools. We then are ready to move into the requirements elicitation part of this phase, which is described in Chapter 7.

6.5 ACTIVE LEARNING EXERCISES

1) Develop a SIPOC to scope your system boundaries from a functionality perspective. Ensure that the suppliers and customers are also included in your stakeholder analysis from your Concept of Operations phase. If they are not aligned, what is the reason for this? Has your scope changed since Phase 1?
2) Develop a Value Chain Model for your system. Ensure that the value chain activities are aligned to your SIPOC process steps or functions.
3) Develop a FDM for your system. Ensure that the first level functions are aligned to your Value Chain Model as well as the SIPOC process steps or functions.
4) Develop use case diagram(s) for your system. Ensure that the uses or functions align to the functional decomposition, typically the first level functions.

BIBLIOGRAPHY

Business Process Management Notation, Object Management Group/Business Process Management Initiative http://www.bpmn.org/, 2010.
Evans, J. R., and Lindsay, W. M. *Managing for Quality and Performance Excellence*, Eighth Edition, South-Western Cengage Learning, Mason, Ohio, 2011, 2008.
Furterer, S., Bonet, M., Venable, C., McKenzie, E., Dunn, M., and Hogg, B, Business Architecture Cookbook, White Paper 2009.
Model element definitions, www.dictionary.com, 2010
Porter, M. E. *Competitive Strategy: Techniques for Analyzing Industries and Competitors*. New York: Free, 1980.
SEBoK Editorial Board, *The Guide to the Systems Engineering Body of Knowledge (SEBoK)*, vol. 2.2, ed. R. J. Cloutier (Hoboken, NJ: Trustees of the Stevens Institute of Technology, 2020), accessed May 25, 2020, www.sebokwiki.org. BKCASE is managed and maintained by the Stevens Institute of Technology Systems Engineering Research Center, the International Council on Systems Engineering, and the Institute of Electrical and Electronics Engineers Computer Society.
TOGAF 8.0, The Open Group, https://pubs.opengroup.org/architecture/togaf8-doc/arch/toc.html, 1999-2006
Use Case Training, Walmart Stores Inc., 2007.
Value Chain, http://en.wikipedia.org/wiki/Value_chain, 2010.

7 Requirements and Architecture Phase 2
Elicit Requirements

In this chapter, we elicit the customer requirements that are part of Phase 2 – Requirements and Architecture.

7.1 PURPOSE

In this chapter, we discuss the elicitation of the customers' requirements, and in the next chapter we discuss the derivation of the systems or technical requirements from the customers' requirements and the derivation of specialty engineering requirements. The models and tools in this phase help to illustrate a view of the high-level functionality to be provided by the system. The models do not yet represent a physical manifestation but a conceptual one. The distinction being that this view does not represent an implementable solution. When I was in software development, I would develop logical requirements for the software developers. I might provide logic related to performing field-level and groups of field-level validation, in the order that a user might enter the data. However, when coding and creating a physical instantiation of the logic, the programmer might pull all of the data and validate the edits in an entirely different order for coding efficiency purposes, while maintaining the same logic and validation requirements. Sometimes the programmers would expect the logic to be provided to them in the most efficient manner, not realizing that my job was to provide the logical requirements, not the coding efficiency parameters and technical specifications, for that was part of the next phase, the design phase. So, these distinctions for understanding what is provided in this phase, separating the customers' requirements from the way that it is implemented as a system solution is important.

7.2 ACTIVITIES

The activities performed and the tools applied in the Requirements and Architecture phase are shown in Table 7.1. We cover the "Elicit Customer Requirements" activities and the associated Requirements tools.

7.3 ELICIT CUSTOMER REQUIREMENTS

Engineers learning design and systems engineering often don't see the need or value of talking to the customers of the system to elicit the "real" requirements. Even if they are working on a real-world project, they still have the tendency to want to make things up. Some quotes from Dannels's (2000)

DOI: 10.1201/9781003081258-9

TABLE 7.1

Activities and Tools in the Requirements and Architecture Phase

Vee Phase	Activities	Tools	Principles

Phase 2: Requirements and Architecture	Develop Logical Architecture
	1. Develop Logical Architecture
	
	Elicit Requirements
	
	2. Elicit customer requirements
	3. Derive requirements
	  • Develop system requirements
	  • Develop specialty engineering requirements
	4. Develop measurement plan
	5. Develop information models
	6. Develop a quality management plan
		Logical Architecture:
		• SIPOC
		• Value Chain and Functional Decomposition
		• Use Case Diagrams
		
		Requirements:
		
		• Process Scenarios via Process Architecture map
		• Use Cases
		• Customer requirements
		• System requirements
		• Specialty engineering requirements: software engineering; environmental engineering; safety and security engineering; Human Systems Integration
		• Data Collection Plan, with operational definitions
		• Class Diagram with information model and hierarchy model
		• Quality management plan
			• System dynamics (behavior, system elements)
			• Cybernetics (information flow)
			• Systems thinking
			• Abstraction
			• Views

real students demonstrate the need for students learning these concepts to understand the value of talking to actual customers to elicit requirements:

- *Laura: We will make up customer needs for this assignment because it is required but actually talking with customers is not real.*
- *Peter: Talking to customers is just a hoop. Just make something up to get the assignment done. It really doesn't matter for our design.*
- *Eric: We just don't have time. There are too many other important design tasks to be done.*
- *Bob: Talking with the end user is not real. We engineers are not responsible for that.*
- *Mike: Just go ask Dr. Woolf. He can act like a customer or go to another design team; they can pretend to be a customer.*

This way of student thinking is detrimental to their learning in this "safe" environment of the university setting but can be catastrophic when these same engineers actually design a system in the real world.

There are many different approaches for eliciting customers' requirements, some of which are:

1) Structured brainstorming workshops
2) Interviews and questionnaires
3) Technical, operational, and/or strategy documentation review
4) Simulations and visualizations
5) Prototyping
6) Modeling
7) Feedback from verification and validation processes

8) Outcomes from systems analysis processes
9) Quality Function Deployment (QFD)
10) Use case diagrams
11) Process Scenarios via Process Architecture Map (PAM)

Developing requirements from process scenarios, via the PAM and use cases, allows for traceability of requirements during the verification and validation phases, because the test cases are easy to develop based upon the use case narratives. This traceability is shown in Figure 7.1.

7.3.1 FACILITATE PROCESS SCENARIOS TO CREATE PROCESS ARCHITECTURE MAPS (PAMS) AND USE CASES

The approach that we will take to elicit requirements is to facilitate process scenarios with the customers, by using the PAM, a tool developed by the author. The information derived from the process scenarios will also be documented in more detailed narrative-based documents called use cases. In this section, we discuss the process scenarios and the application of the PAM and then cover the use cases in the next section. The PAM combines a process map tool with the elements of a process architecture. The process architecture provides the key elements used within a process or system. The traditional process map shows the sequence of activities performed by "actors" or roles within a system. To accurately define a PAM, we must first define a process. A process is a sequence of activities that are performed in coordination to achieve some intended result. A process is also described as a set of activities involved in transforming inputs into outputs. Processes typically cross traditional organizational boundaries or departments (Evans and Lindsey, 2011). The activities jointly realize a *course of action*. Each process is enacted by a single organization but it may interact with processes performed by other organizations, systems, or sub-systems.

The PAM has three primary purposes: communication, knowledge capture, and analysis. In using a PAM for communication, we want to convey a rigorous understanding of the process and the ability to manage change. In knowledge gathering, we are interested in understanding the associations of processes and activities. Finally, in analysis we want to be able to study the process for optimization and transformation purposes.

A process consists of activities. A process exhibits variability. By understanding and controlling the factors that contribute to the variability, we enable the processes to provide competitive advantage. Technical or system requirements are derived from customer requirements. The technical requirements identify process requirements. The process requirements are realized through different processes. This ensures traceability from customers' expectations to customers' requirements to the technical requirements to process requirements and to the processes that enable those requirements.

For the purposes of building a knowledge base, we are concerned with capturing the activities that make up a process and with providing concise and meaningful names to the processes that orchestrate those activities. For the purpose of analysis and communication, a PAM takes the form of a process/workflow diagram and we are interested in the sequencing and constraints of activities.

Traceability of Customer Needs & Requirements

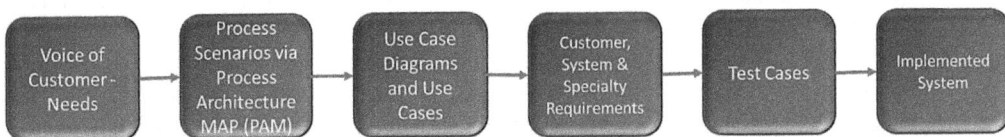

FIGURE 7.1   Traceability of customer needs and requirements.

The steps to develop the process scenarios are:

1. From either the functional decomposition or the use case diagrams, identify the processes that will be mapped, creating a process scenario list. Define the boundaries of the processes to be mapped based on the value chain, SIPOC, and/or Functional Decomposition Model.
2. From the processes to be mapped, the functional decomposition, and the stakeholder analysis, identify the roles that perform the processes. Identify representatives that perform these roles to take part in the process scenario development.
3. Identify the major activities within the process and the sequence of activities performed or to be performed within the process. Identify the process steps and uncover complexities using brainstorming and storyboarding techniques. Brainstorming is a useful tool to have the participants generate ideas in a free form manner, without evaluating the ideas prematurely. Storyboarding is a way to develop different stories of how the system will be used, while generating what the use or process looks like from a user's perspective.
4. Arrange the steps in time sequence and differentiate operations by the particular symbols chosen to represent activities, decisions, connectors, start and end points.
5. Identify who (the actor) performs each activity. Don't use names of people, use roles.
6. Identify the documents or information used for each activity.
7. Identify the system, technology, or method used in each activity.
8. Create a PAM of each process scenario.
9. For a more detailed process architecture, you may identify potential risks to the process, and the control mechanisms. You may then add additional knowledge that is helpful to perform the process.
10. Always observe the actual process as it is performed, if possible, and revise your process map.

The process scenarios in the requirements phase can be documented via the PAM tool and then in more detail in the design phase using a use case or detailed process scenario tool.

A process map is a graphic display of steps, events, and operations that constitute a process. A process map helps us to document, understand, and gain agreement on a process. The process map can be used to document the current "as is" process or a future state or "to be" process to designate the process after it is implemented or improved. A process is a sequence of activities with the purpose of getting work done. A typical process map uses rectangles to represent an activity or task, diamonds to represent decisions, and ovals or circles to represent connectors to either activities on the same page or another page. Flow usually flows from left to right and top to bottom. Process maps help process owners understand how work is accomplished, how all processes transform inputs into outputs, how small processes combine to form bigger processes, and how processes are both connected and interdependent. Process maps also help process owners see how their work affects others and how other roles affect their work, how even small improvements make a difference, and that process improvements require communication and teamwork.

For the PAM specifically, we use the following conventions:

1. **Swim Lanes**, located down the left side of the diagram, are used in our PAM to represent the different elements related to the process, including:
   - the activities or tasks of the process in sequential order;
   - the functional roles or "actors" of the activities taking place in that part of the flow; decisions or delays do not need a role designated.
   - information or documents, forms, or printed reports used in the activity;
   - technology, equipment, information systems, or methods related to the activities. The methods could include indicating a completely automated activity, a completely manual activity, or a combination of a human activity augmented by automation, and;
   - knowledge that is important for training purposes, such as a link to the detailed procedure or work instructions, notes on the process, metrics to be measured, and so on.

- Note that there are three rows of activities, three corresponding rows of roles, information, and technology.
- Knowledge can relate to any row above the knowledge column.

2. **Start** and **End** points in the process flow are designated by an oval.
3. **Process or Activity** (rectangle symbol). General process or a series of activities that make up a process.
4. **Decisions** (diamond symbol) must have at least two arrows exiting and lines should be labeled with the appropriate condition.
5. **Connector** is the oval symbol that connects the sequence or flow between shapes. Connectors are circles that can denote flow on the existing page of the process map or connect to another page. For an on-page connector, the number in the circle starts with the page number and numbers after the period denote the sequential order of the ovals used on that page. For an off-page connector, the first number denotes the page number where the flow continues, and the number after the dot is the sequential order of the off-page connectors. Connectors are used to minimize the number of arrows crossing the page and causing confusion for the reader.
6. **Delay** (D symbol) is used to represent any delay in the process where work is stalled or dependent on a given timeframe. Indicate the amount of time and unit of time when using this symbol, if known (Example: 10 minutes or 4 hours).

Some tips for creating process maps:

- Use roles on the process map, avoid names
- Flow is from top to bottom, left to right
- Connect all symbols, either to the first start oval, the last end oval, or a sequential symbol
- Never (rarely) cross arrows
- Share process maps with other systems engineers developing connected sub-systems
- Control documents, including the process name, date of origination or revision, version number
- Use the wall and flip chart paper or white boards to map out the process with the people involved in the process
- Validate steps/activities/data outside of the meeting
- Don't be afraid to start over
- Don't get stuck on the detail, get stuck on the flow and value
- Walk through the actual existing process, if there is one
- Numbering conventions are recommended
- The activities are inherently referred to by the numbering using the Row (1, 2, and 3) and then the Column numbers (1 through 9).

The author developed a PAM that incorporates process architecture elements, including the role that performs each activity, the information and technology used to perform the activity, additional knowledge related to the process for additional training, a link to procedures, and so forth. The process map for the patient appointment scheduling process in a Women's Healthcare Center is shown in Figure 7.2. The process architecture cover page is shown in Figure 7.3.

The PAM is used as the foundational tool, to document the current process, perform a Lean analysis, identify improvement opportunities, and document the future process after improvements are implemented. Process maps are great tools used to train process owners on the process as well. The process map can be used to generate customer requirements, information systems requirements, and to develop an information model such as a class diagram. The PAM template allows the creation of a cover page that summarizes the high-level process steps from the SIPOC to provide the connection to the system view of where this process fits within the entire system. A system here is referring to the set of processes that interconnect to achieve a work system. The cover page also provides an inventory of the process architecture roles or process stakeholders, information and technology.

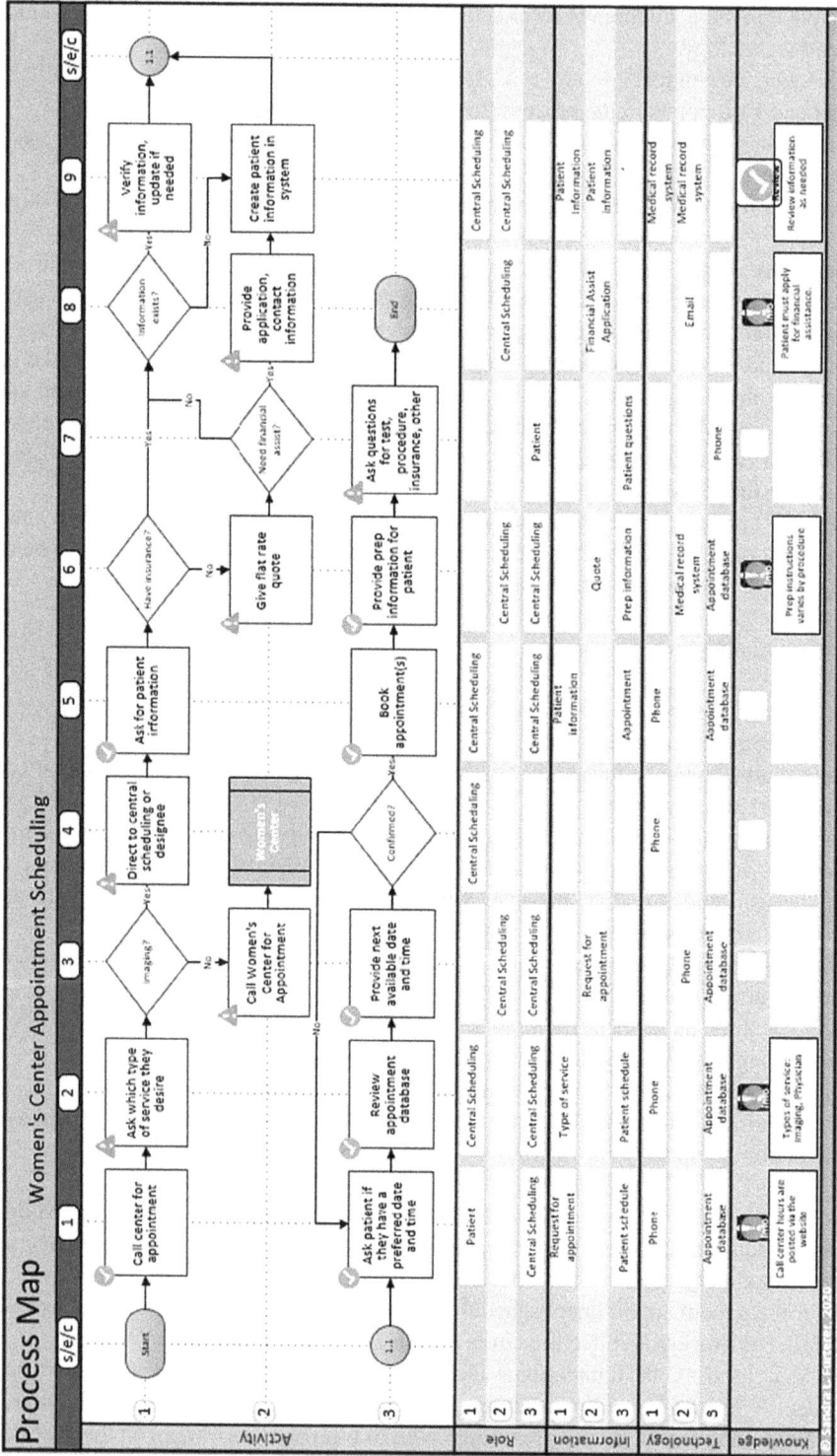

FIGURE 7.2 Women's healthcare center appointment scheduling process map.

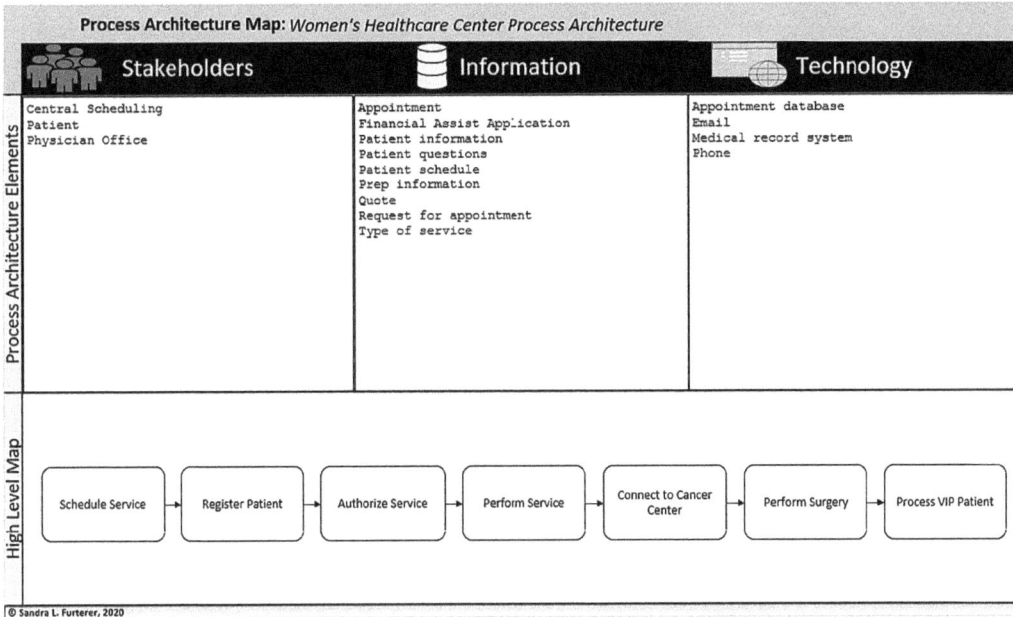

Process Architecture Map: *Women's Healthcare Center Process Architecture*

Stakeholders	Information	Technology
Central Scheduling Patient Physician Office	Appointment Financial Assist Application Patient information Patient questions Patient schedule Prep information Quote Request for appointment Type of service	Appointment database Email Medical record system Phone

High Level Map: Schedule Service → Register Patient → Authorize Service → Perform Service → Connect to Cancer Center → Perform Surgery → Process VIP Patient

© Sandra L. Furterer, 2020

FIGURE 7.3 Process architecture map – process architecture cover page – women's healthcare center.

The PAM is a Visio template with Visual Basic coding and macros to create the Process Architecture cover page and perform a Lean and Waste analysis which is discussed later. Figure 7.4 shows the icons or symbols that are typically used on the process map.

A brief description of the symbols follows. The rectangle is used to describe an activity or task of the process. A diamond represents a decision made in the process. The questions in the decision should be designed to ask a Yes or a No question. If more paths are desired, additional yes/no questions can be asked. The Start/End oval is used to identify where the process starts and ends. The circle connector is used to connect symbols where the arrows may cross the page and make the process map difficult to read. It can be used to connect to other symbols on the page and to other pages. The "D" is used to denote a wait or delay in the process. The external process can be used to connect the process to a process map that is external, separate, but related to this process map. The DRAFT symbol can be used to identify that the process map is a draft and must be reviewed further before it is finalized for use. The Knowledge symbol is used to add additional knowledge text to the knowledge swim lane. The Info icon is placed in the square on the knowledge swim lane to identify that text providing additional knowledge will be placed on the swim lane. The additional symbols are used within the knowledge lane to designate auto – an automated process, review – a review step, quality control – an inspection or quality control step, control – a control mechanism embedded in a process activity, risk – to identify an activity that has inherent risk, where a control may be incorporated. The Limited Value Added, Value Added, and Non-Value Added activity icons are discussed within the Lean Analysis section. As the systems engineer is working with the customers to develop the process scenarios, if ideas for the system arise, they can be noted in the knowledge swim lane and identified with an idea light bulb symbol. If a task is fully manual, identifying that it is a manual task, where there is further opportunity to automate the task, it can be marked in the knowledge swim lane with a manual activity symbol. If there are metrics that are collected at certain points in the process, the metrics icon can be shown in the knowledge swim lane and the metric titles can be entered beneath the symbol.

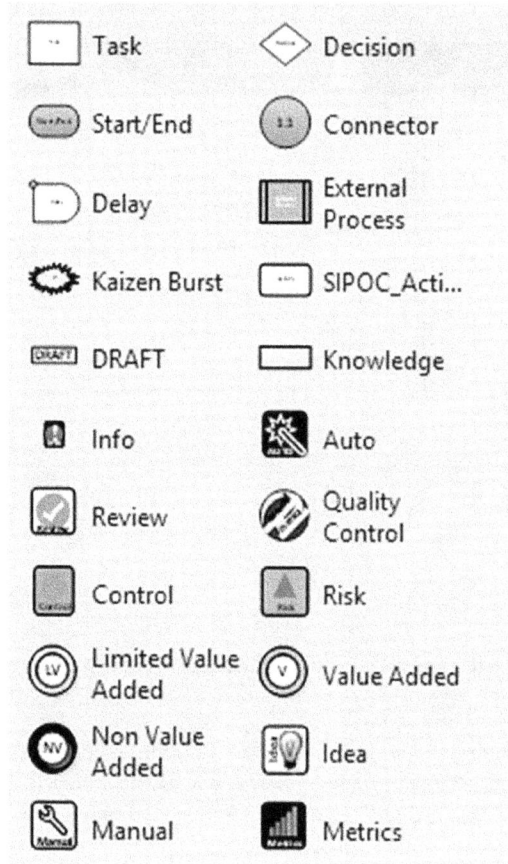

FIGURE 7.4   Process architecture map icons or symbols.

When the systems engineer creates the process map using the PAM, they drag an activity/task from the icon stencil, which opens an activity creation box, as shown in Figure 7.5.

All of the data related to the process map activity can be entered in this simple template. The Value Analysis and Wastes are discussed later.

Following is a process scenario use case checklist that can be used to generate process scenarios and use case documents (TOGAF8.0).

Process Scenario Use Case Checklist

- What is the overall objective of the scenario?
- What are any preconditions that must exist to start the scenario?
- Who is the "actor" of the scenario (who is performing the scenario)?
- What are the inputs (information, etc.) used to start the scenario?
- Walk through the "happy path" steps to get to the end result?
- Walk through alternate steps to get to the end result and "not" to get to end result (failure)?
- Think of who the person would interact with along the way? Who are the suppliers of inputs to the process and which customers receive outputs from the process?

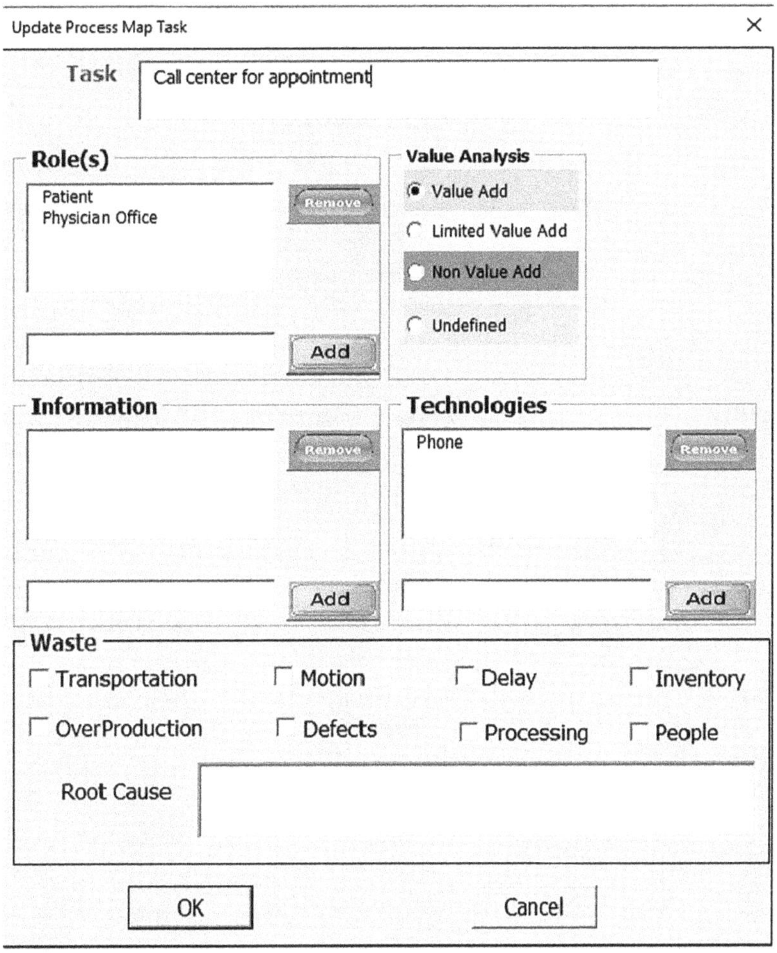

FIGURE 7.5   Process map activity entry or update template.

- Think of what systems, information systems, or technology the person might interact with (generally)?
- Think of assumptions or needs a person might expect to require during the scenario?
- What is the final outcome, result, or output? What is transformed?
- What are the post-conditions, what is the state of the business after the scenario is finished?
- What are the resources, people, and tools needed?
- What is the information, materials, information systems, policies, and procedures used and/or produced?
- How is the process measured today?
- What are the failure points?
- What decisions are made?
- How is work distributed?

The Women's Healthcare Center Process Scenarios Use Cases List is shown in Table 7.2.

TABLE 7.2
Process Scenarios Use Cases List – Women's Healthcare Center

Process Scenarios List

Function	Scenario Name	Purpose	Actors
Schedule Service	Schedule Service 001	To schedule an appointment for services	Scheduler, patient, medical system
Register patient	Register Patient 002	To register patient when they arrive to center	registration, patient, registration system
Perform service	Perform Service 003	Perform service at women's center	Patient, medical staff, equipment
Provide imaging results	Provide Results 004	Provide results to patient	Radiologist, physician, patient, radiology system, medical system
Provide spiritual care	Provide Spiritual Care 005	Provide spiritual care to the patient	Spiritual care, patient
Connect to Cancer Center	Connect to Cancer Center 006	Connect patient to the cancer center, when needed	Medical staff at women's center and cancer center, registration at both
Authorize service	Authorize Service 007	Authorize service prior to appointment with payer	Registration, patient
Process self-referral	Process Self-Referral 008	Process self-referral from patient	Patient, registration
Request records/films	Request Films 009	Request prior film results from imaging library	Patient, imaging library
Perform surgery	Perform Surgery 010	Perform surgery at outpatient facility	Medical staff, patient
Process VIP patient	VIP Processing 011	Provide VIP processing at center	Manager of center, registration system, VIP patient

7.3.2 FOOD SYSTEMS PROCESS SCENARIOS USE CASES LISTS AND PROCESS ARCHITECTURE MAP APPLICATION

The process scenarios for the Food System developed by each team are shown in Tables 7.3 through 7.9.

Team 1's Process Map Architecture for their Advertise to Connect to People process map is shown in Figure 7.6. The process architecture cover sheet is shown in Figure 7.7. This is only one of the process scenarios that the team developed with the stakeholders. However, it gives a good example of the level of detail and the information that is collected during the process scenario building exercises. These will then be used to generate the customer and system requirements.

TABLE 7.3

Process Scenarios Use Cases List – Team 1 Connecting Organizations

Function	Scenario Name	Purpose	Actors
Advertise	Advertise to connect to people	Advertise via digital and non-digital formats	Volunteers, Citywide, IT
Obtain the Form	Obtain the form 001	Obtain form to fill out food order and survey	customer, Citywide
Complete Form	Complete form 002	Fill out information of what foods they would like to purchase and survey food needs	customer
Submit/Return Form	Submit/Return Form 003	Return information to Citywide for passing food order and food requests to the farmer	customer, Citywide, farmer
Kick off Website Process Scenario	Kick off Website Process Scenario 004	Passage of customer information to farmer to complete food orders. Also, Citywide would analyze survey data to identify needs and program gaps	Citywide, farmer
Meet Farmer at Designated location at Designated time	Meet Farmer 005	Create face to face interactions with the farmer and the customer	customer, farmer
Pick up and pay ordered food	Pick up Food 006	For the customer to pick up and pay for their food	customer, farmer

TABLE 7.4

Process Scenarios Use Cases List – Team 2 Creating Food Policy

Function	Scenario Name	Purpose	Actors
Problem Identification	Problem Identification 01	Describes the process of identifying a problem	Community Members, Legislation
Elect a Champion	Elect a Champion 02	Illustrates the process of electing a champion	Community Members, Champion, Commission, Legislation
Gather Support/Data	Gather Support/Data 03	Includes the data to be secured to bring in an envisioned change in the policy	Community, Commission, Government
Present policy change idea to commission	Present policy change idea to commission 04	Describes the process of presenting policy change idea to Commission	Champion, Commission
Commission Discussion	Commission Discussion 05	Describes the process of Commission discussion on the policy change	Commission, Legislation, Community Members

TABLE 7.5
Process Scenarios Use Cases List – Team 3 Kids Growing Food

Function	Scenario Name	Purpose	Actors
Food and Growing Instruction	Instruct microgreens	To enable children to grow microgreens at home	Children, Teacher
Food Production	Produce microgreens	For children to grow microgreens at home	Children, Teacher, Supplier
Food Processing	Harvest microgreens	For children to transfer microgreens back to school in order to be compensated for their crop and the teacher to harvest and clean	Children, Teacher
Food Distribution	Distribute microgreens	For teacher to transfer microgreens to customer (restaurant, market, etc.) and receive compensation	Teacher, Customer

TABLE 7.6
Process Scenarios Use Cases List – Team 4 Food storage

Function	Scenario Name	Purpose	Actors
Storing the food at temporary storage at farm	Temporary storage at farm until collection	To store and keep the food fresh until the day of collection	Farmers Homefull community staff
Collection food from temporary storage	Collection of food from farm to central storage location	To organize food transportation from temporary storage to central storage	Homefull community staff (collectors) Farmers Community storage Maintenance staff
Storing packaged food at central storage location	Food storage at central location	To reduce the food loss and make distribution convenient from central unit	Homefull community storage maintenance staff, community truck driver, community helpers
Distribute the food to storage	Food distribution by electrical bikes	Describes the process of food distribution by electrical bikes	Homefull community staff (delivery person) Customers

TABLE 7.7
Process Scenarios Use Cases List – Team 5 Food Hub Website

Function	Scenario Name	Purpose	Actors
Food Production:			
Harvest Crops/Produce	Farmer Utilization 001	Take farmer goods to the businesses	Take farmer goods to the businesses
Market Produce Availability	Marketing 002	Market demand across all levels of those involved in the system	Entrepreneurs
Promote Efficiency	Efficiency 003	Will increase quality/efficiency of food delivery	Distributors
Food Sales:			
Real Time Costs	Cost Updating 001	Directly relay costs of goods for markets	distribution, produce, and buyers
Food Availability	Food Availability 002	Help customer know what food is available and where	distribution, produce, and buyers

(Continued)

TABLE 7.7   (Continued)

Food Distribution:

Promote Communication	Communication 001	improve communication between all three groups: distribution, produce, and buyers	distribution, produce, and buyers
Promote Delivery	Delivery 002	Will increase efficiency of produce delivery	distribution
Real Time Management	Time Management 003	Show real time distributor tracking	distribution

TABLE 7.8
Process Scenarios Use Cases List – Team 6 GEM City Market

Function	Scenario Name	Purpose	Actors
Obtain food	Improving according to demand (Obtaining the food products)	Meeting the shopper's needs/Wants by improving the products according to the demand	Shoppers Farmers General Manager
Promote pricing	Hybrid Pricing (Promote pricing and managing the store)	Describes the process of implementing hybrid pricing in Gem City Market	General manager
Delivery and pickup	Delivery and Pickup service (Purchasing the food products)	To provide the delivery and pickup services from the Gem City Market to the community people promoting easy, flexible online shopping and also availability to healthy, fresh food options without choosing any other junk alternative	Truck drivers Community Farmers
Marketing of food	Marketing of food products in the market to the customers (Meeting requirements)	Describes the marketing of food products in the gem city market and also to promote the food products in the market to obtain revenue, for people to access healthy and variety of food options	General manager Community Shoppers

TABLE 7.9
Process Scenarios Use Cases List – Team 7 Table 33 Restaurant

Function	Scenario Name	Purpose	Actors
Discuss missing ingredients with Inventory Management Systems	Report missing ingredients	Find out how much ingredients needed for Table 33 to fulfill the menu.	Inventory management, Chief and Restaurant Manager
Reach local farms	Find local farmers	Find suppliers that grow the needed ingredients organically to supply the restaurant for inexpensive price and maintain high quality.	Restaurant Manager and local farmers
Find best way to transport ingredients	Ingredients' transportation	Find a way to transport ingredients easily from farms to Table 33.	Local Farmers and restaurant manager
Move ingredients from inventory to kitchen	Move needed ingredients from inventory to kitchen	Move the needed ingredients from the inventory to kitchen to let the chief make meals.	Chief and Table 33 workers
Deliver meals to customers	Prepare meals	Prepare organic meals for customers using organic ingredients.	Server

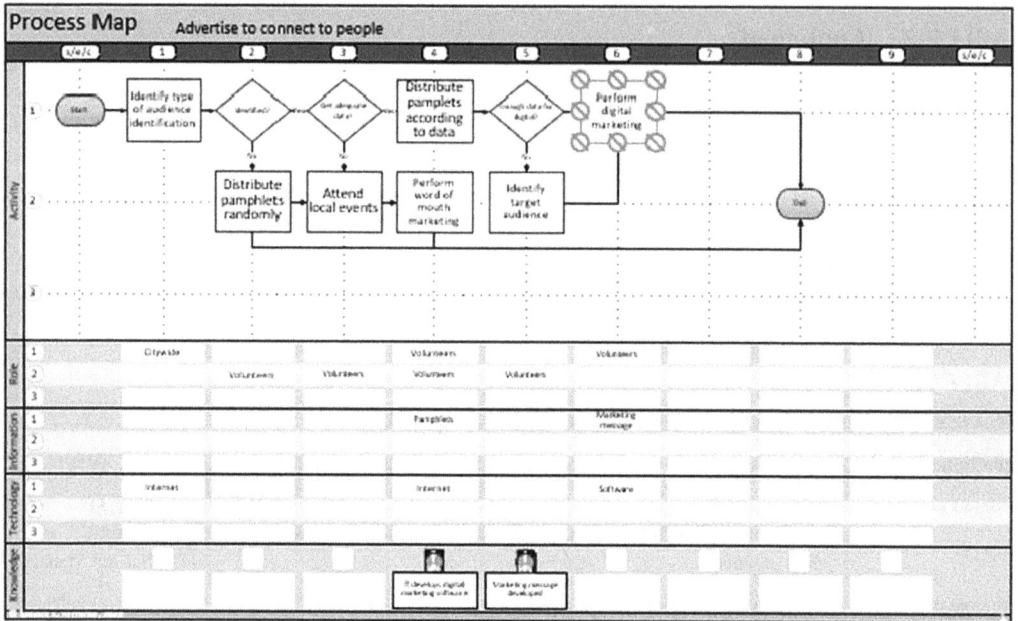

FIGURE 7.6 Process architecture map for advertise to connect to people, team 1 – connecting organizations.

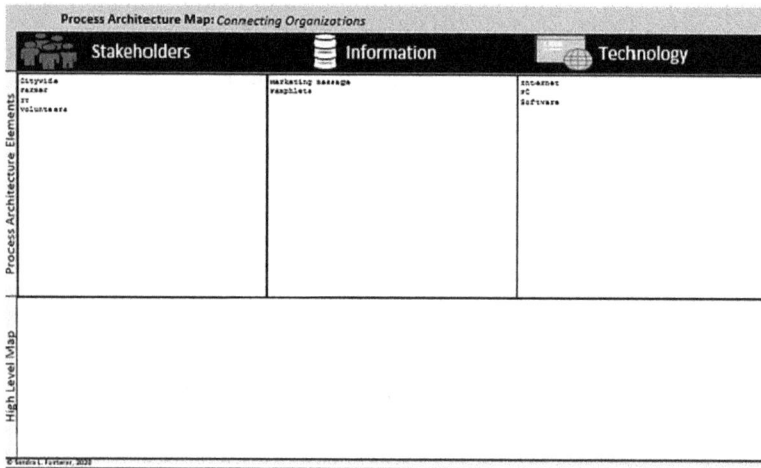

FIGURE 7.7 Process architecture map cover page for advertise, team 1 – connecting organizations.

7.3.3 CREATE USE CASES

Use cases are a way to elicit and organize requirements **from a user's perspective**. All the requirements for a user to accomplish a particular function are gathered together in a single use case. These functions or uses are presented in the use case diagrams in Chapter 6.

Advantages of use case modeling include (Use Case Training, 2007):

- System requirements are placed in context. Each requirement is part of a logical sequence of actions to accomplish a goal of a user.

- Use cases are easy to understand. The model expresses requirements from a user perspective and in the user's own vocabulary. The model shows how users think of the system and what the system should do. Traditional forms of requirements capture need some translation to make them useable. When you translate something, information is often lost or misinterpreted. Use cases require no such translation and therefore provide a more consistent and reliable form of requirement capture.
- Use cases show why the system is needed. Use cases show what users can accomplish by using the system.
- The model is a means to communicate requirements between customers and developers, in order to make sure that the system we build is what the customer wants.

Use-case modeling is the best tool (so far) to capture requirements and put them in context.

There are four main criteria that should hold for a use case to be a valid use case (Use Case Training, 2007):

Value to the stakeholder: There should be value to the stakeholder, user, or customer from the use case or functionality provided. It should help the stakeholder achieve some end result or goals.

- **Avoid system design:** The use case should describe the WHAT the actors and system are doing, not HOW to do it. Avoiding describing how the functionality will be implemented helps you to convey a requirement level of information. For example, for our food system, a use case could describe how a farmer would provide information that would connect their products to a consumer. The use case would describe how a farmer would use a website to enter the product that they produce. It would not describe how the information is placed into a database behind the scenes. The use cases should describe externally observable functionality. The HOW comes into play in the design phase and can be described through specifications, business rules, or design drawings.
- **Complete narrative:** The use case should describe the complete functionality from the users' perspectives through to achieving the end goal and the users' steps from start to finish.
- **Stand alone:** Use cases should make sense on their own. They should describe a reasonable grouping of activities that describe a function. Focusing on what is achieved and what functionality is provided helps to provide a stand-alone use case.

Creating use cases. The best way to create use cases is to interview and/or story board how the system will be used with the stakeholders. For each stakeholder group identified in the stakeholder analysis in Phase 1 –Concept of Operations, identify someone that the team can interview regarding how they envision using the system. Storyboarding is a technique used to create a story of how the user will use the system.

Here are some questions that are useful to help develop the use cases in a system (Use Case Training, 2007):

- What are the goals of each actor?
- Why does the actor want to use the system?
- Will the actor create, store, change, remove, or read data in the system? If so, why?
- Will the actor need to inform the system about external events or changes?
- Will the actor need to be informed about certain occurrences in the system?
- Does the system supply the business with all of the correct behavior?

The use cases consist of the following elements:

1) Name (usually a verb-noun pairing, the verb representing the function from the Functional Decomposition Model or the use from the use case), identifier number for tracking purposes, purpose (description of the use case), and start trigger or stimulus (the event that starts the use case).
2) Customers of the use case, they are also called actors in the use case language.
3) Preconditions that must be in place for the use case to begin.
4) Assumptions that are being made to ensure that the use case can work.
5) Basic flow, which is the most frequent or ideal path through the process or system. It includes step description, inputs, people responsible for the step, outputs, and equipment or technology used in the step.
6) Post conditions are a listing of the possible outcomes of the ideal or basic flow and each alternative flow.
7) Alternate flows can be alternate paths through the process based on need, failure, or error conditions. These are important to describe because much functionality is derived from error conditions and additional uses of the system. The same information is described in the alternate flows, as in the basic flow. Steps from the basic flow can be referenced that are the same in the alternate flows, so that they don't need to be copied and pasted multiple times. This reuse also reduces the need to change the same steps in multiple paths and even use cases.

An example of a use case for the Women's Healthcare center is shown in Figures 7.8 through 7.11.

BUSINESS SCENARIO GENERAL INFORMATION	
Scenario Name	**Schedule Service**
Scenario ID	001
Purpose	Describes the process of a patient scheduling service
Start (Trigger) Stimulus	Patient decides to receive service at Women's Center

CUSTOMERS	
Primary Customers	Imaging Patient
Secondary Customers	Patient
Secondary Customers	Physician office

PRE-CONDITIONS	
1.	Patient is a new patient or existing patient
1.	
1.	
1.	

ASSUMPTIONS	
1.	Basic Flow assumes patient calls central scheduling for an imaging appointment
2.	Imaging Appointments can be for any Women's Center imaging services.
3.	Women's Center physician appointments can be for any physicians in the Women's Center.
4.	Physician Office can request appointments for a patient (35% of appts. Book)
5.	Provide the ability to schedule a physician appointment within the WC.
6.	Be able to schedule physician appointments from Central Scheduling. Direction from Patient Access is to do a warm transfer, central scheduling will not schedule for
7.	Need to run eligibility and authorization to confirm on-line appointment
8.	CHANGE THE SCENARIOS TO INCLUDE A WARM TRANSFER. PATIENT ACCESS WILL NOT SCHEDULE FOR physicians, and Warm transfer includes getting to the live scheduler, giving them the patient information from Meditech, and handing off the patient. Need to add an alternate flow to include what happens if no one picks up the phone.

FIGURE 7.8   Use case initial information for women's healthcare center.

#	BASIC FLOW: Patient Receives Imaging Patient Appointment by Calling Central Scheduling				
	Step Description	Inputs	People Responsible	Outputs	Equipment,v IT, Etc.
1.	Patient calls Central Scheduling and requests an imaging appointment	Phone call	Patient, central scheduling	Request for appointment	Telephone
2.	Ask for name, social security number from patient	Request for appointment	Central scheduling, patient	Name, social security number	Telephone to central scheduling
3.	If Meditech file exists, verify demographic and update if necessary, if doesn't exist, create it.	Demographic and insurance info	Central scheduling, patient	Created/Updated demographic	Meditech
4.	Ask patient which services would like to schedule for, could be multiple appointments	Service requested, patient prescription	Central scheduling, patient	Service requested	Telephone
5.	Ask patient if they have a preferred date and time of day	Preferred appointment date and time	Central scheduling, patient	Preferred appointment date and time	Telephone
6.	Review central schedule database for preferred date and time	Preferred appointment date and time	Central scheduling	Available dates and times	Meditech Outpatient central scheduling
7.	Provide next available date and time	Available dates and times	Central scheduling, patient	Available dates and times	
8.	Confirm with patient, if not repeat steps 6 through 7	Available dates and times	Central scheduling, patient	Confirmed date and time	
9.	Book appointments	Confirmed date and time	Central scheduling	Booked date and time	Meditech
10.	Provide prep information for patient	Prep info for service	Central scheduling, patient	Prep info	Meditech
11.	Ask questions on test, procedure, insurance, other information	Questions for service	Central scheduling, patient	Test questions	Meditech
12.	Reconfirm date, time, remind patient to bring photo id, copay/deductible and script and provide any other instructions	Required info	Central scheduling, patient	Scheduled appointment for service	Meditech

FIGURE 7.9   Use case basic flow for women's healthcare center.

	POST-CONDITIONS		Service Level Goals
	Course	Description	
1.	Ideal	Patient calls central scheduling and receives an appointment or multiple appointments for imaging.	Receive appointment within XX days of request. Same service level for screening vs. diagnostic, vs, biopsy?
2.	Alt-A	Patient is in Women's Center and receives an appointment for imaging	Receive appointment within XX days of request. Same service level for screening vs. diagnostic, vs, biopsy?
3.	Alt-B	Patient calls central scheduling and receives an appointment for physician (with warm transfer from central scheduling to WC physician)	Receive appointment within XX days of request. Same service level for physician?
4.	Alt-C	Patient is in Women's Center and receives an appointment for physician from front desk (with warm transfer from central scheduling to WC physician)	Receive appointment within XX days of request. Same service level for physician?
5.	Alt-D	Patient is in Women's Center and receives an appointment for physician from physician desk.	Receive appointment within XX days of request. Same service level for physician?
6.	Alt-E	Patient changes imaging appoinment by calling central scheduling.	
7.	Alt-F	Patient changes physician appointment by calling central scheduling (with warm transfer from central scheduling to WC physician)	
8.	Alt-G	Patient cancels imaging appointment by calling central scheduling	
9.	Alt-H	Patient cancels appointment by calling central scheduling (with warm transfer transfer from central scheduling to WC physician)	
10.	Alt-I	Physician Office requests an appointment for a patient.	
11.	Alt-J	Patient schedules both imaging and physician appointments from Central Scheduling, (with warm transfer from central scheduling to WC physician)	
12.	Alt-K	Patient schedules both imaging and physician appointments from desk, (with warm transfer from physician to central scheduling)	
13.	Alt-L	Patient cancels physician appointment by calling WC.	
14.	Alt-M	Patient schedules imaging appointment from Medical Group, with warm transfer to central scheduling.	
15.	Alt-N	Be able to view open blocks on-line for possible appointments.	
16.	Alt-O	Patient calls for an appointment, does not have insurance, and needs financial assistance.	
17.	Alt-P	Patient requests an Physician appointment from Central Scheduling or Imaging Scheduler, they attempt to warm transfer, but noone is available.	

FIGURE 7.10   Use case post conditions for women's healthcare center.

#	Step Description	Inputs	People Responsible	Outputs	Equipment, IT, Etc.
Alternate A: Patient is in Women's Center and Receives an Imaging Appointment from Front Desk					
1. A	Patient is in Women's center and requests an imaging appointment	Request for appointment	Check out person	Request for appointment	
1. B	Patient is directed to a designated person in the area	Request for appointment	Designated person	Request for appointment	
	Same steps from 2 through 11 of Basic Flow				
12	Scan and put order into the system.				Efax shared folder

#	Step Description	Inputs	People Responsible	Outputs	Equipment, IT, Etc.
Alternate B: Patient calls central scheduling and receives an appointment for physician					
1.	Patient calls Central Scheduling and requests an physician appointment	Request for appointment	Central scheduling, patient	Request for appointment	Telephone
2.	Do warm transfer to physician scheduler.		physician scheduler.		
3.	Ask for name, social security number from patient	Request for appointment	physician scheduler.	Name, social security number	Telephone to central scheduling
4.	If IDX file exists, verify demographic and update if necessary, if doesn't exist, create it.	Demographic and insurance info	Physician scheduler.	Created/updated demographic	IDX
5.	Steps 4 and 5 are the same as in basic flow				
6.	Review IDX database for preferred date and time	Preferred appointment date and time	Physician scheduler.	Available dates and times	IDX
	Steps 7 and 8 are the same as in basic flow				
9.	Book appointments	Confirmed date and time	Physician scheduler.	Booked date and time	IDX
12.	Steps 10 and 11 are the same as in basic flow Reconfirm date, time, remind patient to bring photo id, copay/deductible and script	Required info	Physician scheduler.	Scheduled appointment for service	IDX

FIGURE 7.11　Use case alternates a and b for women's healthcare center.

7.3.4　Food System Use Case

Following is just one example of a use case for Team 1 – Connecting Organizations for the Distribute Data to Farmers function. All of the teams created use cases, but due to space, they are not included here (Table 7.10).

TABLE 7.10
Example Use Case Distribute Data to Farmers for Team 1 Connecting Organizations
BUSINESS SCENARIO GENERAL INFORMATION

Scenario Name	**Distribute Data to Farmer**
Scenario ID	004-3
Purpose	Provide farmer with what foods are ordered and requested foods by customers
Start (Trigger) Stimulus	Finished data entry and collection at event(s)

(Continued)

TABLE 7.10 (Continued)

CUSTOMERS

Primary Customer	Citywide
Primary Customer	Farmer

PRE-CONDITIONS

All data collected at events entered into the website database

All paper form information received entered into website database

ASSUMPTIONS

1. Data from paper forms was entered in correctly

2. All or most of the paper forms were received from events and community members

BASIC FLOW

#	Step Description	Inputs	People Responsible	Outputs	Equipment, IT, Etc.
	Ensure data entry is complete	Completed paper forms	Citywide volunteer	All forms entered into website database	Computer, internet
	File away paper forms for record	Completed paper forms	Citywide volunteer	Filed forms for record	Filing cabinet
	Accumulate instance of food request in spreadsheet	Data from website database	Citywide volunteer	Organized data of food request instances	Computer, spreadsheet software
4.	Load data into statistical data software (i.e. JMP)	Spreadsheet Info	Citywide volunteer	Spreadsheet data transferred to stat software	Computer, statistical data software
5.	Review statistical data	Output of statistical findings	Citywide volunteer	Understanding of most requested items	Computer, statistical data software
6.	Create list of food needs by Priority	Output data from Statistical software	Citywide volunteer	List of most requested items	Spreadsheet or word doc.
7.	Summarize statistical data and needs	List of prioritized items and stat data output	Citywide volunteer	Summarized info for farmers	Word doc
8.	Send Email and letter to local farmer(s)	Summarized info sheets	Citywide volunteer	Farmers receive to info	Computer, internet

POST-CONDITIONS **Service Level Goals**

Course	Description	
Ideal	Farmers receive summarized information	Farmer receives summarized information of food request to determine if they can help fill orders
Alt-A	There is an error when sending email to farmers	Farmer receives summarized information of food request to determine if they can help fill orders
Alt-B	Farmer has no internet access for Email	Farmer receives summarized information of food request to determine if they can help fill orders

(Coninued)

TABLE 7.10 (Continued)

Alt-C	Letters mailed to wrong address	Farmer receives summarized information of food request to determine if they can help fill orders
Alt-D	Citywide doesn't have address of multiple local farmers	Farmer receives summarized information of food request to determine if they can help fill orders
Alt-E	Statistical software not available to Citywide employees	Complete summarization of food request prioritization
Alt-F	Community member requests for product/food not grown in area	Complete summarization of food request prioritization
Alt-G	Citywide volunteer does not know how to use statistical software	

Alternate A: There is an error when sending email to farmers

#	Step Description	Inputs	People Responsible	Outputs	Equipment, IT, Etc.
1.	Same steps from 1 to 7 of Basic Flow				
2.	Check Internet connection	Summarized info	Citywide volunteer	Determined if connection was lost	Computer, internet
3.	Reboot Computer System	Connection lost	Citywide volunteer	System reset	Computer, internet
4.	Check internet connection	Connection gained	Citywide volunteer	Email system working again	Computer, internet
5.	Resend Email	Summarized info reloaded	Citywide volunteer	Email sent to farmer(s)	Computer, internet

Alternate B: Farmer has no internet access for Email

#	Step Description	Inputs	People Responsible	Outputs	Equipment, IT, Etc.
1.	Same steps from 1 to 7 of Basic Flow				
2.	Send letter of Summarized info only	Summarized info sheets	Citywide volunteer	Farmers receive to info	Computer, internet

(Continued)

TABLE 7.10 (Continued)

Alternate C: Letters mailed to wrong address

#	Step Description	Inputs	People Responsible	Outputs	Equipment, IT, Etc.
1.	Same steps from 1 to 7 of Basic Flow				
2.	Check mailing addresses of farmers	List of addresses	Citywide volunteer	Errors or missing addresses found	List of addresses
3.	Update list of addresses	List of revisions	Citywide volunteer	Updated address list	Computer
4.	Reprint and package letters	New addresses	Citywide volunteer	Prepared envelopes	Paper, printer, envelope
5.	Send letters	Prepared envelopes	Citywide volunteer	Sent letters	Mail box

Alternate D: Citywide doesn't have address of multiple local farmers

#	Step Description	Inputs	People Responsible	Outputs	Equipment, IT, Etc.
1.	Same steps from 1 to 7 of Basic Flow				
2.	Go door to door handing out letters	Letters in envelopes	Citywide volunteer	Meeting new farmer(s)	Paper, envelopes, car
3.	Write down addresses of farmers visited	Visited farmer	Citywide volunteer	Addresses known	Paper, pen clipboard

Alternate E: Statistical software not available to Citywide employees

#	Step Description	Inputs	People Responsible	Outputs	Equipment, IT, Etc.
1.	Same steps from 1 to 3 of Basic Flow				
2.	Summarize spreadsheet data using simple statistical methods	Data in spreadsheet software	Citywide volunteer	Statistical data obtained	Computer, spreadsheet software
3.	Review statistical data	Output of statistical findings	Citywide volunteer	Understanding of most requested items	Computer, statistical data software
4.	Create list of food needs by Priority	Output data from Statistical software	Citywide volunteer	List of most requested items	Spreadsheet or word doc.
5.	Summarize statistical data and needs	List of prioritized items and stat data output	Citywide volunteer	Summarized info for farmers	Word doc
6.	Send Email and letter to local farmer(s)	Summarized info sheets	Citywide volunteer	Farmers receive to info	Computer, internet

(Continued)

TABLE 7.10 (Continued)

Alternate F: Request for product/food not grown in area

#	Step Description	Inputs	People Responsible	Outputs	Equipment, IT, Etc.
1.	Same steps from 1 to 5 of Basic Flow				
2.	Maintain Prioritization of food list	Output data from Statistical software	Citywide volunteer	List of most requested items	Spreadsheet or word doc.
3.	Note food request type for future reference	Non-available food request	Citywide volunteer	List of non-local foods	Spreadsheet or word doc.
4.	Summarize statistical data and needs	List of prioritized items and stat data output	Citywide volunteer	Summarized info for farmers	Word doc
5.	Send Email and letter to local farmer(s)	Summarized info sheets	Citywide volunteer	Farmers receive to info	Computer, internet

Alternate G: Citywide volunteer does not know how to use statistical software

#	Step Description	Inputs	People Responsible	Outputs	Equipment, IT, Etc.
1.	Same steps from 1 to 4 of Basic Flow				
2.	Additional volunteer/ employee trains volunteer on how to use software	Untrained volunteer	Citywide volunteer(s)	Trained on stat software	Statistical software
3.	Volunteer demonstrates understanding of software	Trained volunteer	Citywide volunteer	Confirmation of knowledge of software use	Statistical software
4.	Prioritization of requested goods	Output data from Statistical software	Citywide volunteer	List of most requested items	Spreadsheet or word doc.
5.	Summarize statistical data and needs	List of prioritized items and stat data output	Citywide volunteer	Summarized info for farmers	Word doc
6.	Send Email and letter to local farmer(s)	Summarized info sheets	Citywide volunteer	Farmers receive to info	Computer, internet

REVISION HISTORY

Date	Version	Description	Author
3/22/2019	1.0	Draft for scenario building	Student

7.4   SUMMARY

In this chapter, we elicited the customers' requirements from facilitating process scenario sessions and then documenting them applying the PAM tool and use case documents. In the next chapter we will continue to use these artifacts to derive the customer, system, and specialty engineering requirements, to complete the requirements and architecture phase.

7.5   ACTIVE LEARNING EXERCISES

1) Facilitate a process scenario to create a PAM and a use case.

BIBLIOGRAPHY

Dannels, D. (2000). Learning to be professional: Technical classroom discourse, practice, and professional identify construction. *Journal of Business and Technical Communication*, 14(1), 5–37.

Dick, J., Hull, E., and Jackson, K. (2017). *Requirements Engineering* Fourth Edition, Springer.

Evans, J. and Lindsey, W. (2011). *Managing for Quality and Performance Excellence*, 10th Edition. Cengage Learning.

Fairley, R.E., and Willshire M. J. 2011. "Teaching software engineering to undergraduate systems engineering students." *Proceedings of the 2011 American Society for Engineering Education (ASEE) Annual Conference and Exposition*. 26–29 June 2011. Vancouver, BC, Canada.

IEEE STD 1220-1998 (1998) Standard for application and management of the systems engineering process. IEEE, New York.

INCOSE. 2015. 'Systems Engineering Handbook: A Guide for System Life Cycle Processes and Activities', version

No Author, Use Case Training, Walmart Information Systems Division, 2007.

4.0. Hoboken. NJ, USA: John Wiley and Sons, Inc, ISBN: 978-1-118-99940-0.

Standish Group. (1995), The CHAOS Report 1995, Standish Group International. Inc., Boston MA.

TOGAF. 8.0 The Open Group, Business Scenarios https://pubs.opengroup.org/architecture/togaf8-doc/arch/

TOGAF 8.0. The Open Group, https://pubs.opengroup.org/architecture/togaf8-doc/arch/toc.html, 1999–2006

Wilson, W.M., Rosenberg, L.L., & Hyatt, L.E., 1997, Automated analysis of requirement specifications, in Proceedings of the 19th International Conference on Software Engineering, pp. 161–171.

8 Requirements and Architecture Phase 2
Derive Requirements

8.1 PURPOSE

In Chapter 8, we discuss the derivation of the systems or technical requirements from the customers' requirements and the derivation of specialty engineering requirements. Additionally, in this phase, the team develops a measurement plan, a quality management plan, and a conceptual information model.

8.2 ACTIVITIES

The activities performed and the tools applied in the Requirements and Architecture phase are shown in Table 8.1. We cover the "Elicit Requirements" activities and the associated Requirements tools.

8.3 DERIVE REQUIREMENTS

This section discusses how to derive the requirements from the process maps and use cases. Insight into both customer and system requirements can be derived from the process maps and use cases. The specialty engineering areas can also provide more detailed technical requirements from these documents as well. We'll first discuss customer requirements, system requirements, and then specialty engineering requirements.

8.3.1 CUSTOMER REQUIREMENTS

A study by the Standish Group (1995) discovered that many project failures were related to requirements, some of the reasons that highlight the need for defining good requirements were:

1) Incomplete requirements (13.1%)
2) Lack of user involvement (12.4%)
3) Changing requirements or specifications (8.7%)
4) Didn't need it any longer (7.5%)

Let's first define a requirement:
A requirement definition based on the IEEE standard (IEEE-STD-1220-1998 (IEEE 1998)) is:

Requirement: a statement that identifies a product or process operational, functional, or design characteristic or constraint, which is unambiguous, testable, or measurable, and necessary for product or process acceptability (by consumers or internal quality assurance guidelines).

DOI: 10.1201/9781003081258-10

TABLE 8.1

Activities and Tools in the Requirements and Architecture Phase

Vee Phase	Activities	Tools	Principles
Phase 2: Requirements and Architecture	Develop Logical Architecture 1. Develop Logical Architecture Elicit Requirements 2. Elicit customer requirements 3. Derive requirements • Develop system requirements • Develop specialty engineering requirements 4. Develop measurement plan 5. Develop information models 6. Develop a quality management plan	Logical Architecture: • SIPOC • Value Chain and Functional Decomposition • Use Case Diagrams Requirements: • Process Scenarios via Process Architecture map • Use Cases • Customer requirements • System requirements; • Specialty engineering requirements: software engineering; environmental engineering; safety and security engineering; Human Systems Integration • Data Collection Plan, with operational definitions • Class Diagram with information model and hierarchy model • Quality management plan	• System dynamics (behavior, system elements) • Cybernetics (information flow) • Systems thinking • Abstraction • Views

There are several types of requirements depending upon the taxonomy that the author defines (SEBoK Original)

- Customer requirements: Statements of facts and assumptions that define the expectations of a system in terms of mission objectives, environments, constraints, and measures of effectiveness and suitability.
- Functional requirements: The necessary actions, tasks, or activities that must be performed by the system.
- Performance requirements: How well something a function or task must be done. These are typically measured in terms of quantity, quality, coverage, timeliness, or readiness.
- Usability requirements: Define the quality of a system, from an effectiveness, efficiency, and satisfaction criteria.
- Interface requirements: How the system is required to interact or to exchange material, energy or information with external systems, or how system elements within the system interact with each other, including human elements.
- Operational requirements: Operational conditions or properties that are required for the system to operate or exist. This can include human factors, ergonomics, availability, maintainability, reliability, and security.
- Environmental requirements: Related to the environmental conditions encountered by the system in different operational modes. These can address the natural, induced, and/or self-induced

environmental effects (motion, shock, noise, electromagnetism, thermal) and threats to societal environment (legal, political, economic, social, business).
- Logistical requirements: Includes sustainment, packaging, handling, shipping, transportation.
- Design requirements: Requirements based on how the system will be designed.
- Derived requirements: Requirements derived from higher level requirements.

Requirements engineering is a discipline of systems engineering defined as (Dick, Hull and Jackson, 2017):

Requirements engineering: the sub-set of systems engineering concerned with discovering, developing, tracing, analyzing, qualifying, communicating, and managing requirements that define the system at successive levels of abstraction.

There are several attributes of good individual requirements that should be met to ensure the quality of these requirements (INCOSE. 2015):

- Necessary: Describes an essential capability, characteristic, constraint, and/or quality factor.
- Appropriate: Appropriate to the level of the entity.
- Unambiguous: Can have only one interpretation or meaning.
- Complete: Sufficiently describes the necessary capability, characteristic, constraint or quality factor, without needing other information to understand the requirement.
- Singular: State a single capability, characteristic, constraint, or quality factor.
- Feasible: Can be realized with acceptable risk.
- Verifiable: Its realization can be proven, tested, and have an objective measure.
- Correct: An accurate representation.
- Conforming: Conform to an approved standard template and style for writing requirements.
- Achievable: Able to be met.
- Expressed based on the need: Based on the need, not the solution.
- Consistent: Consistent with other expressed requirements.

For a set of requirements they should have the following characteristics (INCOSE. 2015):

- Complete: Describes the necessary capabilities, characteristics, constraints, and/or quality factor, without needing other information.
- Consistent: Contains individual requirements that are unique, do not conflict or overlap with other requirements in the set.
- Feasible: The requirements set can be realized within entity constraints with acceptable risk.
- Comprehensible: They are clear from an expectation perspective and in relation to the system of which it is a part.
- Able to be validated: Be able to be proven to lead to the achievement of the entity needs.

8.3.1.1   Requirements Terminology
Wilson et al. (1997) provided key words (imperatives) and phrases to be used to define types of requirements, as follows:

- Are applicable: To include, by reference, standards or other documentation as an addition to the requirements being specified.
- Is required to: Provided in specifications statements to be written in the passive voice.

- Must: Used to establish performance requirements or constraints.
- Responsible for: Used in requirements documents that are written for systems whose architectures are predefined.
- Shall: To dictate the provision of a functional capability.
- Should: Not frequently used as an imperative for requirements.
- Will: To cite requirements that are operational to provide the capability being specified.

The requirements should be written using the following format:

The <noun phrase> shall (not) <verb phrase>.

The noun phrase should describe the main subject of the requirement. The shall or shall not represents a required (shall) or prohibited (shall not) behavior. The verb phrase should define the actions of the requirement.

8.3.2 DEVELOP SYSTEM REQUIREMENTS

The system requirements will: *State abstractly what the system will do to meet the stakeholder requirements. Avoid reference to* any particular design (Dick, Hull and Jackson, 2017).

The above definition uses the term stakeholder requirements, where we used the term customer requirements. Once the customer requirements are elicited, the customer requirements should be translated to system requirement. These system requirements are more technical in nature and related to the more detailed definition derived from the customer requirements. This typically results in defining multiple system requirements that relate to and detail one customer requirement. These system requirements are then prescribed more fully in the design phase, where they become physically instantiated, and finally describe the system solution or implementation.

8.3.3 DEVELOP SPECIALTY ENGINEERING DISCIPLINES AND RELATED REQUIREMENTS

Several specialty engineering disciplines are employed within the Vee model alongside the systems engineers, program, and project managers to apply their specialty engineering disciplines, including at a minimum, the following disciplines:

- Software engineering
- Environmental engineering
- Safety engineering
- Security engineering
- Human Systems integration engineering

Each discipline and the types of requirements that are defined are discussed in the following sections.

8.3.3.1 Software Engineering

Many, or shall we dare say, most systems incorporate some type of software or information system application into the system. Whether it's a car or an airplane, a train, or a healthcare system, most use software as a critical way to automate processes and information. The software engineering follows a similar life cycle as the Vee model, developing the concept of the software as the system is being conceptualized, eliciting software requirements within the requirements and architecture phase, designing detailed software specifications within the design phase, then verifying and validating the software code within the verification and validation (V&V) phases.

There are several key concepts that a systems engineer needs to understand about software engineering (INCOSE, 2015), (Fairley and Willshire 2011):

- For the time, effort, and expense devoted to developing it, software is more complex than most other system components.
- Software testing and reviews are sampling processes. Since software is not tangible nor physical, it can be more difficult to test and ensure that the requirements are met. There is only so much time to test different possible scenarios of the ways that the software can be used within the system, so some defects could not be identified prior to system implementation.
- Software often provides the interfaces that interconnect other system components, making it more difficult to test and ensure that defects do not exist, due to the software complexity, and testing capabilities.
- Every software product is unique. They are coded in various different languages, use different technology, and are dependent upon the skills of the software developers.
- In many cases, requirements allocated to software must be renegotiated and re-prioritized. In software development, schedules must be estimated and therefore can slip, causing the need to reprioritize and potentially delay functionality due to schedule, resources, and quality issues.
- Software requirements are prone to frequent change. Many times, systems users do not have a clear understanding or view of the software functionality until it is tested or delivered, requiring the requirements to change.
- Small changes to software can have large negative effects, due to the complexity and nature of software development and requirements elicitation.
- Some quality attributes for software are subjectively evaluated, based on the users' needs, the requirements elicited, the test criteria and humans as the users of the system.
- The term *prototyping* has different connotations for systems engineers and software engineers. Prototyping for a systems engineer may produce a useable working product, while a prototype of software still is not tangible and may only contain a portion of the requirements.
- Cyber security is a present and growing concern for systems that incorporate software, requiring the need for cyber security experts to be involved in system design and development.
- Software growth requires spare capacity, to ensure that additional functionality can be added after the initial requirements are delivered. The capacity can consist of people resources, software developers, users to test as well as physical capacity of equipment, computer hardware, and networks, to name a few resources.
- Several Pareto 80-20 distributions apply to software, especially when it comes to coding, testing, and defects. Eighty percent of the resources and effort may apply to 20% of the requirements, code, testing, and defects discovered.
- Software estimates are often inaccurate. There are multiple techniques used to estimate how long it takes to code and deliver software, none of which are perfect or 100% accurate. Since software requirements and code is unique, so are the estimates for development.
- Most software projects are conducted iteratively, especially with incorporation of agile software development that groups requirements into small sprints that can be delivered within weeks or months, instead of years. This iterative software development approach may not align with a waterfall or even Vee product development life cycle.
- Teamwork within software projects is closely coordinated to design, develop, test, and deliver software.
- V&V of software should preferably proceed incrementally and iteratively, to ensure that requirements are met, and defects are discovered early in the design and development processes.
- Risk management for software projects differs in kind from risk management for projects that develop physical artifacts. Risk management in software not only includes risk related to schedule, requirements, and resources but also include risks related to software development processes, including collaboration and communication issues related to the software development team.

- Software metrics include product measures and process measures and some of these are unique to software development and do not always apply to physical components, elements, and systems.
- Progress on software projects is sometimes inadequately tracked and can be difficult to do so, due to fact that software code can progress, but until it is tested and delivered, the completeness is difficult to assess.

8.3.3.2 Environmental Engineering

Environmental engineering requirements address four issues related to the system design and operation. They include:

1) Design for a given operating environment
2) Environmental impact
3) Green design
4) Compliance with environmental regulations.

Areas to be included for environmental engineering requirements include (INCOSE, 2015):

- Air pollution. Any gaseous, liquid, toxic, or solid material suspended in air that can result in health hazards to humans.
- Water pollution. Any contaminating influence on a body of water brought about by the introduction of materials that will adversely affect the organisms living in that body of water.
- Noise pollution. The introduction of industrial noise, community noise, and/or domestic noise that will result in harmful effects on humans.
- Radiation. Any natural or human-made energy transmitted through space that will result in harmful effects on the humans.
- Solid waste. Any garbage and/or refuse that will result in a health hazard.

The U.S. Environmental Protection Agency (EPA) defines green engineering as:

The design, commercialization, and use of processes and products, which are feasible and economical, while minimizing (1) generation of pollution at the source and 2) risk to human health and the environment.

(EPA 2012)

The following are principles of green engineering, provided by the EPA (2011):

- Engineering processes and products holistically, use systems analysis, and integrate environmental impact assessment tools.
- Conserve and improve natural ecosystems while protecting human health and well-being.
- Use life-cycle thinking in all engineering activities.
- Ensure that all material and energy inputs and outputs are as inherently safe and benign as possible.
- Minimize depletion of natural resources.
- Strive to prevent waste.
- Develop and applying engineering solutions, while being cognizant of local geography, aspirations, and cultures.
- Create engineering solutions beyond current or dominant technologies; additionally, improve, innovate, and invent (technologies) to achieve sustainability.
- Actively engage communities and stakeholders in development of engineering solutions.

8.3.3.3   Safety Engineering Requirements

The goal of safety engineering is to minimize hazards that can result in a mishap with an expected severity and with a predicted probability. This discipline focuses on identifying hazards, their causal factors, and predicting the severity and probability of occurrence. Hazards are potential risk events that could result in injury, death, or damage to equipment or property. The purpose is to reduce or eliminate the severity and/or probability of the hazards, resulting in minimizing the risk and severity of the hazards (INCOSE, 2015). The requirements for safety engineering are to incorporate features, processes, or system elements that will reduce the risks of safety hazards.

8.3.3.4   Security Engineering Requirements

Security engineering focuses on building systems that remain secure despite malice or error. Security engineering requirements incorporate elements or systems that proactively and reactively mitigate system vulnerabilities. There are many disciplines included in developing security engineering requirements, including cryptography, computer security, tamper-resistant hardware, applied psychology, supply chain management, and law (INCOSE, 2015). Cryptography focuses on disguising information to ensure the integrity, confidentiality, and authenticity of information during communication and storage. Developing and deploying business continuity and disaster recovery plans are part of security engineering processes. Security engineering requirements should address the following areas (INCOSE, 2015):

- Physical
- Personnel
- Procedural
- Emissions
- Transmissions
- Cryptographic
- Operations
- Computer security

8.3.3.5   Human Systems Integration Requirements

Human Systems Integration (HIS) is:

> *An interdisciplinary technical and management process for integrating human considerations with and across all system elements.*

> (INCOSE, 2015)

The disciplines that are commonly included in HIS are (INCOSE, 2015):

- Manpower and Personnel: Manpower determines the number of people resources that will be assigned to jobs and tasks, to complete them efficiently, effectively, and safely. Personnel defines the types of skill sets used for jobs and tasks.
- Training: Identifying the knowledge, skills, and abilities required to properly operate, maintain, and support the systems.
- Human Factors Engineering: Involves the understanding of human capabilities from a cognitive, physical, sensory, and team dynamic perspective and integration of these capabilities into the systems' requirements and design.
- Occupational Health: This discipline incorporates design features whose purpose is to minimize the risk of injury, acute or chronic illness, and disability. Areas can include addressing noise, chemical exposures and skin protection, atmospheric hazards, vibration, shock, acceleration and motion protection as well as chronic disease and discomfort.

- Habitability: Addressing issues related to system living and working conditions including lighting and ventilation, adequate space, availability of medical care, food and/or drink services and suitable sleeping quarters, sanitation and personal hygiene, and fitness/recreation facilities.
- Safety: The focus is to minimize risk of accident or mishaps to those operating, maintaining, and supporting the system, including issues such as safety of personnel, walking/working surfaces, emergency egress pathways, and personal protection devices, pressure and temperature extremes, prevention and control of hazardous energy release.
- Survivability: Addressing issues related to reducing the susceptibility of the total system to mission degradation or termination, injury or loss of life, and partial or complete loss of the system or its elements.
- Environment: Focusing on environmental considerations that can affect operations and requirements related to human performance.

8.3.3.6 Women's Healthcare Center Requirements
The customer and system requirements for the Women's Healthcare Center is shown in Table 8.2.

8.3.3.7 Food System Requirements
The students developed process scenarios with the stakeholders, created the Process Architecture Maps and process architecture, and then extracted the requirements by type for their sub-system. The requirements tables for each team are included in Tables 8.3 through 8.5, Figures 8.1 through 8.4.

TABLE 8.2
Customer and System Requirements for the Women's Healthcare Center

Requirements		
Requirement Type	**Customer Requirement**	**System Requirement**
Non-functional	Women's Center shall have a comfortable environment	Center shall have ergonomic and comfortable furniture in waiting rooms
Non-functional	Women's Center shall have ease of parking	WC shall have valet parking
Non-functional	Rooms with be aesthetic and geared psychologically to women	Rooms shall have state of art design features
Non-functional	WC will have creative access points to modern facilities	Center shall have multiple entrances from parking lots
Non-functional	Women will be guided through health issues in a nurturing, relaxing environment	Center shall have a patient navigator
Functional	Services shall be provided efficiently, with speed	Center shall have optimized processes
Functional	Services shall be provided with ease	Center shall have optimized processes
Functional	System shall be easy to navigate	Center shall have optimized processes
Functional	Services shall be provided with time-saving and convenience	Center shall have optimized processes
Functional	Shall have the ability to have combined service appointments	Scheduling shall allow combined appointments
Functional	Shall have ability to receive same day results	Center shall add additional radiologists to provide same day results

(Coninued)

TABLE 8.2   (Continued)

	Requirements	
Requirement Type	Customer Requirement	System Requirement
Functional	Shall have caring and competent professional staff	Center shall train staff in customer service
Functional	Shall have customer focused amenities	Center shall have spa services
Functional	Shall have advanced technology as an enabler of superior care	Center shall have state of art imaging technology
Functional	Services provided as One Visit, One Stop	Center shall allow combined appointments
Functional	Patients shall be guided through health issues in a nurturing, relaxing environment	Center shall have optimized processes
Functional	Women's Center services shall be a seamless integration of service components	Center shall have optimized processes
Functional	Services shall be provided for all ages and life phases	Center shall have services for all ages
Functional	Services shall be a comprehensive range of services	Center shall have services for all ages
Functional	Services shall be provided with coordination of medical and health concerns and treatments, multi-disciplinary team	Center shall incorporate team-based case management
Functional	Services shall be integrated between physicians, diagnostics, and ancillary	Center shall have optimized processes
Functional	Women's Center shall have functional medicine that combines traditional and integrative	Center shall incorporate team-based case management
Functional	Women's Center shall have gender-specific medicine	Center shall provide gender specific services
Functional	Women's Center shall provide holistic care	Center shall provide holistic services
Functional	Women's Center services shall be aligned with core values	Center shall have spiritual care
Human Factors	Imaging technology is easy to use	Human factors design is considered in the imaging technology interface
Human Factors	Process technology dashboard interface is easy to use	Dashboards in BPMS system are easy to use
Human Factors	Process technology dashboard interface identifies alerts when patients are delayed	Dashboards in BPMS system identify alerts when patients are delayed, more than 10 minutes for any service point
Human Factors	Process technology dashboard interface identifies alerts when patient's results are delayed	Dashboards in BPMS system identify alerts when patient's results are not delivered in less than a day
Information	BPMS reporting is complete	All identified reports are provided
Information	BPMS reporting is easy to use	User can pull a report in less than 10 minutes
Performance	System process design reduces delays from current system design by 30%	Future system process design reduces service wait times for each process to less than 10 minutes.
Performance	Patient results are delivered on the same day	Patient results are provided to patient's physician same day
Performance	Valet parking is available	Valet parking for patients requesting is available at peak hours.

(Continued)

TABLE 8.2 (Continued)

	Requirements	
Requirement Type	Customer Requirement	System Requirement
Safety	Patient walkways are available from parking lots to ensure pedestrian safety	Covered patient walkways are available from parking lots to ensure pedestrian safety.
Safety	Women's center is secure and safe.	Security cameras are available and operational in all critical areas.
Safety	Women's center is secure and safe.	Panic buttons are available and operational in all patient areas in the women's center
Security	All patient medical and personal identifiable information is secure and hack-proof.	State of art information security is in place.
Security	All patient medical and personal identifiable information is secure and hack-proof.	Information audits are in place to ensure no unauthorized access of medical records is being done.
Software	Create dashboard for patient delay alerts	Incorporate patient delay alerts in the BPMS
Software	Allow VIP patients to be identified when they arrive in the Women's Center	Incorporate VIP tracking and alerts.
Software	Allow tracking of same day results delivery	System shall track same day results
Software	Allow tracking of same day results delivery	System shall provide alerts when patient's results are not delivered to physician same day
Environmental	Construction materials shall be disposed in environmentally friendly manner	Adhere to ISO standard for environmental disposal of construction materials
Environmental	Medical waste shall be disposed in an environmentally friendly and legal manner	Adhere to environmental standards for disposal of medical waste

TABLE 8.3
Requirements for Team 1 – Connecting Organizations

	Requirements	
Requirement Type	Customer Requirement	System Requirement
Software	Interface to interact with the website	The website shall have platform that is easy to navigate and self-intuitive
Software	Website shall be easy to navigate	Website shall contain drop down lists for user to select the desired section of website to go to
Software	Data shall be accumulated in a single database	Website shall have a means for community members and volunteers to input data into the website
Software	Data shall be accumulated in a single database	Website shall have means of means of data collection internal to the website
Software	Data shall be accumulated in a single database	Website shall have algorithm to file sort and organize data in database
Software	Website shall be safe for users	Website shall have a means of cyber security to protect user information

(Coninued)

TABLE 8.3   (Continued)
Requirements

Requirement Type	Customer Requirement	System Requirement
Software	Website shall allow for many users at once	Website shall have enough server space for expansion of the website and prevention of website shutdowns
Software	Website shall allow for many users at once	Website shall allow for a minimum of 300 persons on the server at any given time
Software	Shall have means of communication for all persons involved	Website shall have a forum or other electronic means of communication in website to allow members, farmers, advertisers, and citywide to converse with each other
Software	Shall have means to contact Citywide	Website shall have contact information and a means to contact Citywide directly
Software	Citywide shall be able to view website traffic	The website shall have a counter that continually monitors the number of visitors to the webpage
Software	Shall have a means notification when an order is placed	Website shall have a quick response to customer of when their order is placed
Software	Shall have a means notification when an order is placed	Website shall send out an automatic reply to customer email of online purchase order
Software	Shall have a means of obtaining customer input	Website shall have a section where customers can fill out a service survey provide feedback for the program
Software	Shall have a way for farmers to advertise their products	Website shall have sections with high server traffic flow where farmers products can be advertised
Software	Shall have a way for community partners to advertise	Website shall have sections with high server traffic flow where community partners can be recognized and can advertise
Software	Website shall have a language option that is easy to change	Website shall include a dropdown menu where a user can easily select their language of choice and the page will translate accordingly
Software	Shall have a means of notifying customers pick up time and locations	Website shall have a dashboard display list of next pick-up times and locations for a minimum of a week prior
Software	Allow tracking of number of orders requested	Website shall monitor and tally number of online orders placed
Software	Allow tracking of number of orders filled	Website shall monitor and track number of orders full filled
Software	Citywide shall have means to prioritize foods	Citywide shall on database or spreadsheet software offline to collect and store survey data
Software	Citywide shall have means to prioritize foods	Citywide shall have means to conduct statistical studies for prioritization of foods and customer requests
Human Factors	Citywide volunteers shall always have access to internet	Citywide volunteers shall be equipped with necessary technology to conduct their work inside and outside office
Human Factors	Farmers have way to complete transaction	Farmers shall own means of accepting payment via cash, check, and/or electronic transfer
Human Factors	Website easy to navigate	Website shall be organized for ease of locating online function required

(Continued)

TABLE 8.3 (Continued)

Requirements

Requirement Type	Customer Requirement	System Requirement
Human Factors	Website shall include an administrative moderator to ensure forum civility	Website volunteer reads through and moderates discussions in the website forum
Information	Website alerts members when a pick-up time is delayed	Website volunteers send user wide notifications to alert members when an event or pick up time is changed
Information	Citywide shall provide weather updates and cancelation notices	Citywide shall use all forms of communication to send members and farmers cancelations due to climatic weather
Information	Provide information to farmers for food orders	Citywide shall sends updates immediately to participating farmers on what food orders have been made
Information	Provide farmers with information of data from surveys	Citywide shall provide a weekly newsletter or summarization of survey data and food requests to all local farmers
Functional	Provide means to fill out orders and surveys without a computer	Citywide shall provide a volunteer/employee to input paper documented information into website database daily
Functional	Provide a means to store paper documented information	Citywide shall provide a volunteer/employee to input paper documented information into website database daily
Functional	Provide members means to obtain event, pick-up information off line	Citywide shall work with community partners to post and promote program to communities
Functional	Discussion with farmers when they are available for pick-up days	Citywide shall discuss best dates and times for community member and farmers to meet in the communities
Functional	Meeting locational shall be within community orders are placed for	Event locations shall be public area accessible by all patrons
Functional	Pick up locations shall vary from on community to another	Farmers shall be notified of which community pick-up is scheduled for routinely
Security	Website is secure from leak or hacking of member information	Website shall have state of the art cyber security
Security	Website is secure from leak or hacking of member information	Website shall have two-part identification for login
Environmental	Farmers and Citywide responsible for cleanup of pick-up location after event	Farmers and Citywide shall ensure event location is maintained and cleaned up after event
Environmental	Space not be overcrowded and provide accessible parking	Citywide responsible for scouting of appropriate pick-up location for events
Environmental	Event area shall be handicap accessible	Event area shall include ramps, parking level ground, and additional handicap accommodations
Safety	customers shall feel safe and comfortable at market or event	local police shall be contacted of brought in to monitor or patrol event
Safety	Shall not have major disruption of traffic flow	Citywide shall work with local authorities to ensure traffic flow through area is fluent and safe

TABLE 8.4

Requirements for Team 2 – Creating Food Policy

	Requirements	
Requirement Type	**Customer Requirement**	**System Requirement**
Non-functional	Commission center shall have ease of parking	Commission center shall have valet parking
Functional	Problem shall be well identified	Community members and the champion shall identify the problem properly
Functional	Candidates shall be decided	Community members shall decide candidates through discussion and election
Functional	A champion shall be selected	Community members shall elect a champion from candidates through discussion and election
Functional	Enough support shall be gathered	Support groups who are impacted by the problem shall provide enough support
Functional	Data shall be collected and processed	The champion and support groups shall process data
Functional	Policy change files shall be created	Policy change files shall be made through data processing
Functional	Policy change files shall be present to commission center	The champion shall be able to present policy change files to commissioners in commission center
Functional	Commission report shall be created	Commission members shall come up with the commission report through discussion and processing based on policy change files
Functional	Drafted legislation shall be created	Commission members shall come up with a drafted legislation
Functional	The legislation shall be passed	Mayor and commission members shall pass the legislation based on the discussion and decision meeting
Security	Policy change files shall be well protected	Commissioners, the champion shall keep files safe
Security	All community members election information is secure	Information audits are in place to ensure no unauthorized access of election records is being done
Safety	Community center and commission center is secure and safe	Security cameras are available and operational in all critical areas
Safety	Champion walkways are available from parking lots to ensure pedestrian safety	Covered walkways are available in commission center
Human Factors	Calendar is easy to be seen	Calendar in Excel is easy to get
Human Factors	Scheduling system is easy to use	Scheduling system in commission center is easy to use
Human Factors	Election system is easy to use	Election system during champion selection or decision-making is easy and clear
Software	Create calendar sheet for community meeting and commission meeting	System shall provide calendar sheet using MS software

(Continued)

TABLE 8.4 (Continued)

	Requirements	
Requirement Type	**Customer Requirement**	**System Requirement**
Software	Calendar information is displayed on official website	System shall provide official website
Software	Scheduling system is correct for making appointment and meeting schedule	System shall provide proper scheduling system
Environmental	Draft waste shall be put into trash center	Available trash bins and trash processing

TABLE 8.5
Requirements for Team 3 – Kids Growing Food

	Requirements	
Requirement Type	**Customer Requirement**	**System Requirement**
Software	Table 33 and teacher need proof of purchase	Software capable of delivering a "carbon copy" receipt detailing date, time, weight, price, and signature of both parties
Software	Student and teacher need proof of sale	Software capable of delivering a "carbon copy" receipt detailing date, time, weight, price, and signature of both parties
Software	Teacher needs a monthly statement of cash flow	Software capable of producing a monthly cash flow report
Functional	Children shall receive proper instruction and training in growing microgreens	Microgreen program shall use qualified teacher
Functional	Children shall grow high-quality microgreens	Microgreen program shall supply microgreen seeds, tray with lid, medium, and watering apparatus
Functional	Children shall transfer microgreens back to school for compensation	Microgreen program shall have a standard method of evaluating quality for compensation
Functional	Teacher shall harvest, clean, and bag microgreens	Teacher shall have adequate supplies to perform processing tasks
Functional	Teacher shall transport microgreens to Table 33 for compensation	Teacher shall have adequate means of transportation for a timely transport
Environmental	Waste from microgreen program shall be recycled	Microgreen program shall provide access to a place to compost
Environmental	Microgreen program shall use environmental friendly materials where possible	Microgreen distribution bags shall be composed of recycled material
Non-functional	Microgreen program shall attempt to engage children wherever possible	Microgreen program training shall contain fun deliverables and activities to learn about microgreens and their health benefits

(Coninued)

TABLE 8.5   (Continued)

Non-functional	Microgreen program shall encourage exceeding baseline quality standards	Microgreen program shall have a standard method of evaluating quality for compensation
Safety	Safety shall be a key component in the microgreen program	Microgreen program shall utilize non-toxic medium
Safety	Microgreens shall not be exposed to unsanitary conditions	Children shall use program distributed tray with lid to grow microgreens
Performance	Children shall be compensated for microgreens on the same day they transfer back to school	Microgreen program shall have adequate starting funds
Information	Payment records shall be thorough and complete	Microgreen program shall use Excel workbooks to detail payments
Security	All student and teacher records shall be stored securely	Microgreen program shall use password protected files for storing records
Human Factors	Microgreen program training shall be easy to understand and implement	Microgreen program training shall include deliverables that are easy to comprehend and available to refer back to
Human Factors	Standard quality evaluation chart shall be easy to use and understand	Microgreen program shall utilize a clear and well-organized chart for the teacher to correctly compensate children and the children to understand how they are being compensated

The reader will notice the difference in level of detail, completeness, and number of requirements across the teams. Some of this is due to the different project and system scopes, while some of it corresponds to the level of detail derived during the process scenarios and requirements elicitation sessions. This highlights the extreme criticality of the process scenarios to develop robust, accurate, and complete requirements.

8.4   DEVELOP MEASUREMENT PLAN – DATA COLLECTION PLAN, WITH OPERATIONAL DEFINITIONS

8.4.1   DATA COLLECTION PLAN

8.4.1.1   Critical to Satisfaction Criteria

A data collection plan should be developed to identify the data to be collected that relate to Critical to Satisfaction criteria. Critical to Satisfaction is a characteristic of a product or service which fulfills a critical customer requirement or a customer process requirement. CTSs are the basic elements to be used in driving process measurement, improvement, and control.

CTSs are the basic elements to be used in driving process measurement, improvement, and control. It is critical to ensure that our selected CTS accurately represent what is important to the customer. These were already discussed in Chapter 5.

The data collection plan ensures:

- Measurement of critical to satisfaction metrics.
- Identification of the right mechanisms to perform the data collection.
- Collection and analysis of data.
- Definition of how and who is responsible to collect the data.

Requirement Type	Customer Requirement	System Requirement
Safety	shall need the food to be less preserved and fresh	Food will be stored only until the desired period varying on different foods
Safety	Shall require food bags to be appropriate for long term storage	Use of Mylar bags for long term food storage
Safety	Food needs to be healthy to consume	Keep food away from harmful preservative chemicals during storage
Safety	Shall expect to consume only organic fresh food	avoid adultration of the food at the markets or storage
Safety	Shall demand cleanliness at the storage	Helpers will maintain cleanliness
Functional	Shall require food to be undamaged at the farm	On time pick up of food from farms to not let it damage with improper storage
Functional	Food availability for purchasing at all days	Delivery of food on time to markets to keep up with the demand for it
Functional	Food is available at all seasons for consumption	Keep a track of demand and produce of the food to make it available at all seasons
Human Factors	Shall assume the staff is maintaining hygiene methods	The staff is wearing hygiene gears at work
Information	Customers would want to consume healthy food	Maintaining realtime records or expiry dates
Information	Shall expect to consume only organic fresh food	Will maintain on time pickups from farm without delay and also mentioning of dates on food to state the expiry period
Performance	Customers would want to consume fresh food	Will have skilled workers at the Storage place to operate the temperatures for different food accordingly to maintain it in a fresh condition
Human Factors	Food available to consume at all seasons	The staff at the system will keep up with time and be punctual with work to deliver food on time
Environmental	Food available to consume at all seasons	Farmers will learn methods to grow food on all seasons which is taught by homefull staff
Functional	Food available to consume at all seasons	The food is stored at desirable temperature for various foods making it available throughout the year
Safety	Food shall be transported with safe methods	The transportation is maintained with safety for food, by providing a cooling system in the vehicle
Non-functional	Food should not be mishandelled	The storage of food should be like avoid placing food one on top of the other
Security	Food should be handelled with care to maintain it fresh	staff should keep monitoring the storage place to avoid any failure of refrigerators during storage
Environmental	Food should be handelled with care to maintain it fresh	Storage staff will monitor the temperature and adjust it according to the different food varieties
Information	staff shall be skilled at handling different food	The staff must be taught about each and every aspect of storing for diffferent foods
Information	Shall expect the food grower be experienced in growing	Farmers or community grower should be aware of what methods of maintaning the food
Information	Customer expects the community to provide food reaching the demand	The inventory should be kept up to date and thorough with details of food coming in the storage place and going out
Information	Food shall be undamaged	The food collector should be provided with good information of the quantity of food to be transported
Security	Food needs to be fresh to consume	Food will be kept safely at the farm avoiding damage from the insects or others creatures
Security	Food needs to be fresh to consume	Food should be kept safe from any fungus formation
Environmental	Expects the storage of food is clean	By maintaining containers used at the storage very clean
Information	Food shall always be refrigerated while storing	Be preplanned with the food coming to central storage and setting space for it binstead of not having space
Safety	Shall demand cleanliness at the storage	The staff should wash hands and use of sanitizers after every activity
Information	Food shall be undamaged	Food to be transported shall not exceed the limit of carriage on the vehicle to avoid overloading
Functional	Food needs to be fresh to consume	the food collector should collect the food from the farms where are colse to the central storage location
Safety	Good Food Handling Practices	It is important that transporters and distributors establish sanitary food handling practices for carrier vehicles, storage areas.
Safety	Proactive Shipping and Receiving Programs	It is important to pack food proprierly and maintain desired temperature of food before delivering it to customers.
Safety	Good Cleaning and Sanitation Practices	personnel can be trained in proper cleaning and sanitation procedures.The distributor must establish these good practices
Safety	Expects the containers to be clean	Staff shlould clean the containers in order to store the food inside them.

FIGURE 8.1   Requirements for team 4 – food storage.

Requirements		
Requirement Type	Customer Requirement	System Requirement
Functional	Need standard farmers within the system	Farmers shall have a contract between businesses and distributors in order to do business within the system
Information	Visibility for entry level customers and entrepeneurs	Website will show foods available at each local businesses
		Website shall show real time prices of produce
		Website shall show contact information for businesses within the system
Safety	Website needs provisions against hacking	Website shall be examined for weak points
		Website shall fortified to avoid malware and hacking
Environmental	Farmers cannot overuse land	Farmers shall only provide produce as is normal to growing pattern, that is beneficial/normal to the environment
		Farmers shall only provide produce that is beneficial/normal to the environment
Functional	Transportation company capable of delivering produce (either farmers, buyer, or middle-man responsible)	The system shall have a set of reliable, go to distributors of whom are contracted to do dealings with businesses within the system
Software	Business Owners can see live Delivery updates	Website shall be capable of giving real time delivery updates
Software	Customers and Business Owners can see produce availability	Website shall be capable of displaying food availability
Performance	Business Owners can see produce costs including shipping	Website shall be capable of updating costs in real time
Software	All involved in the system	Website shall be capable of allowing live communication
Performance	Business Owners need to access and update the website	Website shall be user-friendly through utlizing drop down menus
		Website shall have clearly marked buttons, etc.
Non-functional	Alert customer of price drops	Website shall allow customers to sign up and receive e-mails when business owners note produce price changes within the network of

FIGURE 8.2 Requirements for team 5 – food hub website.

Table 8.6 shows a data collection plan and Table 8.7 shows the related operational definitions for the Women's Healthcare Center.

The steps for creating a data collection plan are:

1. Define the Critical to Satisfaction
2. Develop Metrics
3. Identify data collection mechanism (s)
4. Identify analysis mechanism (s)
5. Develop sampling plans
6. Develop sampling instructions

A description of each step in the data collection plan development follows.

Requirements		
Requirement Type	Customer Requirement	System Requirement
Non-functional	To have a well organized store	Having proper sign boards in the store
Non-functional	To have access to healthy food resource	To provide each and every healthy product with proper inspection
Non-functional	Have enough parking spaces for the customers	100 parking space will be provided
Non-functional	The store should be secure and safe	Having 24/7 surveillance cameras in the store
Functional	Having enough cashiers	Coustomers don't have to wait more than 10 min
Environmental	high quality food products from the Farmers	To have a proper harvesting techniques to grow crops
Functional	Having enough checkout counters to process payments	workers with technical knowledge of handling billing
Functional	Having different modes of billing options like cash, credit, debit, gift cards etc	To accept all the modes of payments done by the customer
Software	Having a good online shopping website	website easy to navigate
Non-functional	To have a well organized store	Have a good management in the store for organizing the food products
Safety	The store should be safe for the customers	Having a good security team
Functional	Delivering products to the customers	Having a delivery system that drops shopped items to the customers address
Human Factors	To have fesible prices which are afforadble for the customers	Following hybrid pricing in the store

FIGURE 8.3   Requirements for team 6 – gem city market.

1. Define the Critical to Satisfaction (CTS) criteria (as discussed in Chapter 5).
2. Develop metrics: In this step, metrics are identified that help to measure and assess improvement related to the identified CTS. Some rules of thumb for selecting metrics are to (Evans and Lindsey, 2011):
   - Consider the vital few versus the trivial many.
   - Metrics should focus on the past, the present, and future.
   - Metrics should be linked to meet the needs of stakeholders, customers, and employees.

It is vital to develop an operational definition for each metric, so it is clearly understood how the data will be collected by anyone that collects it. The operational definition should include a clear description of a measurement, including the process of collection. Include the purpose and metric measurement. It should identify what to measure, how to measure it, and how the consistency of the measure will be ensured. A summary of an operational definition follows.

8.4.1.2   Operational Definition

The operational definition consists of 1) defining the measure, the definition, 2) the purpose of the metric, and 3) a clear way to measure the process.

Requirement Type	Customer Requirement	System Requirement
Software	Customer must be able to order any meal from the menu	Missing ingredients shall be tracked using softwares such as Excel to determine exactly the type and amount of missing ingredients
Software	Customer must be able to order any meal from the menu	Create a website or application to connect restaurant with local farmers to make the process of supply restaurant with ingredients easier
Environmental	Customer must be able to order any meal from the menu	Table 33 manager shall find the best local farms available, who can provide ingredients during all seasons.
Safety	Customer must have the best quality meals	Table 33 manager and chief shall test ingredients carefully before purchasing ingredients
Software	Customer must have the best quality meals	Table 33 manager and chief shall use devices such as greentest, to help them check the quality of ingredients
Human Factors	Customer must be able to order meals anytime	Table 33 manager shall schedule ingredients delivery properly
Human Factors	Customer must be able to order meals anytime	Table 33 manager must hire the appropriate number of resources, who can move the ingredients from the inventory to the kitchen

FIGURE 8.4   Requirements for team 7 – table 33 restaurant.

TABLE 8.6
Data Collection Plan for Women's Healthcare Center

Data Collection Plan

Required

Identify Metrics to Measure and Assess Improvement That Relate to the CTSS From the Define Phase.

Critical to Satisfaction (CTS)	Metric	Operational Definition	Data Collection Source	Analysis Mechanism	Sampling Plan (size, frequency)	Process to Collect and Report
Timeliness	Procedure time	Time from when patient goes into procedure room to when they leave	Medical information system	Statistical analysis	July through September	IT Director will provide reports
	Wait time for procedure	Time from when the patient is registered to when they get into the procedure room				
	Time to register patient	Time from when patient arrives in the facility to when registration is complete	Registration system			
	Time to receive results	Time to receive results of exam, from when you finished the procedure	Imaging system			
	Wait to get an appointment	Time to wait until get an appointment	Scheduling system			

TABLE 8.7
Operational Definitions for Women's Healthcare Center

Operational Definition

Define a clear and concise description of the measurement and the process by which it is collected.

Operational Definition	Defining Measure	Purpose	Clear Way to Measure
Time from when patient goes into procedure room to when they leave	Procedure time	Identify issues with resources and process issues	From medical information System report
Time from when the patient is registered to when they get into the procedure room	Wait time for procedure		From medical information system report
Time from when patient arrives in the facility to when registration is complete	Time to register patient		From registration system report
Time to receive results of exam, from when you finished the procedure	Time to receive results		From imaging system report
Time to wait until get an appointment	Wait to get an appointment		From scheduling system report, implementing new fields to capture desired appointment time compared to actual appointment time

Defining the Measure, Definition:

A clear, concise description of a measurement and the process by which it is to be collected.

Purpose: Provides the meaning of the operational definition, to provide a common understanding of how it will be measured.

Clear way to measure the process:

- Identifies what to measure
- Identifies how to measure
- Makes sure the measuring is consistent

3. Identify data collection mechanism(s): Next you can identify how you will collect the data for the metrics. Some data collection mechanisms include: customer surveys, observation, work sampling, time studies, customer complaint data, emails, websites, and focus groups.
4. Identify analysis mechanism(s): Before collecting data, consider how you will analyze the data to ensure that you collect the data in a manner that enables the analysis. Analysis mechanisms can include the type of statistical tests or graphical analysis that will be performed. The analysis mechanisms can dictate the factors and levels for which you may collect the data.
5. Develop sampling plans: In this step you should determine how you will sample the data and the sample size for your samples. Several types of sampling are:
   - Simple Random Sample: Each unit has an equal chance of being sampled
   - Stratified sample: The N (population size) items are divided into subpopulations or strata and then a simple random sample is taken from each stratum. This is used to decrease the sample size and cost of sampling

- Systematic sample: N (population size) items are placed into k groups. The first item is chosen at random, the rest of the sample selecting every kth item.
- Cluster Sample: N items are divided into clusters. This is used for wide geographic regions.
6. Develop sampling instructions: Clearly identify who will be sampled, where you will sample and when and how you will take your sample data.

8.4.1.3   Food System Data Collection Plans

Team 1s data collection plan is shown in Table 8.8, with the related operational definitions in Table 8.9.

The data collection plans for Teams 2 through 6 are shown in Table 8.10, 8.11 and Figures 8.5 through 8.8.

8.5   DEVELOP INFORMATION MODELS

Figure 8.9 displays the information model from Chapter 3 that shows the information elements for managing information used by the system. The information elements in this model can help us to understand software engineering requirements.

The purpose of Information Management processes is to generate, obtain, confirm, transform, retain, retrieve, disseminate, and dispose information, to designated stakeholders. The information includes technical, project, organizational, agreement, and user information. We further discuss the technical or system information. There are two tools that we'll use to develop the instantiations of the information models: 1) the information hierarchy model and 2) the class diagram with information model. An information model is a representation of concepts and the relationships, constraints, rules, and operations to specify data semantics for a chosen system domain. If you have a complex system, you can develop a hierarchy of the conceptual information and then create the conceptual information model – class diagrams. You may have multiple class diagrams, one for each domain or sub-system. To develop the information hierarchy model, the Process Architecture Map cover page, listing the information derived from the process scenarios and maps can be a starting point. This should provide a good list of the information or documents used by the system from a functionality perspective. The relationships of the information and groupings of the information into themes is shown in the hierarchy information model. An example of the hierarchy information model for the Women's Healthcare Center is shown in Figure 8.10.

This information inventory list is then used to develop the class diagram information model that shows the relationship of the information elements. The Class Diagram is an Unified Modeling Language (UML) model. It defines the system elements (people, information, materials, etc.) and their relationships to each other. To develop the class diagram information model, follow the steps:

1. Subject Matter Experts (SMEs) brainstorm the system elements or use the information from the Process Architecture Maps cover page and the Hierarchy Information Model.
2. Identify how the elements are related.
3. Draw the class diagram.
4. Validate the class diagram with the SMEs.

An example is shown in Figure 8.11 for the Women's Healthcare Center. The conceptual class diagram provides the listing of elements in our system, both physical and informational, and the relationships of these elements. The elements can be derived from the roles, information, technology, and knowledge provided in the Process Architecture Map.

TABLE 8.8
Data Collection Plan for Team 1 – Connecting Organizations

Critical to Satisfaction (CTS)	Metric (short title)	Operational Definition (metric description)	Data Collection Source	Analysis Mechanism	Sampling Plan (size, frequency)	Process to Collect and Report
Access to food	Number of orders placed	The number of people who have filled out a form (online/paper) to gain access to fresh food	Website database	Counter on the website	Orders per month	Use website database to draw performance variables/ indicators
Access to food	Number of orders filled/ completed	The number of people who actually got the food	Website database	Counter on the website	Monthly	Use website database to draw performance variables/ indicators
Access to food	Number of Farmers recruited	Actual number of farmers who are working on the website to serve people	Website database	Counter on the website	Monthly	Use website database to draw performance variables/ indicators
Access to food	Time required to fulfill order	The time (in days) between people ordering the food and getting it	Website database/ customer feedback	Tracker on the website	Weekly	Use either website order tracker or customer feedback to estimate average time of order delivered
Ease of website use	Time required to fulfill a task using the website	Time it takes for a user to find and complete a form or survey on the website	Time volunteers for the task	Built in timer on the website	Average time less than 10 minutes	Bring in volunteers to conduct study
Access to food	Survey	Food will be provided at locations near effected areas (local markets). In this way survey will be conducted to check if people are satisfied	Website survey forms	Statistical	Every month	Use of statistical tools and graphs
Engaging youth in Food Justice system	Age range of websites users	Survey will ask what age range a user falls into	Website database	Statistical	Running total	Use of website survey to collect user data

TABLE 8.9

Operational Definitions for Team 1 – Connecting Organizations

Operational Definition	Defining Measure	Purpose	Clear Way to Measure
The number of people who have filled out a form (online/paper) to gain access to fresh food	Number of people who filled the form	It is used to check how many people are getting involved in this business	There is a counter on the website which would measure how many people actually filled out the form
The number of people who actually got the food	Number of people who got the food	It can be used to measure the efficiency of our system	There is a "food delivered" option on the website which would count those orders
Actual number of farmers who are working on the website to serve people	Number of farmers recruited on the website	Website needs to continually recruit farmers so this would be used to measure how well the website is doing	Website has a counter to count number of farmers on the system
The time (in days) between people ordering the food and getting it	Time to measure if people get the food on time	It is used to measure efficiency of the business	There is a tracker on the website which would track all of this
Time it takes for a user to find and complete a form or survey on the website	Time it takes for people to fill out the form to check its user friendliness	It is actually used to improve the system. The system always needs to be user friendly	Counter will begin when a user starts to fill the form and complete it
Food will be provided at locations near effected areas (local markets). In this way survey will be conducted to check if people are satisfied	Conducted surveys to improve the system	It is used as a measure to improve the system	There should be paper or E-forms which would be filled out by customers
Survey will ask what age range a user falls into	Conducted surveys to track age groups of customers	It is used to track what age groups' people are involved in the system so that more youth would be involved	E-forms or paper forms will be filled out by customers which would mention the age

TABLE 8.10

Data Collection Plan for Team 2 – Creating Food Policy

Data Collection Plan

Identify metrics to measure and assess improvement that relate to the CTSs from the Define Phase.

Critical to Satisfaction (CTS)	Metric (short title)	Operational Definition (metric description)	Data Collection Source	Analysis Mechanism	Sampling Plan (size, frequency)	Process to Collect and Report
Timeliness	Procedure time	Time from when the community members start discussing a problem to when that problem gets addressed	Champion	Statistical analysis	After every Policy implementation	Champion will provide the details
Timeliness	Community Members response	Time from when the request is made to community to when the required support received	Champion	Statistical analysis	After every request made to community	Champion will provide the details
Timeliness	Commission response	Time from when the request is made to commission to the response given by the commission team	Champion	Statistical analysis	After every request made to commission	Champion will provide the details
Timeliness	Time to pass the law	Time to pass the law since when it was approved	Champion	Statistical analysis	After every policy implementation	Champion will provide the details
Authenticity	Reliability of the data collected	Feedback received on the data once it is validated	Champion and Data collection team	Statistical analysis	After every policy implementation	Champion will provide the details
Cost	Able to get the required funds	Ability to get the required funds from Community and other benefactors to drive the request forward	Champion	Statistical analysis	After every policy implementation	Champion will provide the details

TABLE 8.11
Data Collection Plan for Team 3 – Kids Growing Food

Data Collection Plan

The purpose of the data collection plan is to identify measurable aspects of the system that are critical to the system fulfilling its purpose so that these aspects may be modified in order to improve the system.
This tool allowed the team to identify measurable components of the system and think of how to best analyze them to ensure the system's success.

Critical to Satisfaction (CTS)	Metric (short title)	Operational Definition (metric description)	Data Collection Source	Analysis Mechanism	Sampling Plan (size, frequency)	Process to Collect and Report
Quality	Visual appearance of microgreens	How the microgreens look in terms of color, height, vitality	Visual observation	Ranked in comparison to standard	Every time student turns in microgreens to teacher	Teacher collects and evaluates student's microgreens at school after each growing cycle. Teacher stores in spreadsheet.
Quality	Taste	How the microgreens taste to an experienced microgreen user	Taste test	Ranked in comparison to standard	Every time student turns in microgreens to teacher	Teacher tastes at least one microgreen from student's crop. Teacher stores in spreadsheet.
Safety	Cleaning process	Microgreen is cleaned with hydrogen peroxide	Yes or no test	Yes or no	Every time student turns in microgreens to teacher	Teacher records when crop was cleaned, including time and date.
Cost	Payment amount	Table 33 must pay the teacher at least as much money as the teacher pays the children for the microgreens	Payment records from both Table 33 and children	Compare sale price to children and Table 33	Evaluate every payment exchange	Teacher records sale price to children and Table 33 in payment records.
Timeliness	Time between harvesting and delivery to Table 33	The time taken by the teacher between harvesting the microgreens and delivering the microgreens to Table 33	Teacher kept time-stamped harvesting records	Ensure time taken between harvesting and delivering microgreens is within agreed upon standard	Every harvest and delivery	Teacher records harvesting time and Table 33 certifies time of delivery
Critical to Process	Number of student participants	Number of students participating in program in a given growing cycle	Teacher kept records	Counting number of participating students	Weekly analysis	Count and log number of student participants

Data Collection Plan

Identify metrics to measure and assess improvement that relate to the CTS's from the Define Phase.

Critical to Satisfaction (CTS)	Metric (short title)	Operational Definition (metric description)	Data Collection Source	Analysis Mechanism	Sampling Plan (size, frequency)	Process to Collect and Report
Timeliness	Time to gather the food in carts	Time to take the food from the field to the carts	Farmer	Statistical analysis	Throughout the year	Community staff
	Carts moved to inventory storage	Time to transport the food from the field to the temporary storage location	Farmer	Statistical analysis	Throughout the year	Community staff
	Time for packaging	Time to bag each kind of food in a suitable bag	Farmer	Statistical analysis	Throughout the year	Community staff
	Time to transport food	Time to transport the food from the temporary storage to the center	Food collector	Statistical analysis	Throughout the year	Community staff
	Final transportation	Time to transport the food from the center to the distributors	Food transporter	Statistical analysis	Throughout the year	Community staff
Safety	Fresh food	Storing the food in the suitable temperature and condition	Farmer	Physical analytics	Throughout the year	Community staff

FIGURE 8.5 Data collection plan for team 4 – food storage.

Data Collection Plan

Identify metrics to measure and assess improvement that relate to the CTS's from the Define Phase.

Critical to Satisfaction (CTS)	Metric (short title)	Operational Definition (metric description)	Data Collection Source	Analysis Mechanism	Sampling Plan (size, frequency)	Process to Collect and Report
Timeliness	Produce Delivery	Produce is Delivered on time to correct Businesses	Business Owners and Distributors keep records of delivery times, produce pick up times, and so forth.	Compare amount of time it takes to deliver to estimated driving time from a map service multiplied by a factor to accommodate traffic.	All businesses currently in the DDP	Ask all distributors and businesses in the current network to keep an excel spreadsheet of the necessary times
	Produce is Fresh	Check to see if produce is fresh from time of delivery to time of use/sale	Business Owners will track how much produce they throw away due to spoil or damage	Examine produce for damage or signs of spoil before placing in foods or for sale	All businesses currently in the DDP	Business Owners will track in an excel spreadsheet how many of each produce they received and how many they had to throw out due to spoil or damage
Investment	Business Profits	See by what margin businesses are profiting per month	Businesses report their gross profits per month	Analyze profit margins from inclusion in the system compared to those outside	All businesses currently in the DDP	Business Owners within (and those willing outside the network) report earnings for comparative analysis on business profits.
	New Investors	How many new investors are investing in the system	Poll business owners to see if they have new investors	Poll	All businesses currently in the DDP	Poll businesses about whether they've had new investors in the past 3 months.Also, ask otherbusinesses outside if they've had newinvestors in equal time frame.
	Property Values	See if property values near businesses have increased	Survey areas surrounding businesses	Ask realtors of local properties	Local Realtors	Ask realtors of local properties for updated property value information and compare to prior postings.

FIGURE 8.6 Data collection plan for team 5 – food hub website.

Data Collection Plan

Identify metrics to measure and assess improvement that relate to the CTS's from the Define Phase.

Critical to Satisfaction (CTS)	Metric (short title)	Operational Definition (metric description)	Data Collection Source	Analysis Mechanism	Sampling Plan (size, frequency)	Process to Collect and Report
Affordable products	Pricing of products	Products are affordable to the average/low income community people	Management and surveys	Statisticsal analysis	May 2020 (Open Date) through May 2021	Management will provide the reports
Healthy food	Healthy products	The Products provided are healthy and affordable	Surveys	Statisticsal analysis	May 2020 (Open Date) through May 2021	Management will provide the reports
Proximity to residence	Location of the market	The Market is close to the community people who are suffering in the food desert	Surveys	Statisticsal analysis	May 2020 (Open Date) through May 2021	Management will provide the reports
Quality of products	Quality of food	The products would have a range from high to low quality	Management and surveys	Statisticsal analysis	May 2020 (Open Date) through May 2021	Management will provide the reports
Human interaction	Interaction	The workers would learn how to deal with the customers in a nice and polite way	Surveys	Statisticsal analysis	May 2020 (Open Date) through May 2021	Management will provide the reports

FIGURE 8.7 Data collection plan for team 6 – gem city market.

Data Collection Plan

Identify metrics to measure and assess improvement that relate to the CTS's from the Define Phase.

Critical to Satisfaction (CTS)	Metric (short title)	Operational Definition (metric description)	Data Collection Source	Analysis Mechanism	Sampling Plan (size, frequency)	Process to Collect and Report
Food quality	Farmers ways of growing ingredients	Quality of ingredients will be measured by the chief and manager	Collected from the farm	Food quality will be tested by checking farms certificates in growing food and by using chief skills to measure the quality of ingredients	Based on the Inventory Management systems need, some ingredients are needed daily while some are needed weekly	Manager, chief and inventory system management are responsible for finding organic local ingredients from local farms whenever needed
Food quality	Chief skills in making food and customer opinion	Measuring food quality is done first by the chief who is skilled in making food and then by asking the customer about how food looks like and taste like.	Collected by waitress who is responsible to ask customers about their opinion on the food	Data will be collected by asking the customers directly	For each customer and should be collected while the customer is eating	Waitress should ask the customer directly
Ingredients Cost	Local farmers prices	Cost of ingredients will be measured by manager and point of sales	Collected from farmers	By comparing prices from multiple local farms	Whenever the prices change	Use a table or software to compare all the prices given by local farms
Ingredients supplement	Inventory Management Systems report	Number of missing ingredients will be measured by Inventory Management Systems	Collected by Inventory Management Systems	It will be measured by the number of ingredients in the inventory over the number of ingredients needed	Whenever missing ingredient is found	Using a daily/weekly list of needed ingredients or by using a software such as Excel

FIGURE 8.8 Data collection plan for team 7 – table 33 restaurant.

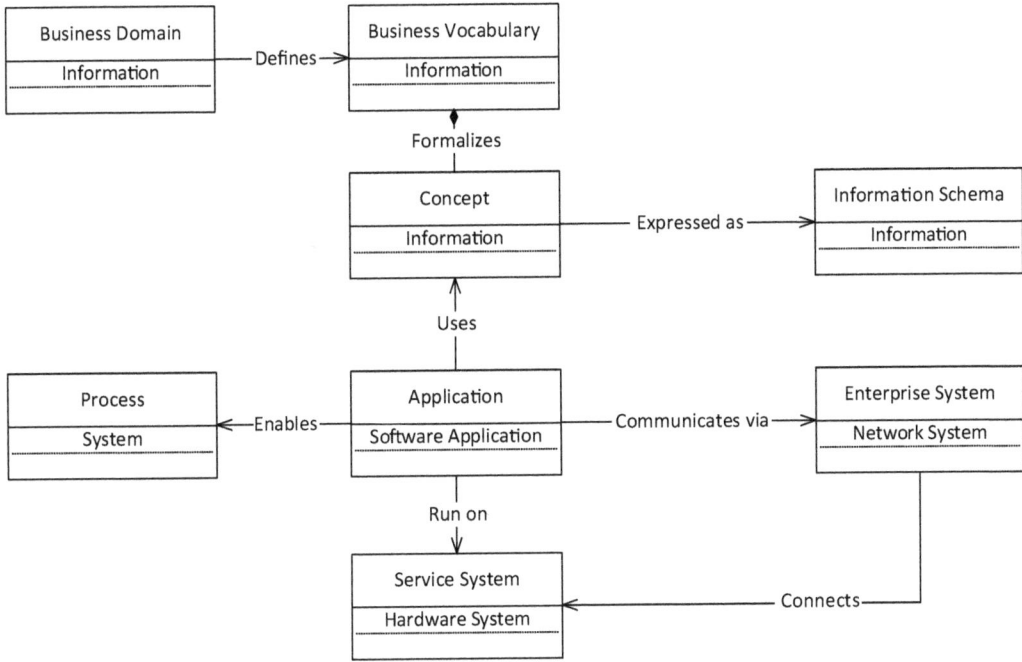

FIGURE 8.9   Information model.

8.6   FOOD SYSTEM INFORMATION MODELS

The class diagram information models will be shown for the seven teams, but only the hierarchy information model for Team 1 – Connecting Organizations is shown in Figure 8.12.

    The class diagram information models are shown for the seven teams in Figure 8.13 through 8.19.

8.7   DEVELOP QUALITY MANAGEMENT PLAN

The quality management plan is covered as part of the Systems Engineering Management Plan chapter.

8.8   SUMMARY

In this chapter, we discussed how to elicit the customer and system requirements for our system. This is a critical phase to ensure that we capture what the users of the system want from the system. It is important to meet with the customers to understand their needs based upon the desired elicitation technique chosen. We discussed using the process scenarios and documenting the requirements with the Process Architecture Map tool, then deriving these requirements in this chapter.

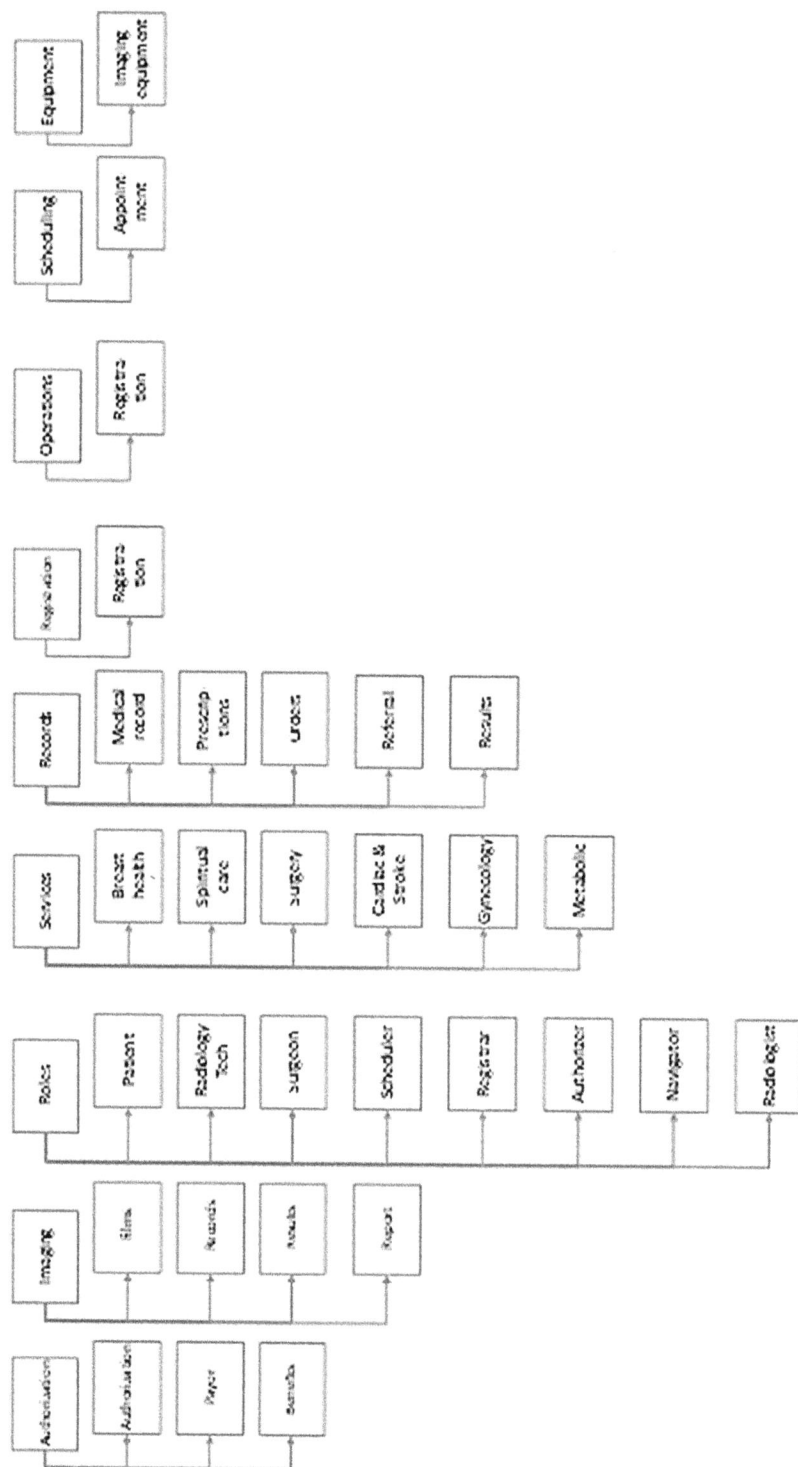

FIGURE 8.10   Hierarchy information model for the women's healthcare center.

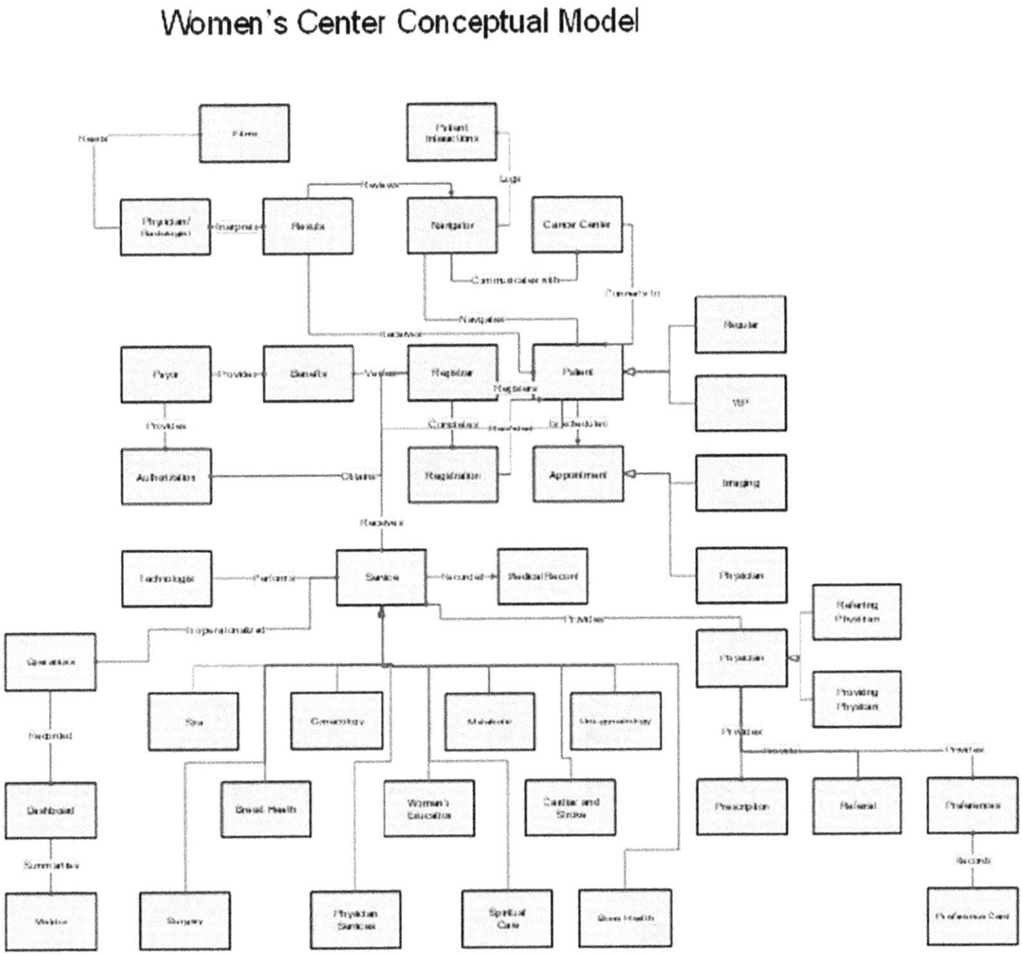

Women's Center Conceptual Model

FIGURE 8.11 Class diagram information model for the women's healthcare center.

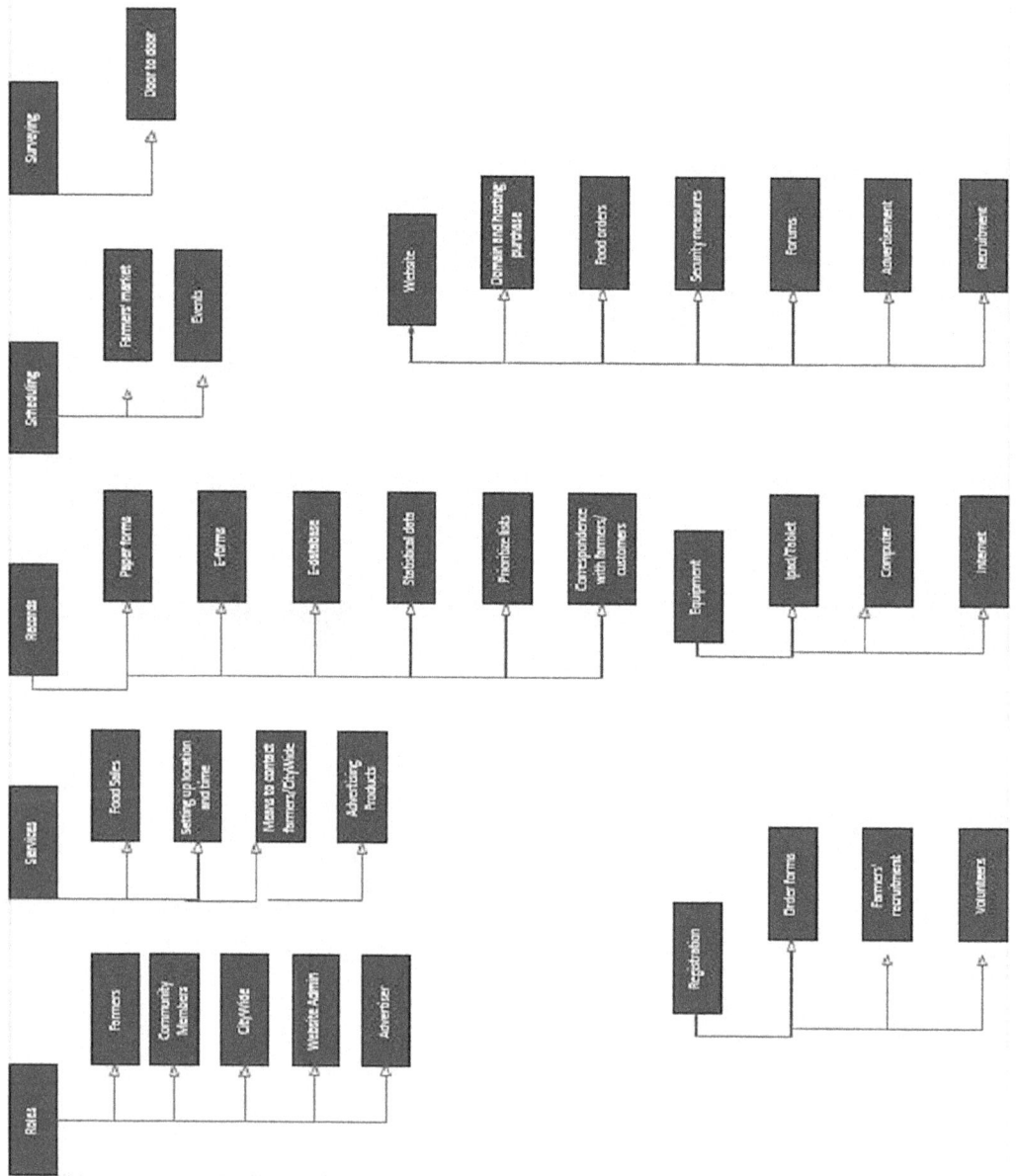

FIGURE 8.12 Hierarchy information model for team 1 – connecting organizations.

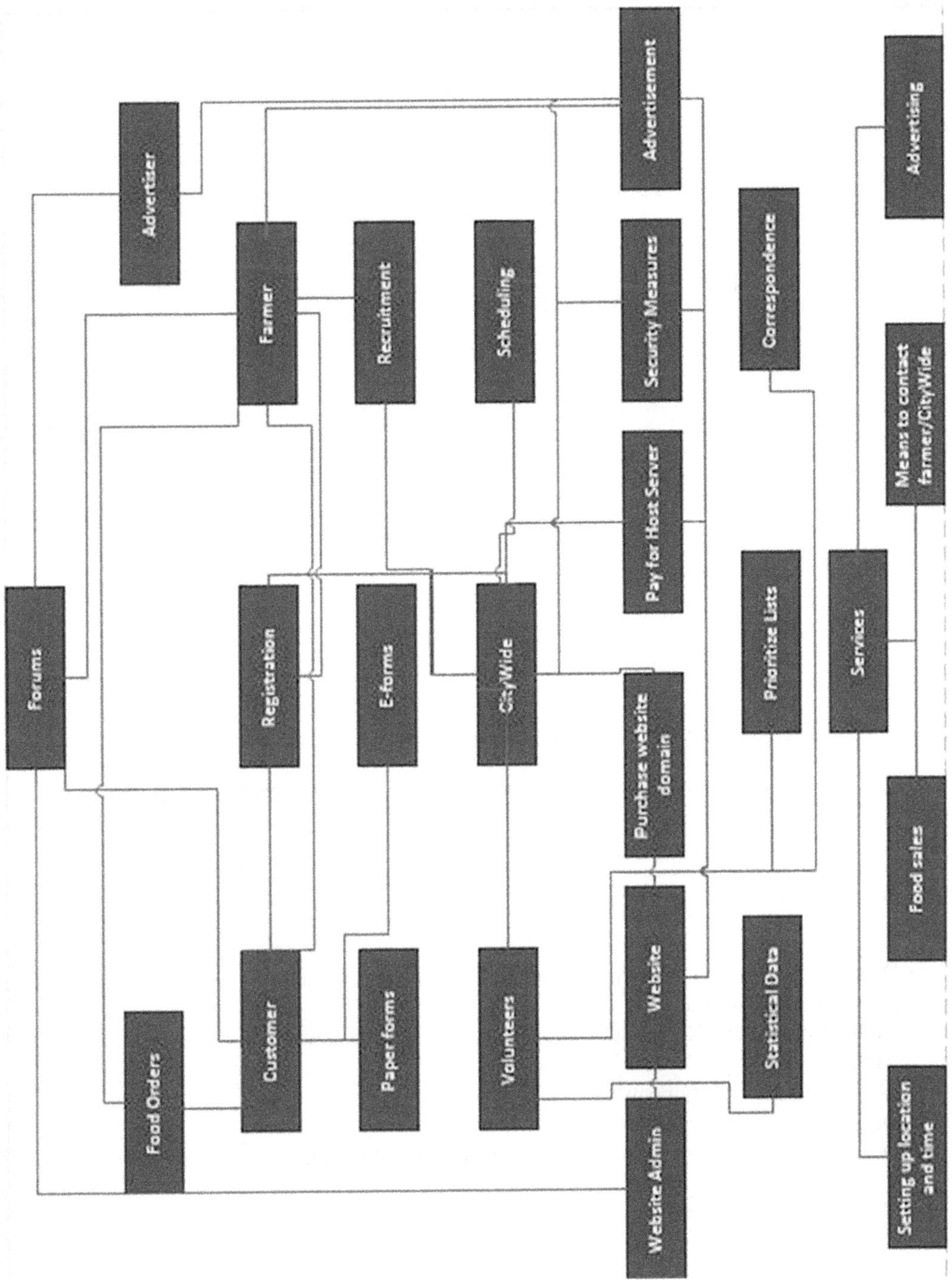

FIGURE 8.13 Class diagram information model for team 1 – connecting organizations.

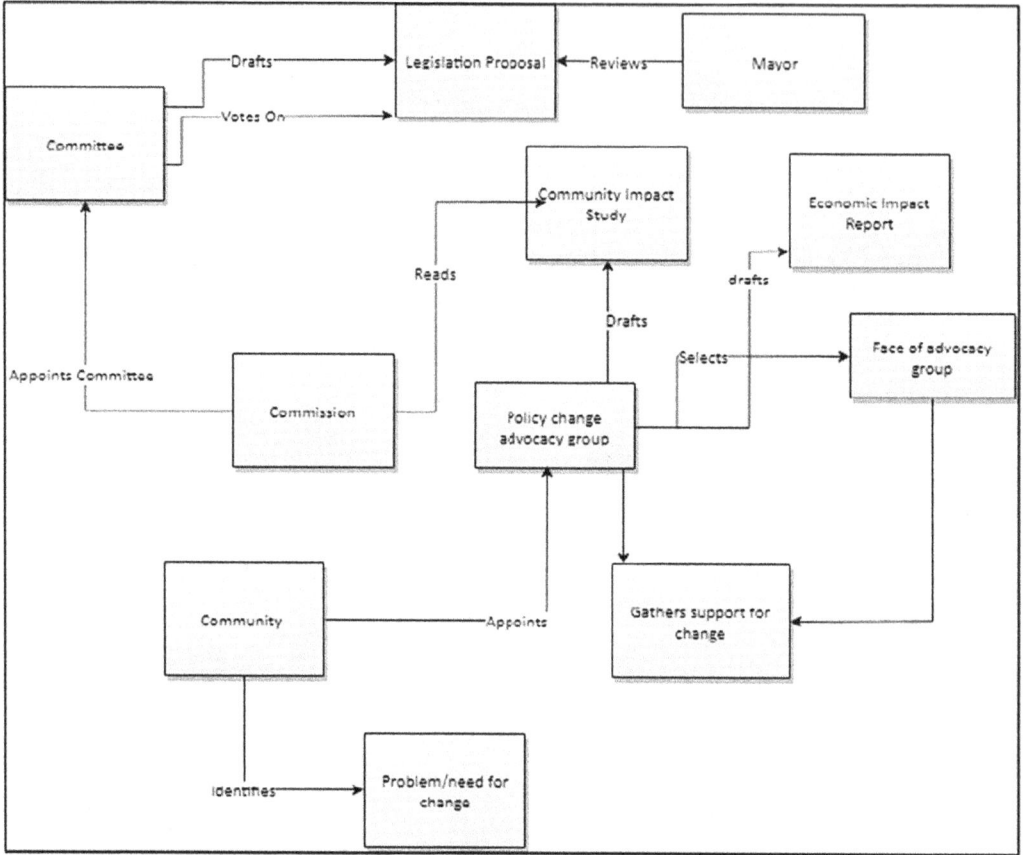

FIGURE 8.14 Class diagram information model for team 2 – creating food policy.

8.9 ACTIVE LEARNING EXERCISES

1) Develop questions for your system that would help you to hold process scenarios. Create a process scenarios list.
2) Hold process scenarios sessions with your stakeholders, document these via the Process Architecture Map.
3) Elicit customer requirements.
4) Derive system requirements.
5) Develop specialty engineering requirements for the following specialty engineering areas:
   - Software engineering
   - Environmental engineering
   - Safety engineering
   - Security engineering
   - Human Systems Integration
6) Develop a measurement system plan using the data collection plan and operational definitions templates.
7) Develop an information model for the following models:
   - Information Hierarchy Model
   - Class Diagram Information Model

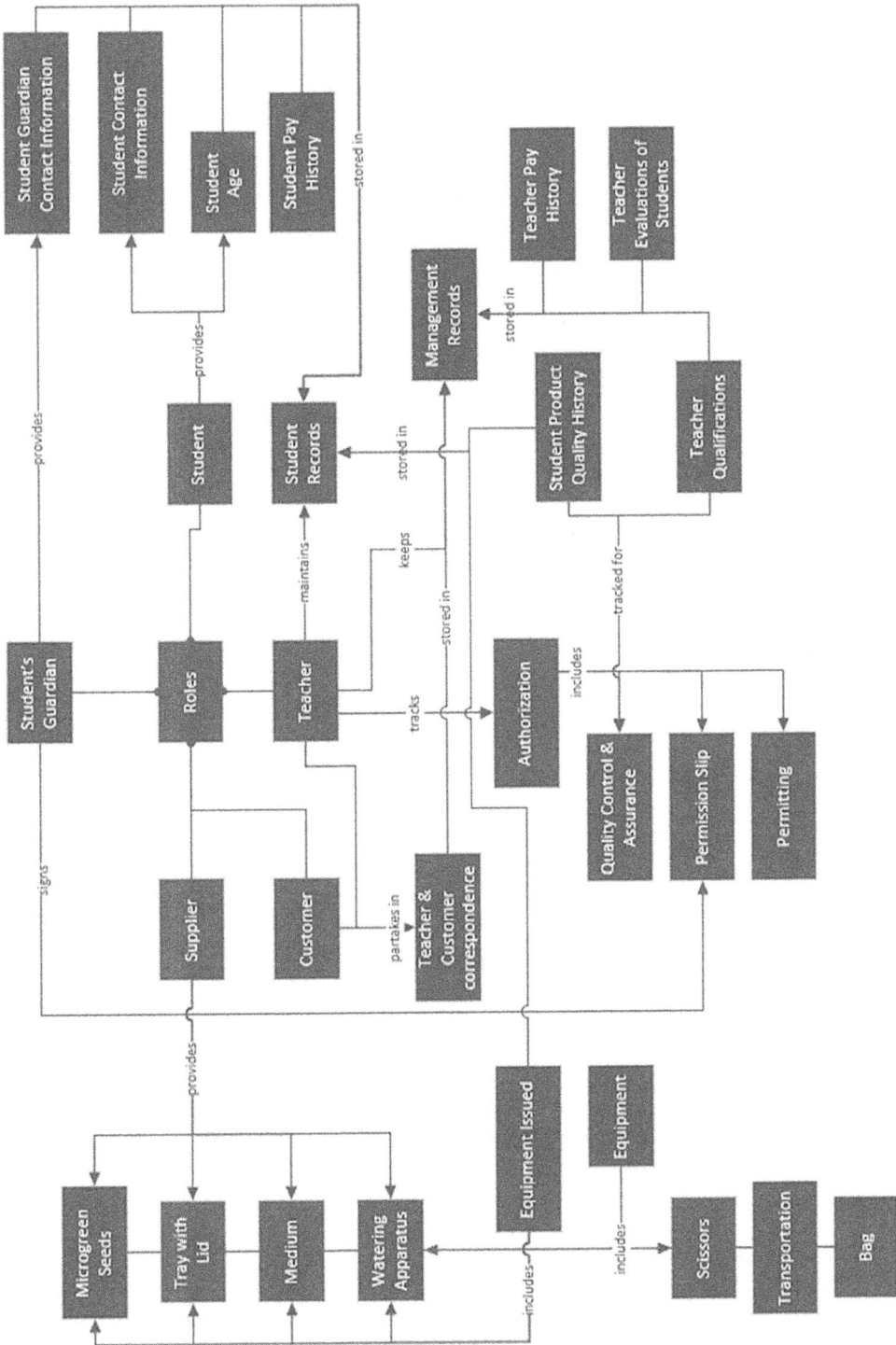

FIGURE 8.15 Class diagram information model for team 3 – kids growing food.

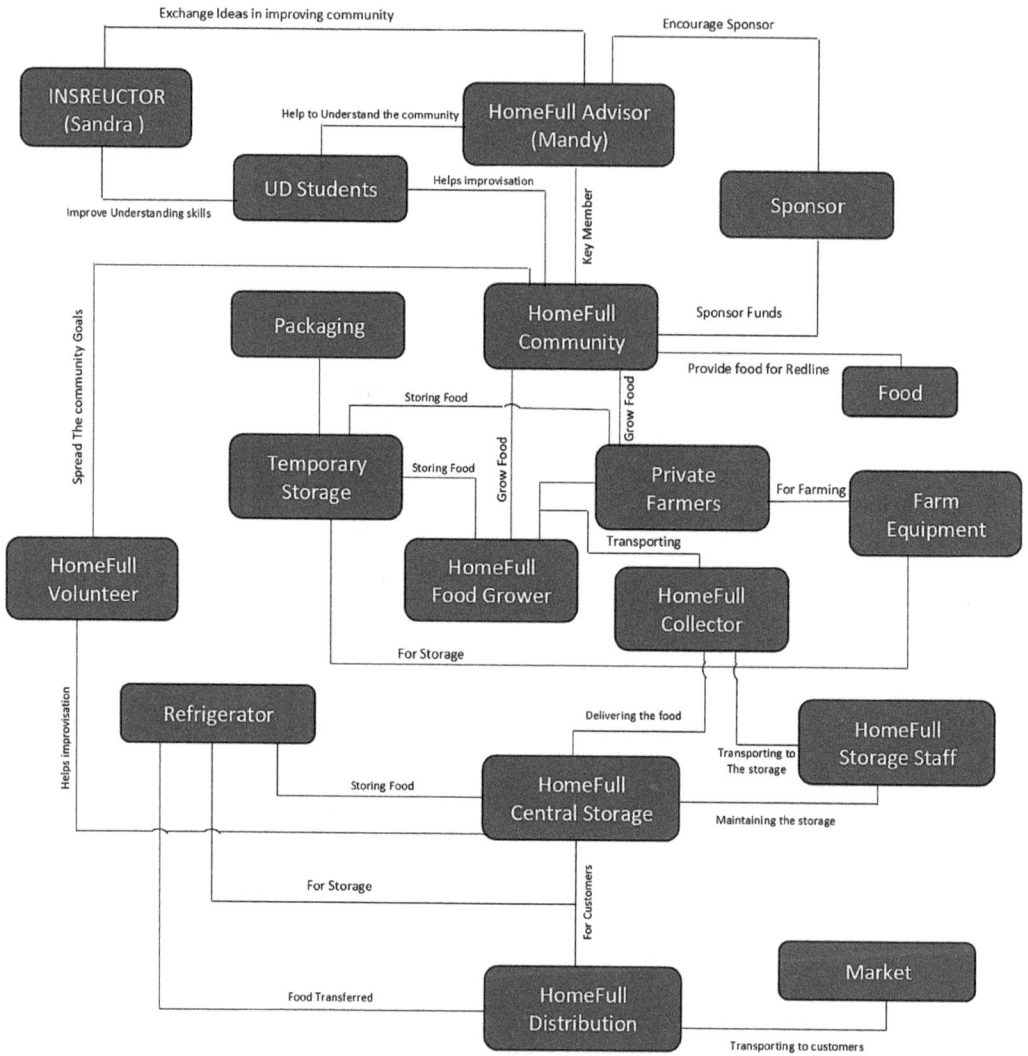

FIGURE 8.16   Class diagram information model for team 4 – food storage.

8) Review the examples for each type of model in this chapter, for one of the teams. Reflect on what the team did well and what could be improved.
   - Process scenarios list
   - Process Architecture map
   - Customer requirements
   - System requirements
   - Specialty requirements
   - Measurement system plan – data collection plan and operational definition templates
   - Information models

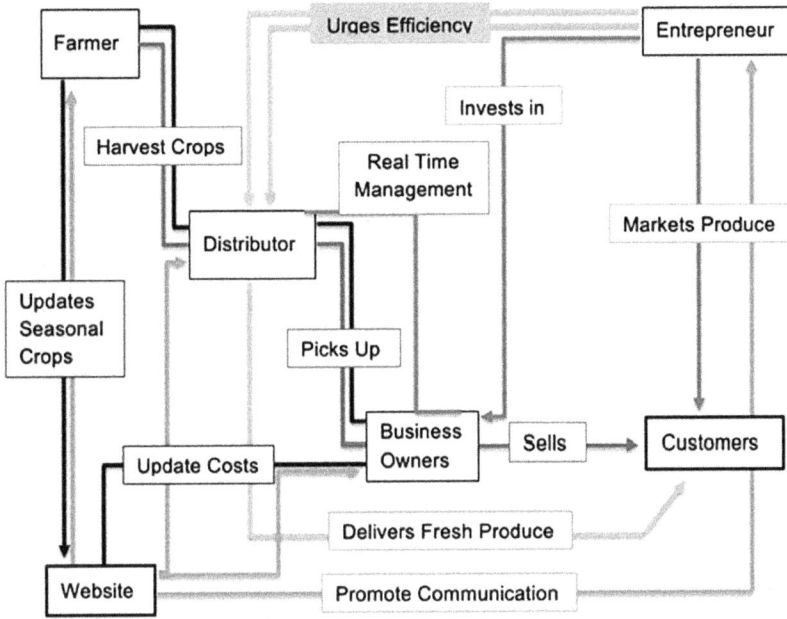

FIGURE 8.17 Class diagram information model for team 5 – food hub website.

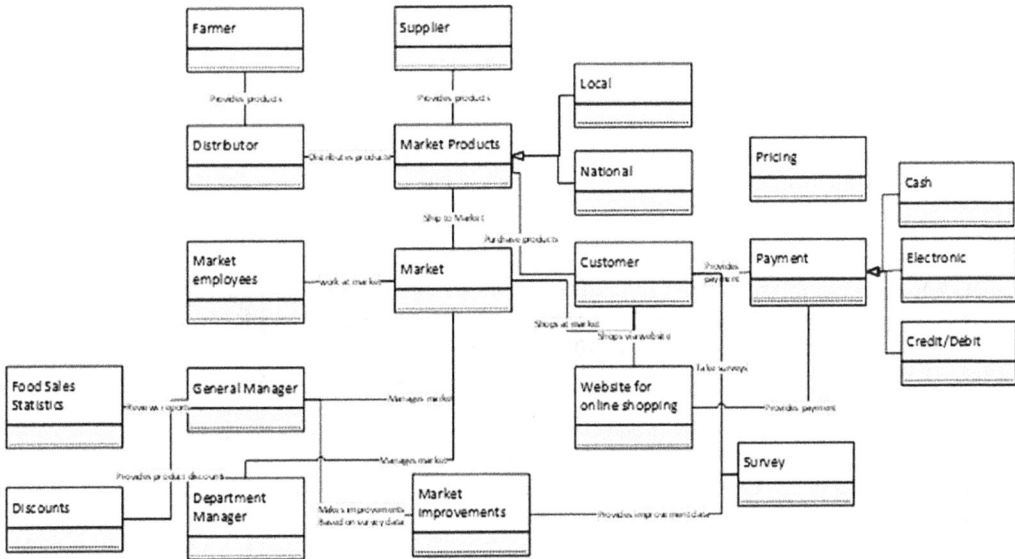

FIGURE 8.18 Class diagram information model for team 6 – gem city market.

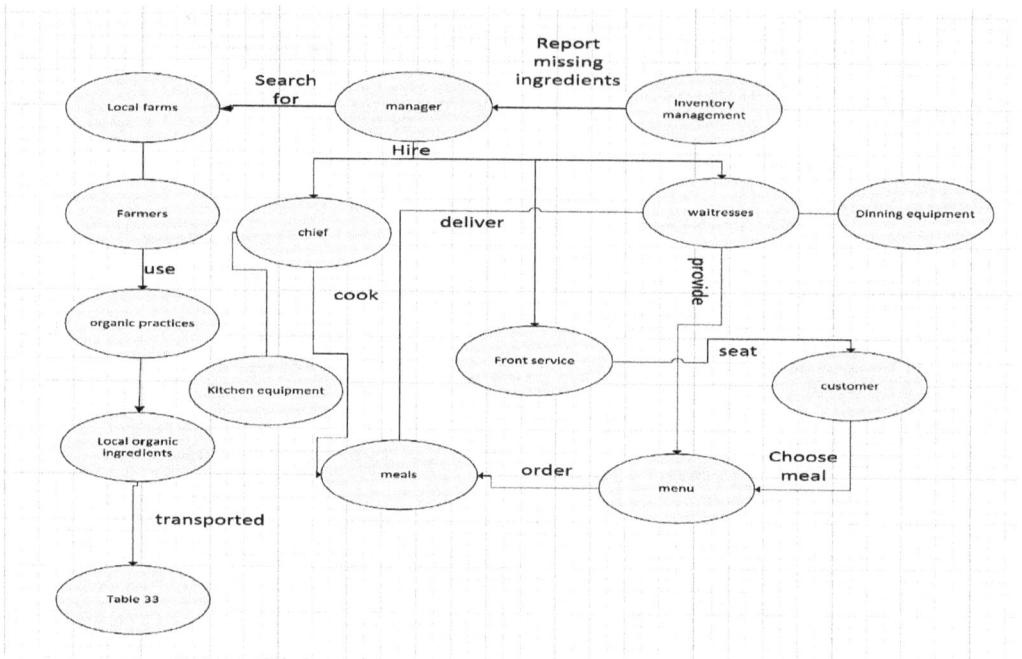

FIGURE 8.19   Class diagram information model for team 7 – table 33 restaurant.

BIBLIOGRAPHY

Dannels, D. (2000). Learning to be professional: Technical classroom discourse, practice, and professional identify construction. *Journal of Business and Technical Communication*, 14(1), 5–37.

Dick, J., Hull, E., Jackson, K. (2017). *Requirements Engineering*, Fourth Edition, Springer.

EPA. 2012. "Green Engineering." United States Environmental Protection Agency. Accessed on 11 September 2012. Available at http://www.epa.gov/oppt/greenengineering/

Evans, J. and Lindsey, W. (2011). *Managing for Quality and Performance Excellence*, 10th Edition. Cengage Learning, 2011.

Fairley, R.E. and M.J. Willshire. (2011). "Teaching software engineering to undergraduate systems engineering students." *Proceedings of the 2011 American Society for Engineering Education (ASEE) Annual Conference and Exposition*. 26-29 June 2011. Vancouver, BC, Canada.

IEEE STD 1220-1998 (1998) Standard for application and management of the systems engineering process. IEEE, New York

INCOSE. 2015. 'Systems Engineering Handbook: A Guide for System Life Cycle Processes and Activities', version 4.0. Hoboken, NJ, USA: John Wiley and Sons, Inc, ISBN: 978-1-118-99940-0.

Standish Group (1995), The CHAOS Report 1995, Standish Group International. Inc., Boston MA.

TOGAF 8.0 The Open Group, Business Scenarios https://pubs.opengroup.org/architecture/togaf8-doc/arch/

TOGAF 8.0, The Open Group, https://pubs.opengroup.org/architecture/togaf8-doc/arch/toc.html, 1999–2006

Wilson, W.M., Rosenberg, L.L., & Hyatt, L.E., 1997, *Automated analysis of requirement specifications*, in *Proceedings of the 19th International Conference on Software Engineering*, pp. 161–171.

9 Detailed Design Phase 3

In this chapter, we perform the detailed design activities to create the system specifications for the elements, their interfaces, and processes. The phase is called the Detailed Design phase.

9.1 PURPOSE

In this chapter, we perform the detailed design and systems analysis activities. To perform the detailed design, we'll develop use cases that provide additional detail describing the system. We will also develop physical architecture models, physical block diagrams, SysML block definition diagrams, and/or SysML internal block diagrams to show the system elements, their interconnectivity, and their behavior. We will use Quality Function Deployment (QFD) to ensure that the system (technical) requirements align to the customer requirements. In the systems analysis, the systems engineer identifies selection criteria to make trade-off analyses. We describe an effective analysis using Lean analysis tools. A cost analysis can be used to make cost trade-offs. We update our risk analysis focusing on technical risks for our system.

9.2 ACTIVITIES

The activities performed and the tools applied in the Detailed Design phase are shown in Table 9.1. We cover the "Perform detailed design" and "Perform systems analysis" activities and the associated Detailed Design and Systems Analysis tools. The system elements and design phase model are shown in Figure 9.1.

9.3 PERFORM DETAILED DESIGN

9.3.1 Physical Architecture Model Hierarchy

The physical architecture model shows the physical elements, how the elements relate to each other, and the physical interfaces of the elements of the system. The system elements are discrete parts of the systems. The physical interface is the connection between physical elements. Table 9.2 (SEBok Original) shows the types of systems elements and potential physical interfaces.

The Physical Architecture Hierarchy Model is similar to a Functional Decomposition Model where it cascades and decomposes the system elements, instead of the functions as was done in the FDM. This model provides a view of the systems, their sub-systems and the elements, and their interfaces.

The Physical Architecture Model for the Women's Healthcare Center is shown in Figure 9.2.

The Physical Architecture Hierarchy Models for the Food System are shown in Figures 9.3 through 9.9.

9.3.2 Physical Architecture Model: Physical Block Diagram, and/or SysML Block Definition Diagram, and/or SysML Internal Block Diagram

The physical architecture models include the following models, with slightly different purposes:

1) Physical Block Diagram (PBD) – to show the system, sub-systems, elements, their relationships, and interfaces.

DOI: 10.1201/9781003081258-11

TABLE 9.1

Activities and Tools in the Detailed Design Phase

Vee Phase	Activities	Tools	Principles
Phase 3: Detailed Design	1. Perform detailed design 2. Perform systems analysis	Detailed Design: • Physical Architecture Model Hierarchy • Physical Architecture Model: Physical Block Diagram; and/or SysML Block Definition Diagram; and/or SysML Internal Block Diagram • System Elements • QFD • Simulation (not required) Systems Analysis: • Selection criteria • Trade-off analysis; and/or Effectiveness analysis; and/or Cost analysis; and/or technical risk analysis • Justification report	• Systems analysis • Wholeness and interactions

TABLE 9.2

Types of System Elements and Physical Interfaces

Element	Product System	Service System	Enterprise System
System Element	• Hardware elements • Operator roles • Software pieces	• Processes, databases, procedures • Operator roles • Software applications	• Corporate, division, department, project, team, leader • IT components
Physical interface	• Hardware parts, protocols, procedures	• Protocols, documents	• Protocols, procedures, documents

(Adapted from SEBoK Original.)

2) SysML Block Definition Diagram (BDD) – to provide a black box or external representation of a system block, with the hierarchy of its composite block. The blocks can include any types of blocks, including software, hardware, components, etc.
3) SysML Internal Block Diagram (IBD) – to provide the white box or internal view of a system block.

The physical block diagram for the Women's Healthcare Center is shown in Figure 9.10

Most of the following figures are physical block diagrams for the teams' food systems, shown in Figures 9.11 through 9.17.

9.3.3 System Elements

After the physical architecture models are developed, the list of system elements can be derived. The template that the class used included a system name, the system elements related to each

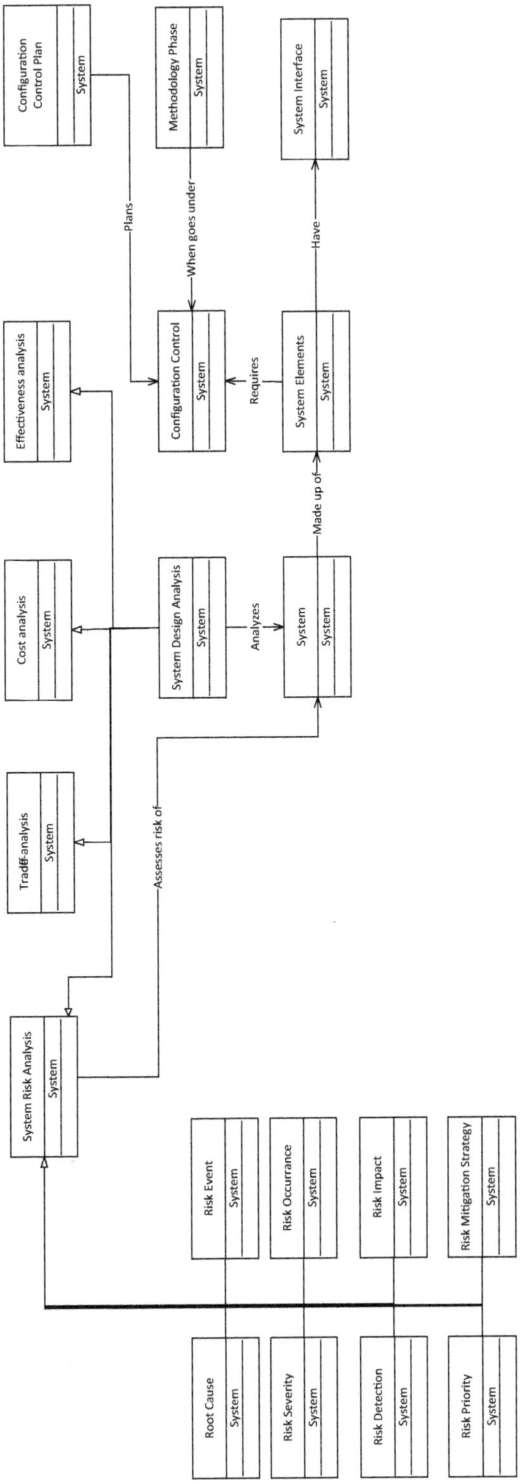

FIGURE 9.1 System detailed design model.

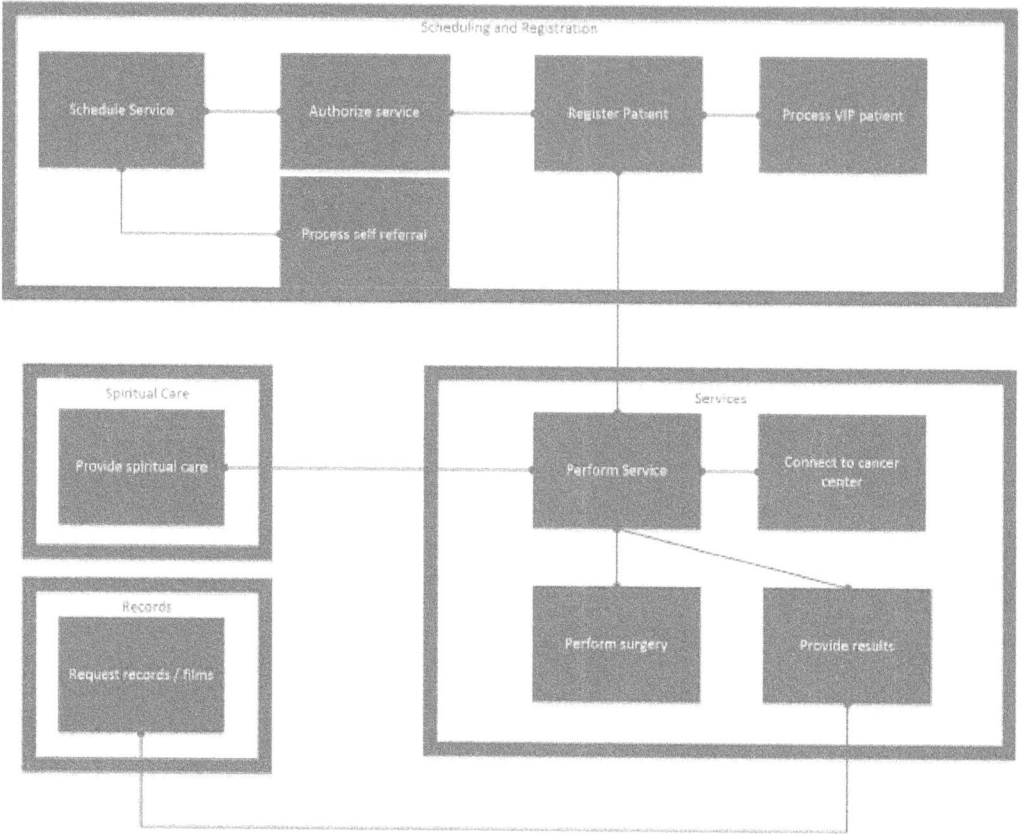

FIGURE 9.2 Physical architecture model for women's healthcare center.

system, the type of system, whether the system element should be under configuration control, and the phase that the system element goes into configuration control. Configuration control is a process for controlling the documents related to the system elements. We discuss configuration control as part of the system engineering planning chapter. The typical system or system element types are:

- Database
- Document
- Equipment
- Facility
- Hardware
- Materials
- Operator roles
- Organizational unit
- Procedure
- Policies
- Process
- Product
- Project
- Protocol
- Services
- Software

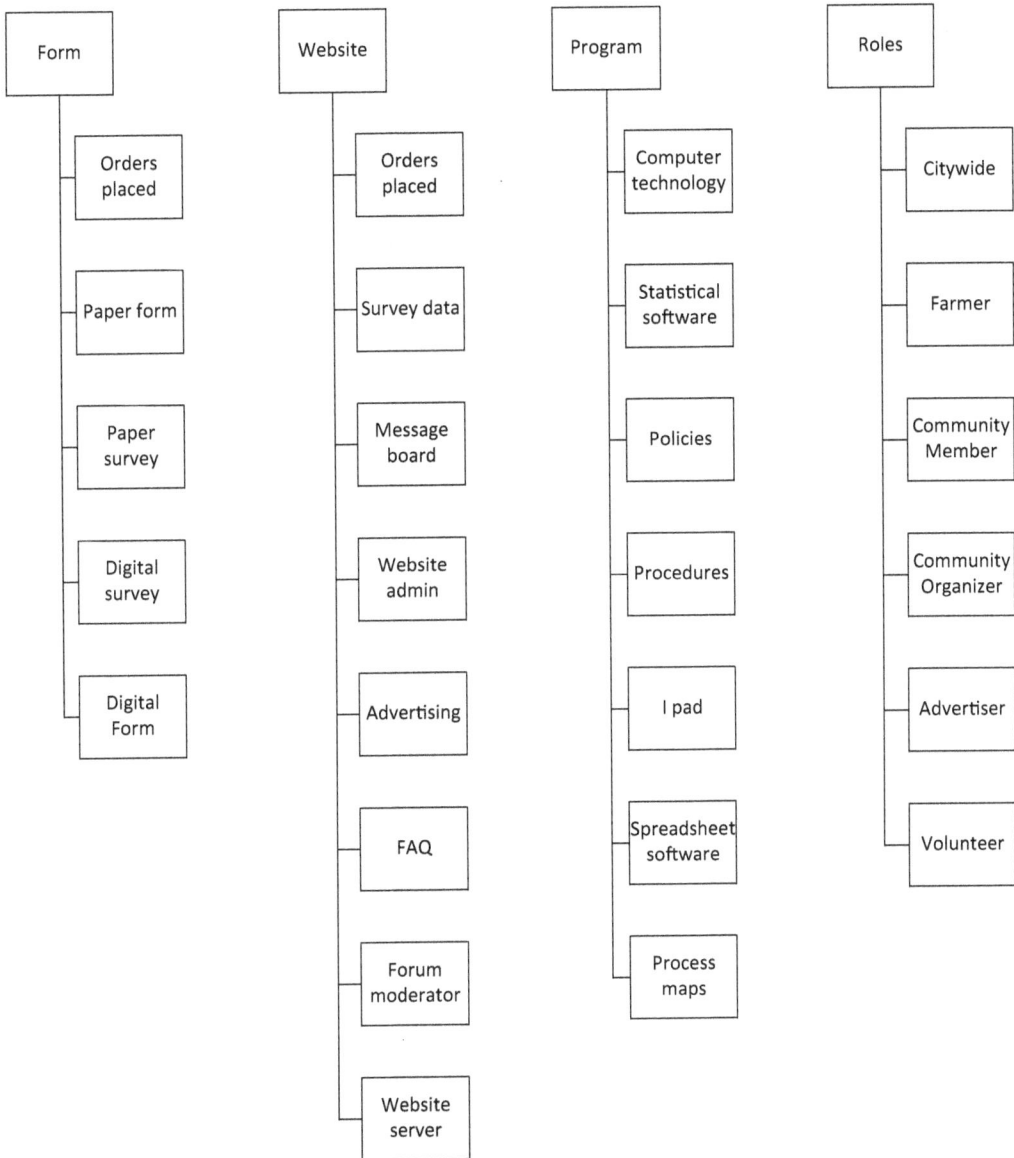

FIGURE 9.3 Physical architecture hierarchy model, team 1 – connecting organizations.

- Supplies
- Telecommunication
- Other

The systems and their elements are shown in Table 9.3 for the Women's Healthcare Center.

The systems and their elements for the Food System are shown in Tables 9.4 and 9.5, and Figures 9.18 through 9.22.

9.3.4 QUALITY FUNCTION DEPLOYMENT (QFD)

QFD and the House of Quality is an excellent tool to help translate the customer requirements from the Voice of the Customer into the technical or system requirements of your product, process, or

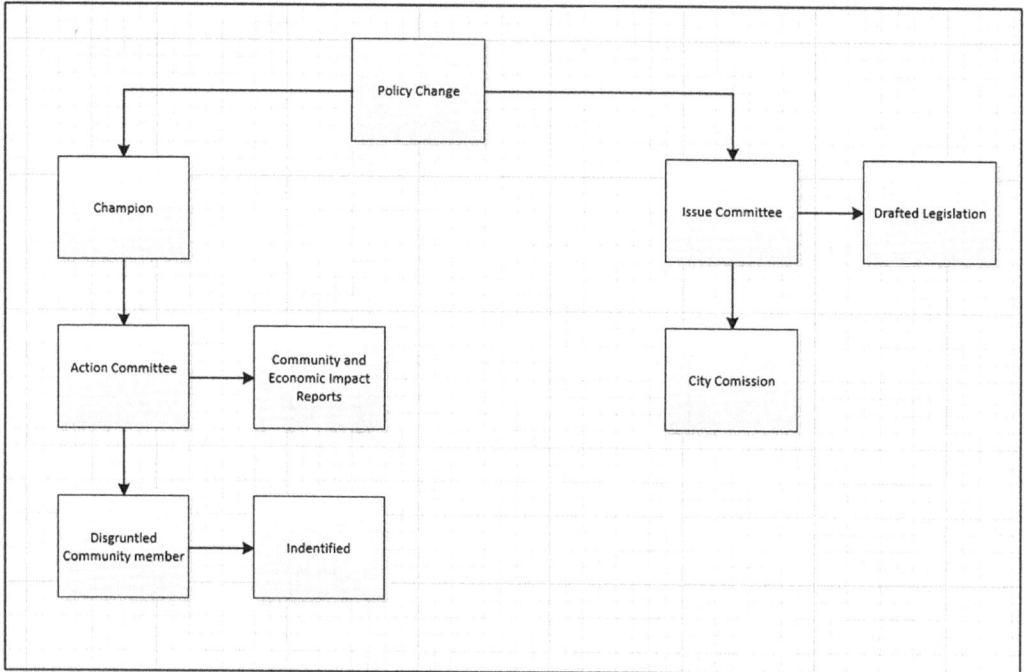

FIGURE 9.4 Physical architecture hierarchy model, team 2 – creating food policy.

service. Although many QFD users only apply the first House of Quality, relating the customer and technical requirements, there are three additional houses that can incorporate traceability from the following, as shown in Figure 9.23:

1) Customer requirements to technical (system) requirements (House of Quality 1);
2) Technical (system) requirements to component characteristics (House of Quality 2);
3) Component characteristics (system elements) to process operations (House of Quality 3);
4) Process operations to quality plan (House of Quality 4).

Figure 9.24 shows the format for the first House of Quality and how each house is applied.
 The steps for creating a House of Quality are (Evans and Lindsay, 2011):

1. Define the customer requirements or CTS (Critical to Satisfaction) characteristics from the Voice of the Customer data. The customer can provide an importance rating for each CTS.
2. Develop the technical requirements with the organization's design team.
3. Perform a competitive analysis, having the customer's rank your product, process, or service against each CTS to each of your competitors.
4. Develop the relationship correlation matrix by identifying the strength of relationship between each CTS and each technical requirement. Typically, a numerical scale of 9 (high strength of relationship), 3 (medium strength of relationship), 1 (low strength of relationship), and blank (no relationship).
5. Develop the trade-offs or relationship between the technical requirements in the roof of the House of Quality. You can identify a positive (+) relationship between the technical requirements, as one requirement increases the other also increases; no relationship (blank) or a negative (-) relationship, there is an inverse relationship between the two technical

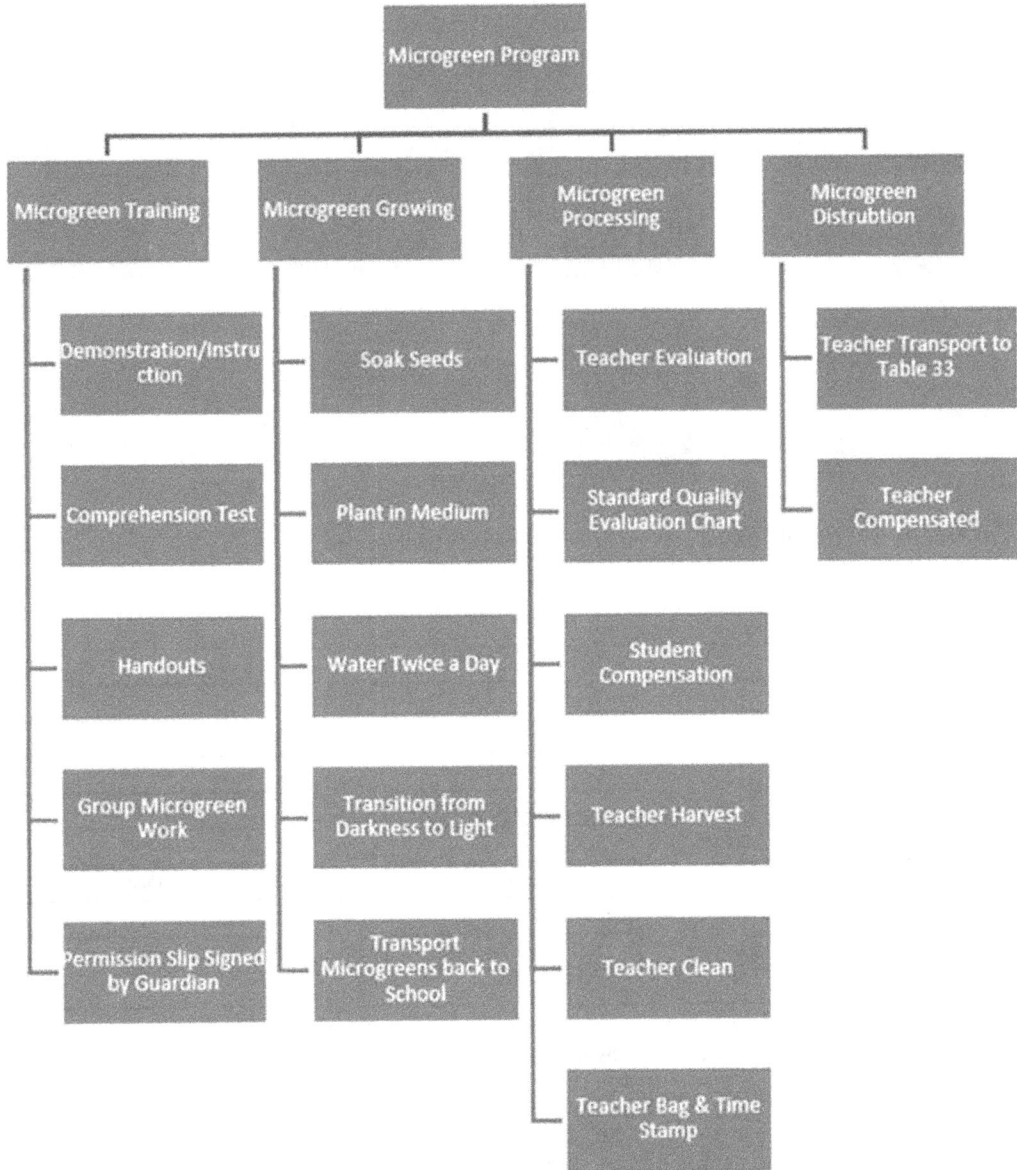

FIGURE 9.5   Physical architecture hierarchy model, team 3 – kids growing food.

requirements. An example of a positive relationship can be illustrated in the design of a fishing pole. The line gauge and tensile strength both increase as the other increases. A negative relationship can be illustrated by line buoyancy and tensile strength. As tensile strength of the line increases, the buoyancy will be less.

6. The priorities of the technical requirements can be summarized by multiplying the importance weightings of the customer requirements by the strength of the relationships in the correlation matrix. This helps to identify which of the technical requirements should be incorporated into the design of the product, process, or service first.

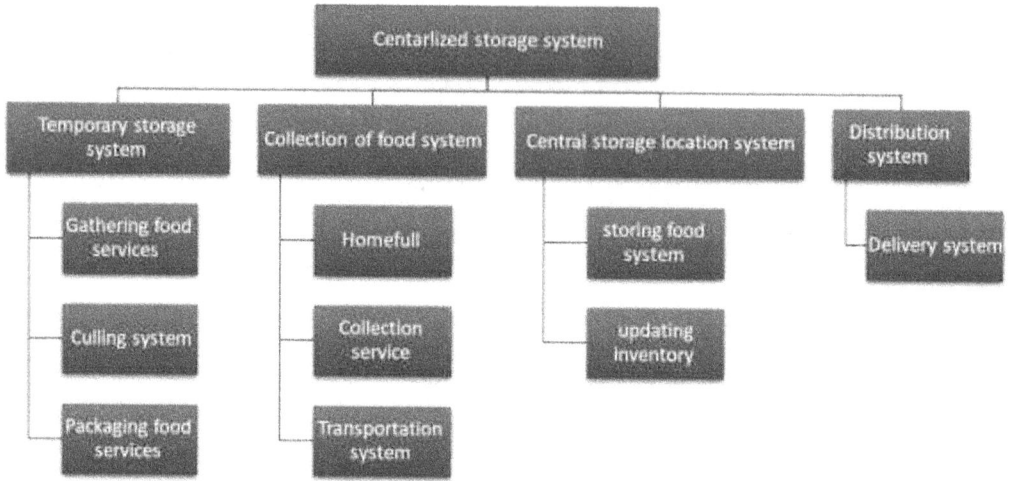

FIGURE 9.6 Physical architecture hierarchy model, team 4 – food storage.

FIGURE 9.7 Physical architecture hierarchy model, team 5 – food hub website.

There are many benefits to using QFD, including:

- Improved communication and teamwork between the systems engineering team members and stakeholders
- Enhances designing quality into the system by collecting and incorporating the voice of customers' needs
- Translates customers' needs into the technical requirements
- Incorporates systems analysis of the system elements
- Identifies processes to manufacture the system or deliver the services
- Enables verification and validation through traceability of Houses of Quality
- Enables competitor evaluation from the customer and technical perspectives
- Can reduce the development life cycle time
- Minimize design changes
- Enhances cross-functional understanding of processes and the system

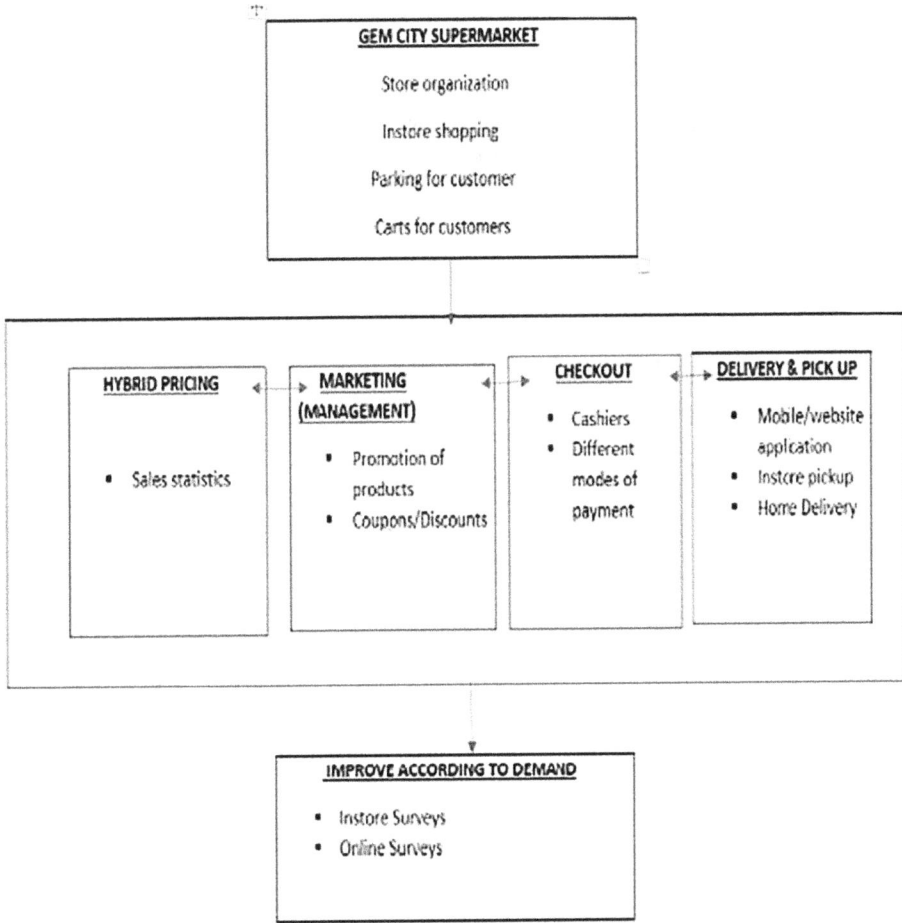

FIGURE 9.8 Physical architecture hierarchy model, team 6 – gem city market.

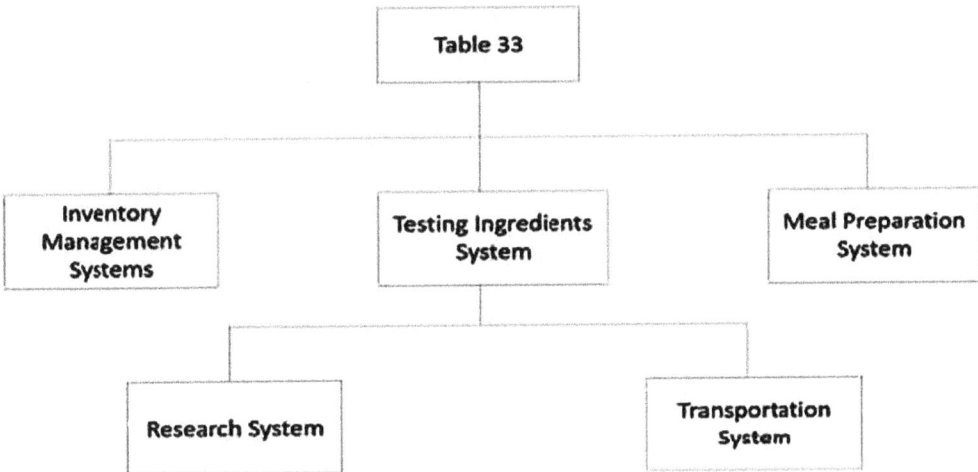

FIGURE 9.9 Physical architecture hierarchy model, team 7 – table 33 restaurant.

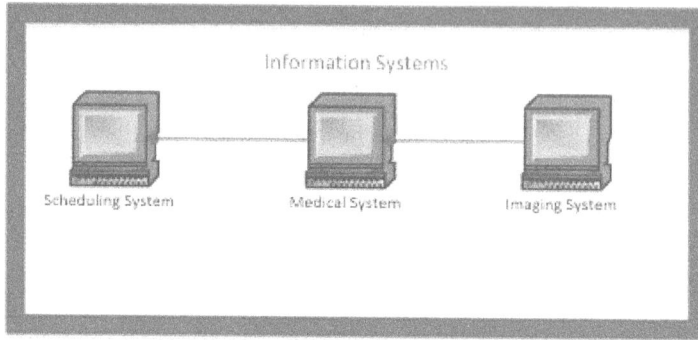

FIGURE 9.10 Physical block diagram for the women's healthcare center.

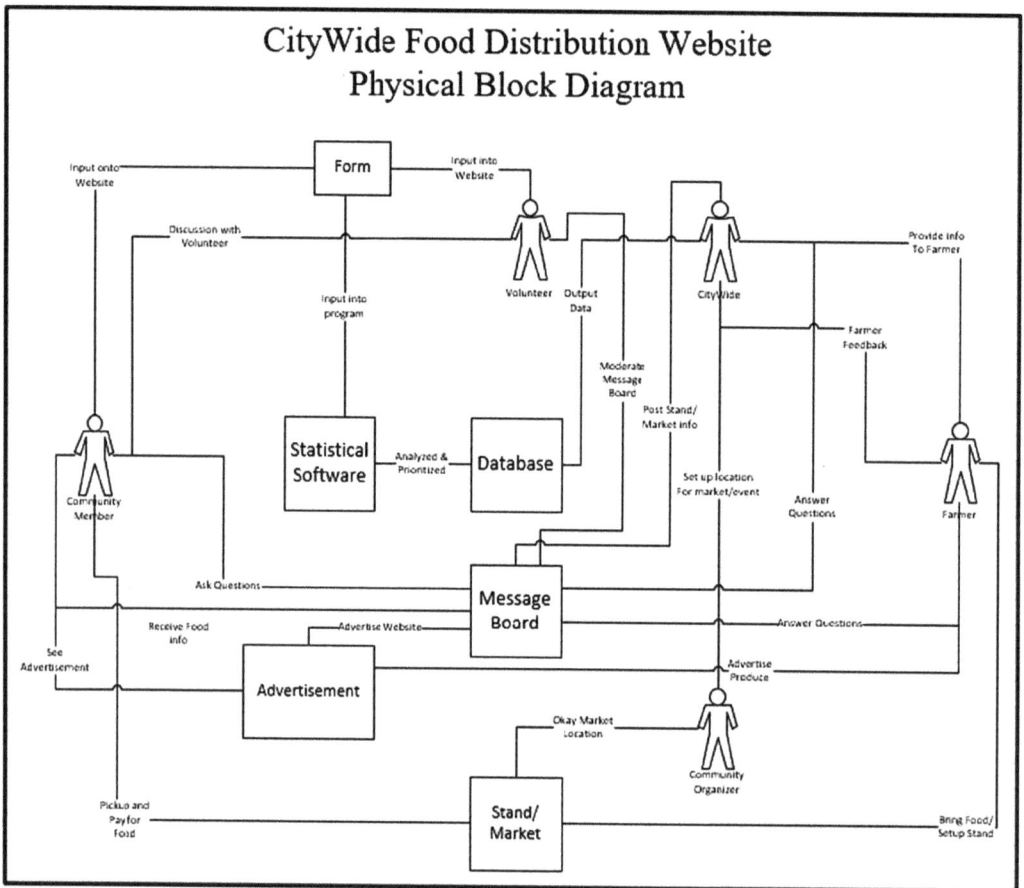

FIGURE 9.11 Physical architecture model – physical block diagram, team 1 – connecting organizations.

9.3.5 Food System QFD House of Quality 1

The food systems' QFD House of Quality 1 are shown in Figures 9.25 through 9.32. As you may notice, the level of detail of the requirements differs by team.

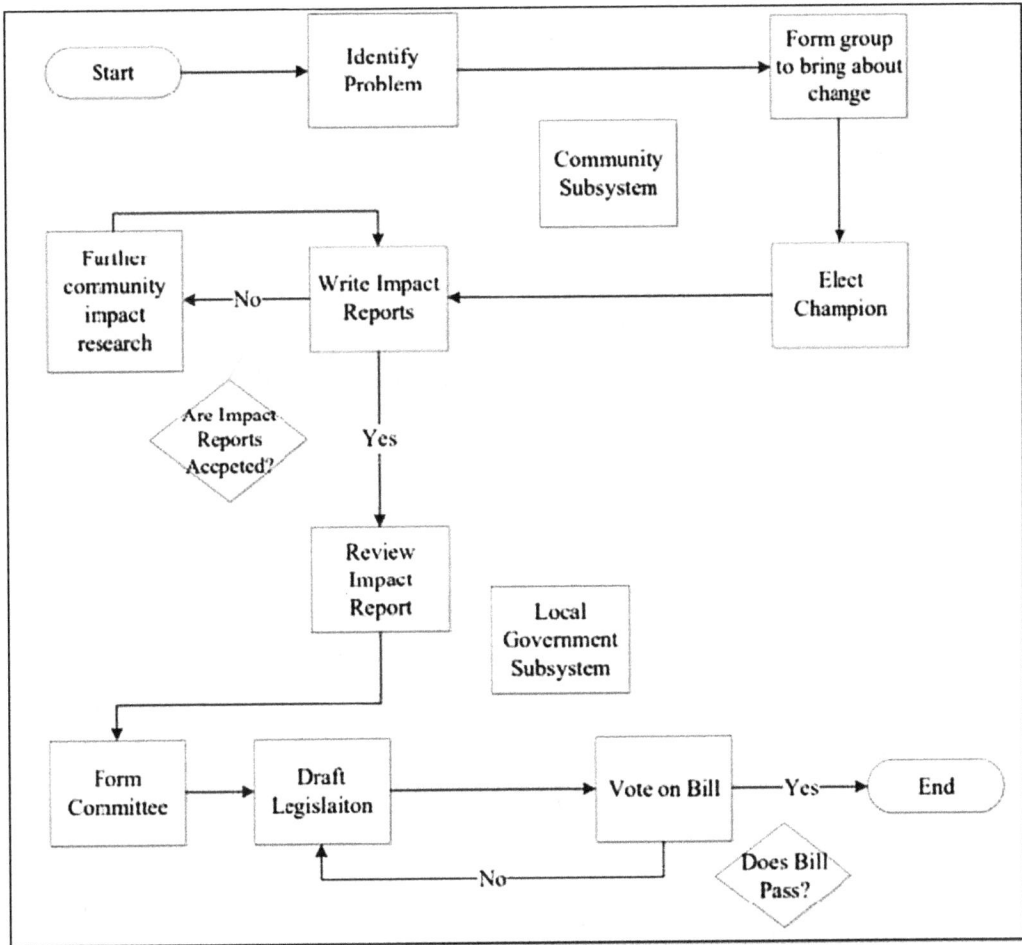

FIGURE 9.12 Physical architecture model – physical block diagram, team 2 – creating food policy.

9.4   SYSTEM DESIGN

System design includes activities to specify the system elements to achieve the system and customer requirements. The system elements are selected that fit the system architecture and comply with the trade-off of the system requirements. The system design includes detailed models, properties, characteristics, and business rules that describe the system for implementation.

Examples of generic design characteristics in the mechanics of solids include (SEBoK 8.0):

- Shape
- Geometrical patterns
- Dimension
- Volume
- Surface
- Curves
- Resistance to forces
- Distribution of forces
- Weight
- Velocity of motion
- Temporal persistence

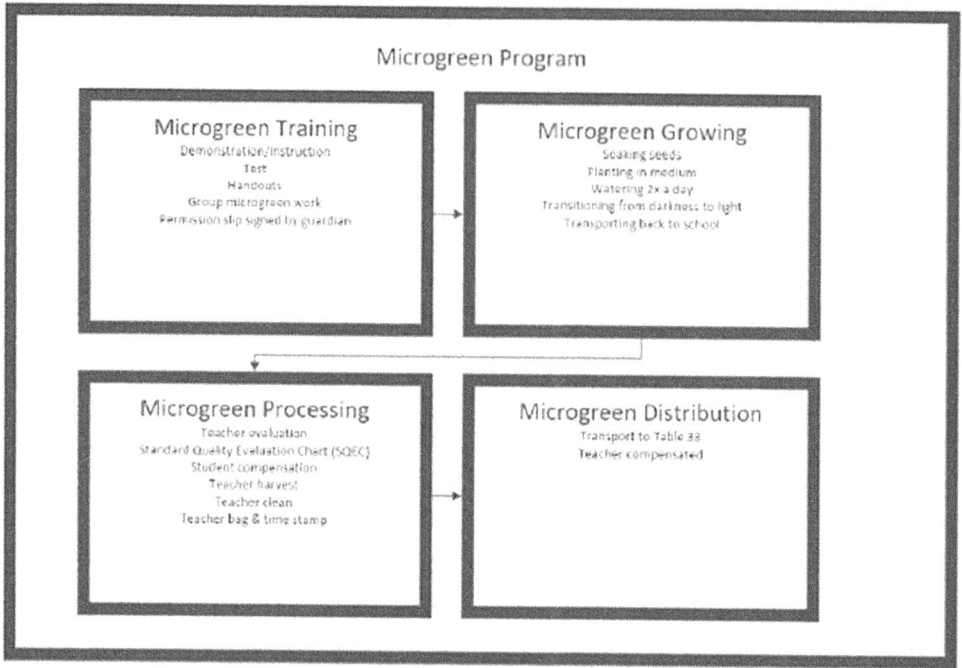

FIGURE 9.13 Physical architecture model – physical block diagram, team 3 – kids growing food.

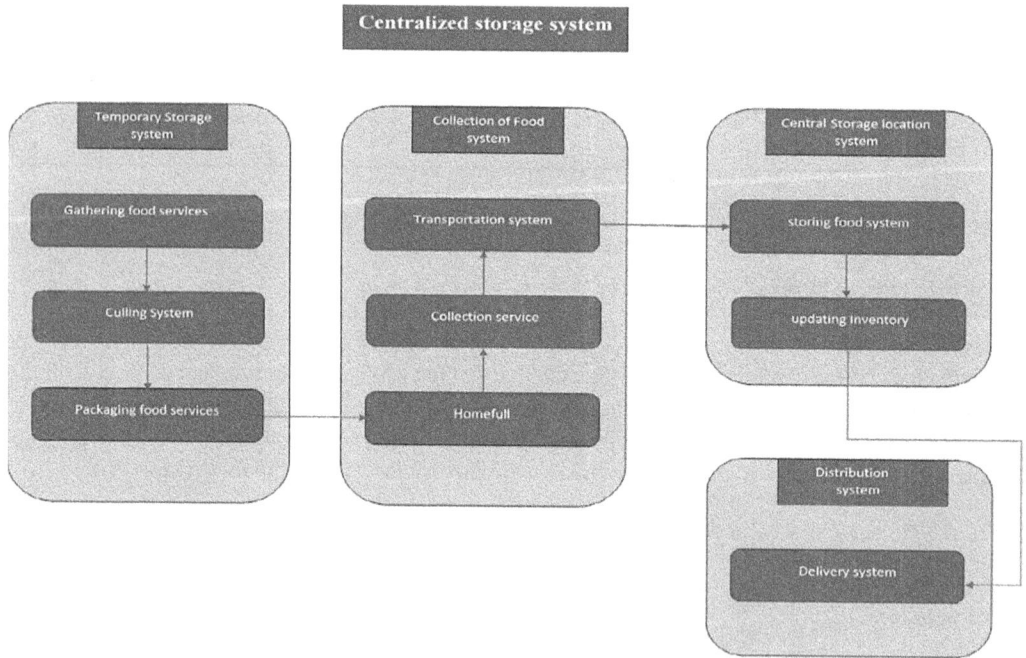

FIGURE 9.14 Physical architecture model – physical block diagram, team 4 – food storage.

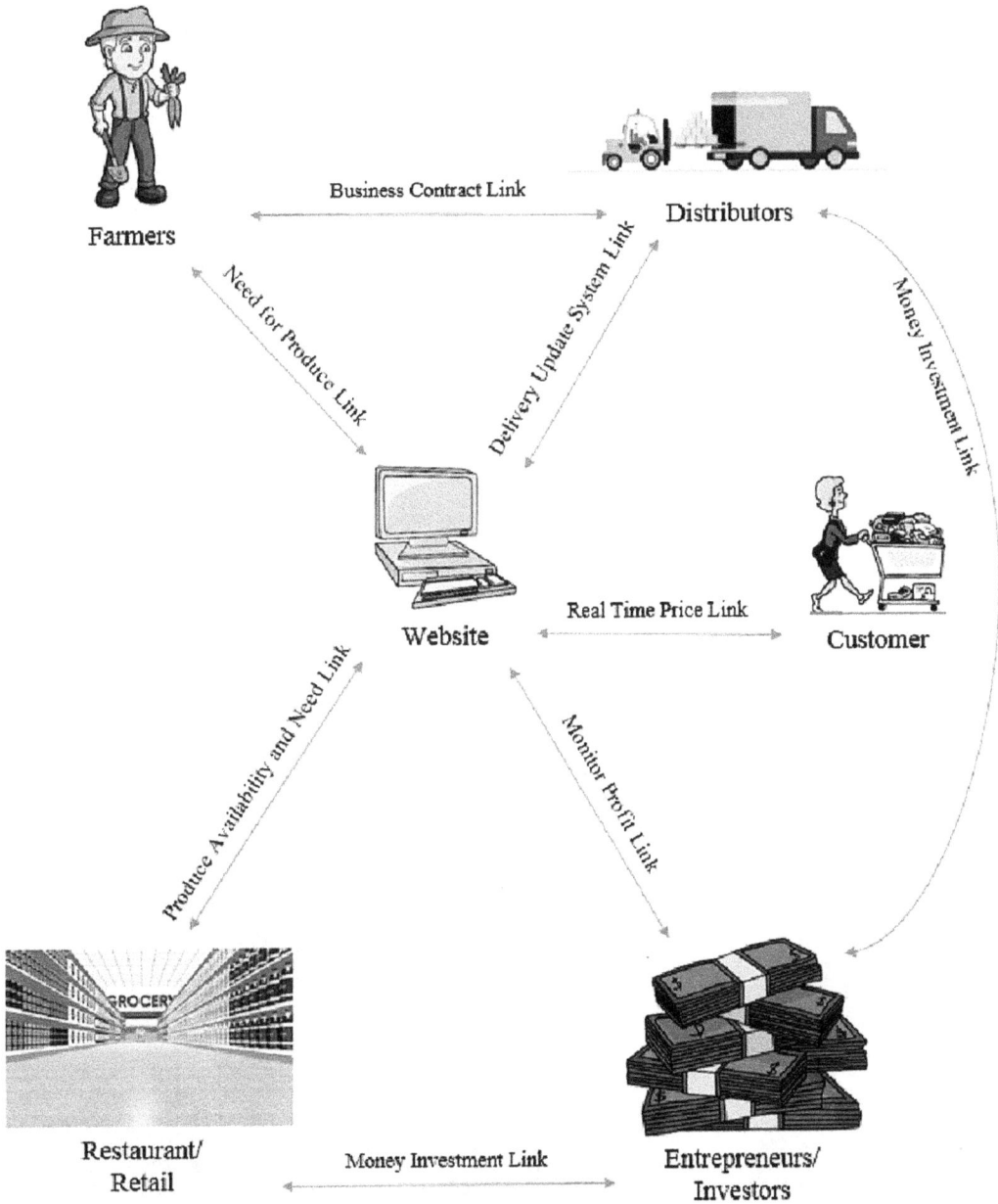

FIGURE 9.15   Physical architecture model – physical block diagram, team 5 – food hub website.

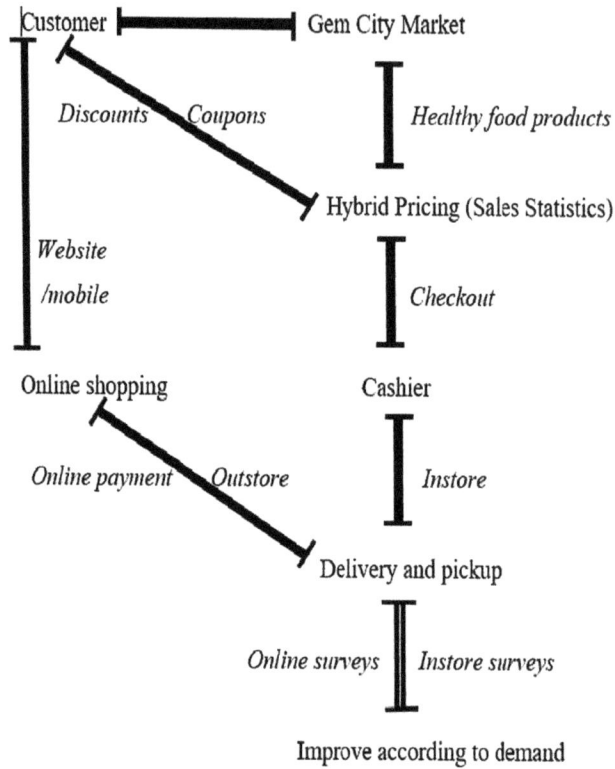

FIGURE 9.16 Physical architecture model – physical block diagram, team 6 – gem city market.

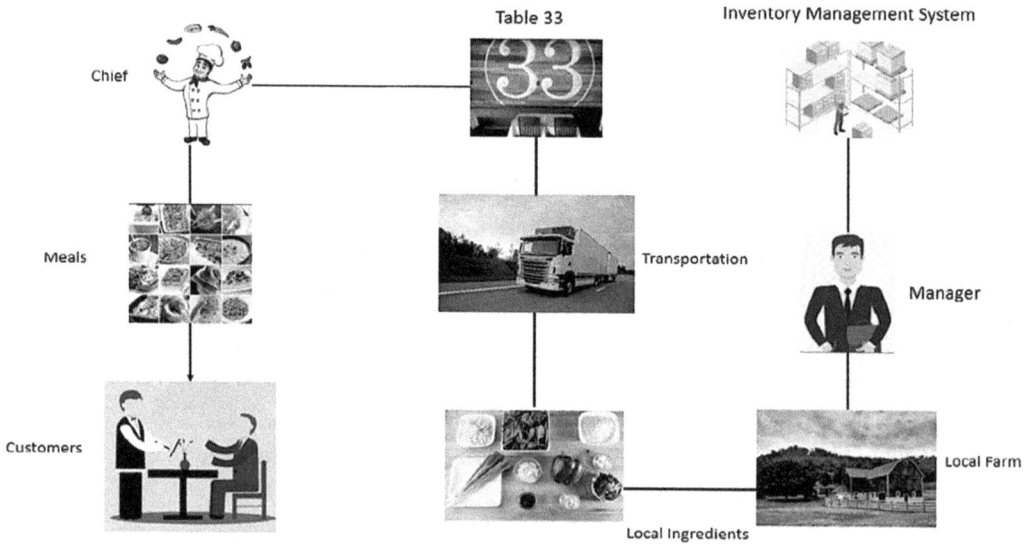

FIGURE 9.17 Physical architecture model – physical block diagram, team 7 – table 33 restaurant.

TABLE 9.3
Women's Healthcare Center System Elements

	System Elements			
System	System Elements	Type	Under Configuration Control?	Phase Goes into Config. Control
Connect to Cancer Center	Patient Navigator	Operator roles	No	
Provide results	Radiologist	Operator roles	No	
Construction of Facility	Facility	Operator roles	Yes	5 – Op Maintenance
Women's Center	Services	Operator roles	Yes	5 – Op Maintenance
Women's Center	Physician	Operator roles	No	
Perform Service	Technician	Operator roles	No	
Spiritual Care	Spiritual care resources	Operator roles	No	
Schedule Service	Resources	Operator roles	No	
Women's Center	Concierge	Operator roles	No	
Women's Center	Volunteers	Operator roles	No	
Process VIP Patient	VIP service members	Operator roles	No	
Women's Center	Process dashboard	Software	Yes	4 – System validation
Perform Service	Imaging technology	Equipment	Yes	5 – Op Maintenance
Women's Center	Process maps	Process	Yes	6 – Op Maintenance
Women's Center	Policies	Policies	Yes	6 – Op Maintenance
Women's Center	Procedures	Procedure	Yes	6 – Op Maintenance
Women's Center	Facility plans/ drawings	Document	Yes	3 – Design
Women's Center	Test plans	Document	Yes	4 – Integration Test

Examples of generic design characteristics in software include (SEBoK 8.0):

- Distribution of processing
- Data structures
- Data persistence
- Procedural abstraction
- Data abstraction
- Control abstraction
- Encapsulation
- Creational patters
- Structural patterns

TABLE 9.4
System Elements, Team 3 – Kids Growing Food

	System Elements			
System	System Elements	Type	Under Configuration Control?	Phase Goes Into Config. Control
Growing	Microgreen Seeds	Materials	NO	
Growing	Tray with lid	Equipment	NO	
Growing	Medium	Materials	NO	
Growing	Watering Apparatus	Equipment	NO	
Processing	Scissors	Equipment	NO	
Processing	Bags	Supplies	NO	
Processing	Hydrogen Peroxide	Supplies	NO	
Distributing	Transaction Software	Software	NO	
Processing	Standard Quality Evaluation Chart	Document	NO	
Distributing	Transportation	Other	NO	
Processing	Compost	Protocol	NO	
Distributing	Starting Funds	Materials	NO	
Processing	Excel Workbook	Database	YES	4 - Integration Test
Training	Training Deliverables	Document	YES	6 - Op Maintenance
Training	Training Demonstration	Protocol	NO	
Processing	Management Records	Database	YES	4 - Integration Test
Processing	Student Records	Database	YES	4 - Integration Test
Training	Qualified Teacher	Operator roles	NO	
Growing	Interested Students	Operator roles	NO	
Distributing	Monthly Cash Flow Report	Document	NO	
Training	Permission Slip	Document	NO	
Training	Student Comprehension Assessment	Document	NO	
Training	Sponsor	Operator roles	NO	
Processing	Money	Materials	NO	
Growing	Fully Grown Microgreens	Product	NO	
Distributing	Customer	Operator roles	NO	
Training	Student Guardian	Operator roles	NO	
Training	Permit	Document	NO	
Processing	Cleaning Procedure	Procedure	YES	6 - Op Maintenance
Processing	Microgreen Evaluation Procedure	Procedure	NO	

(Continued)

TABLE 9.4   (Continued)

System Elements

System	System Elements	Type	Under Configuration Control?	Phase Goes Into Config. Control
Processing	Taste Test	Protocol	NO	
Distributing	Carbon Copy Receipt	Document	NO	
Training	Classroom	Facility	NO	
Growing	Student's Home	Facility	NO	

TABLE 9.5
System Elements, Team 6 – GEM City Market

System Elements

System	System Elements	Type	Under Configuration Control?	Phase Goes into Config. Control
Supermarket	Store operation	Operator roles	No	
Supermarket	Parking for customers	Policies	Yes	6 – Op Maintenance
Supermarket	Carts for customers	Operator roles	No	
Supermarket	Volunteers	Operator roles	No	
Supermarket	In store shopping	Operator roles	No	
Hybrid pricing	Sales statistics	Operator roles	No	
Marketing (management)	Promoting food products	Operator roles	No	
Marketing (management)	Discounts	Operator roles	No	
Checkout	Cashiers	Process	Yes	6 – Op Maintenance
Checkout	Different modes of payment	Process	Yes	
Delivery and pickup	Application (Mobile/ website)	Operator roles	No	
Delivery and pickup	In store pickup	Operator roles	No	
Improve according to demand	In store surveys/feedback	Database	Yes	4 – Integration Test
Improve according to demand	Online Surveys/feedback	Database	Yes	4 – Integration Test

The system design process and activities include the following (SEBoK 8.0):

1. Initialize design definition
   - Plan for technology management.
   - Develop design strategy.

2. Establish design characteristics and design enablers related to each system element
   - Define system elements.
   - Define interfaces.
   - Document and provide rationale of major design decisions.

System	System Elements	Type		Under Configuration Control?		Phase Goes Into Config. Control	
Form	paper food order form	Document	▾	No	▾		▾
Form	paper survey	Document	▾	Yes	▾	5 - System validation	▾
Form	electronic food order form	Database	▾	No	▾		▾
Form	Electronic survey	Database	▾	Yes	▾	5 - System validation	▾
Website	Orders placed	Database	▾	No	▾		▾
Website	Survey data	Database	▾	Yes	▾	3 - Design	▾
Website	Advertising	Database	▾	Yes	▾	3 - Design	▾
Website	Message Board	Software	▾	No	▾	5 - System validation	▾
Website	Frequently asked questions	Database	▾	No	▾		▾
Program	Computer technology	Equipment	▾	Yes	▾	6 - Op Maintenance	▾
Program	Ipad	Equipment	▾	Yes	▾	6 - Op Maintenance	▾
Program	Statistical software	Software	▾	Yes	▾	5 - System validation	▾
Website	Website Server	Software	▾	Yes	▾	5 - System validation	▾
Program	spreadsheet software	Software	▾	Yes	▾	5 - System validation	▾
Website	Forum moderator	Operator roles	▾	No	▾	5 - System validation	▾
Program	Citywide	Operator roles	▾	No	▾		▾
Website	Website Admin	Operator roles	▾	No	▾		▾
Roles	Community Organizer	Operator roles	▾	No	▾		▾
Roles	Farmer	Operator roles	▾	No	▾		▾
Roles	Advertiser	Operator roles	▾	No	▾		▾
Roles	Volunteers	Operator roles	▾	No	▾		▾
Roles	Community Member	Other	▾	No	▾		▾
Program	Policies	Policies	▾	Yes	▾	6 - Op Maintenance	▾
Program	Process Maps	Process	▾	Yes	▾	6 - Op Maintenance	▾
Program	Procedures	Procedure	▾	Yes	▾	6 - Op Maintenance	▾

FIGURE 9.18   System elements, team 1 – connecting organizations.

3. Assess alternatives for obtaining system elements
   • Identify existing implemented system elements and alternatives.
   • Assess design options for the system element, using selection criteria that are derived from the design characteristics.
   • Select the most appropriate alternatives.

4. Manage the design
   • Document rationale for all selections.
   • Establish and maintain traceability between requirements and architecture.

The different engineering disciplines will perform their detailed design activities, based on the type of system provided. Some of the engineering disciplines and their types of design are shown in Table 9.6.

A more detailed design discussion is beyond the scope of this book.

9.5   PERFORM SYSTEMS ANALYSIS

Systems analysis in a set of activities that allows the systems engineering team to perform assessments of the systems and the system elements based on set criteria. It also enables the selection of an efficient system architecture. Assessments should be performed when technical choices or decisions are to be made to assess compliance with the system requirements. Some analyses that are performed include (SEBoK, 8.0):

   • Trade-off studies: Comparing the characteristics of each system element to each candidate system architecture using the assessment criteria. Modeling and simulation tools may be used to perform trade-off studies.

System Elements				
System	System Elements	Type	Under Configuration Control?	Phase Goes into Config. Control
Community	Policies	Policies	Yes	2-Requirements & Architecture
Community	Meeting Rooms	Equipment	Yes	2-Requirements & Architecture
Community	MS Word, Excel	Software	Yes	2-Requirements & Architecture
Community	Email Service	Software	Yes	2-Requirements & Architecture
Community	Survey	Other	Yes	1-Concept
Community	Community Residents	Operator Roles	No	2-Requirements & Architecture
Community	Voting Machines	Equipment	Yes	2-Requirements & Architecture
Community	Champion	Operator Roles	No	1-Concept
Community	Stakeholders	Operator Roles	No	2-Requirements & Architecture
Appointment Scheduler	Scheduling Tool	Software	Yes	2-Requirements & Architecture
Appointment Scheduler	Champion	Operator Roles	No	2-Requirements & Architecture
Appointment Scheduler	Clerk	Operator Roles	No	2-Requirements & Architecture
Appointment Scheduler	Telephone	Equipment	Yes	2-Requirements & Architecture
Commission	Rules	Document	Yes	5 - System Validation
Commission	Commissioners	Operator Roles	No	
Commission	Appointment Scheduler	Service	Yes	
Commission	Computer	Equipment	Yes	
Commission	Procedures	Procedure	Yes	5 - System Validation
Legislation	Laws	Document	Yes	6 - Op Maintenance

FIGURE 9.19   System elements, team 2 – creating food policy.

- Cost analysis: An assessment of the cost of the system for the overall life cycle.
- Technical risks analysis: The technical risk analysis should be updated from those performed in the Concept of Operations phase discussed in Chapter 5. Any additional risk events of a technical nature should be identified as well as assessing the likelihood of occurrence and impact or consequences of the initially established risk events from the Concept of Operations phase.

System Elements

System	System Elements	Type	Under Configuration Control?	Phase Goes Into Config. Control
gathering food services	farmer grower	Operator roles	Yes	2 - Requirements & Arch.
gathering food services	community grown food	Operator roles	Yes	3 - Design
culling system	culling food	Equipment	Yes	4 - Integration Test
gathering food services	food's cart	Process	Yes	3 - Design
packaging food services	cooler	Process	Yes	3 - Design
packaging food services	weighing food	Hardware	Yes	2 - Requirements & Arch.
packaging food services	sutable packing	Document	Yes	2 - Requirements & Arch.
packaging food services	dating	Operator roles	Yes	4 - Integration Test
collection services	food collector	Equipment	Yes	2 - Requirements & Arch.
transportaion system	truck with cooler trail	Software	Yes	1 - Concept
transportaion system	GPS	Operator roles	No	2 - Requirements & Arch.
transportaion system	drivers	Software	Yes	2 - Requirements & Arch.
homefull	communication	Services	Yes	4 - Integration Test
packaging food services	farm helper	Document	Yes	3 - Design
homefull	collecting request	Document	Yes	4 - Integration Test
collection services	food records	Document	Yes	4 - Integration Test
transportaion system	collection plan	Process	No	1 - Concept
collection services	arrangement process	Operator roles	Yes	2 - Requirements & Arch.
storing food system	skilled worker	Document	Yes	2 - Requirements & Arch.
inventory	document	Document	Yes	2 - Requirements & Arch.
inventory	resources	Procedure	Yes	3 - Design
storing food system	labeling	Procedure	Yes	3 - Design
storing food system	weighing	Procedure	Yes	4 - Integration Test
storing food system	packing	Procedure	Yes	2 - Requirements & Arch.
storing food system	refrigerator	Equipment	Yes	3 - Design
inventory	temprature system	Hardware	Yes	3 - Design
delivery system	distrebutor	Operator roles	Yes	4 - Integration Test
delivery system	community goods	Hardware	Yes	4 - Integration Test
delivery system	redline region	Operator roles	Yes	3 - Design
delivery system	transported plan	Document	No	1 - Concept
delivery system	truck with cooler trail	Equipment	Yes	3 - Design

FIGURE 9.20   System elements, team 4 – food storage.

System Elements

System	System Elements	Type	Under Configuration Control?	Phase Goes Into Config. Control
Food Production	Farmers	Operator roles	No	
Food Production	Farm Equipment	Equipment	No	
Food Production	Raw Materials	Equipment	Yes	
Food Distribution	Distributors	Operator roles	No	
Food Distribution	Vehicles	Equipment	No	
Food Distribution	Produce	Product	Yes	
Food Distribution	Website Use - planning shipment	Procedure	Yes	
Food Sales	Entreprenuer	Operator roles	No	
Food Sales	Business Owners	Operator roles	No	
Food Sales	Stores	Facility	No	
Food Sales	Restaurants	Facility	No	
Food Sales	Communication	Telecommunication	Yes	
Food Sales	Profit Tracking	Software	Yes	
Food Production	Website Availability	Software	Yes	

FIGURE 9.21   System elements, team 5 – food hub website.

- Effectiveness analysis: Assessing the effectiveness of the engineered system with respect to performance, usability, dependability, manufacturing, maintenance or support, environment, etc. For our service system we will assess the value-added activities and wastes related to the processes and activities.

System	System Elements	Type	Under Configuration Control?	Phase Goes Into Config. Control
System Elements				
Inventory Management system	Manager	Operator roles	No	
	Inventory (place to store ingredients)	Facility	Yes	6 - Op Maintenance
	Ingredients	Supplies	No	
	Inventory data base	Database	Yes	4 - Integration Test
	Data collection software (Excel)	Software	Yes	5 - System validation
	Data Analyser	Operator roles	No	
Research System	Manager	Operator roles	No	
	Internet search	Database	No	
	Computer/Smart devices	Hardware	No	
	Website/Application to connect farmers with Table 33	Database	No	
	Missing ingredients report	Document	Yes	4 - Integration Test
Testing ingredients System	Manager	Operator roles	No	
	Chief	Operator roles	No	
	Ingredients tester	Operator roles	No	
	Greentest	Equipment	No	
Transportation System	Manager	Operator roles	No	
	Local Farmers	Operator roles	No	
	Transportation Vehicle	Equipment	No	
	Driver	Operator roles	No	
Meal preparation System	Chief	Operator roles	No	
	Chief Assitants	Operator roles	No	
	Cooking Equipments	Equipment	Yes	6 - Op Maintenance
	Local Ingredients	Supplies	No	
	Kitchen	Facility	Yes	6 - Op Maintenance

FIGURE 9.22   System elements, team 7 – table 33 restaurant.

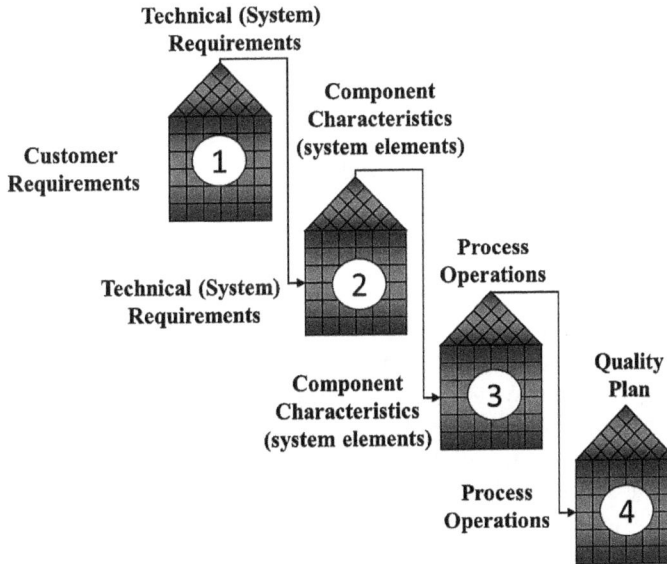

FIGURE 9.23   Quality function deployment houses of quality.

9.5.1 Trade-Off Analysis

The systems engineering team should decide where trade-off analysis can be useful to decide between different candidate designs that can meet the system requirements.

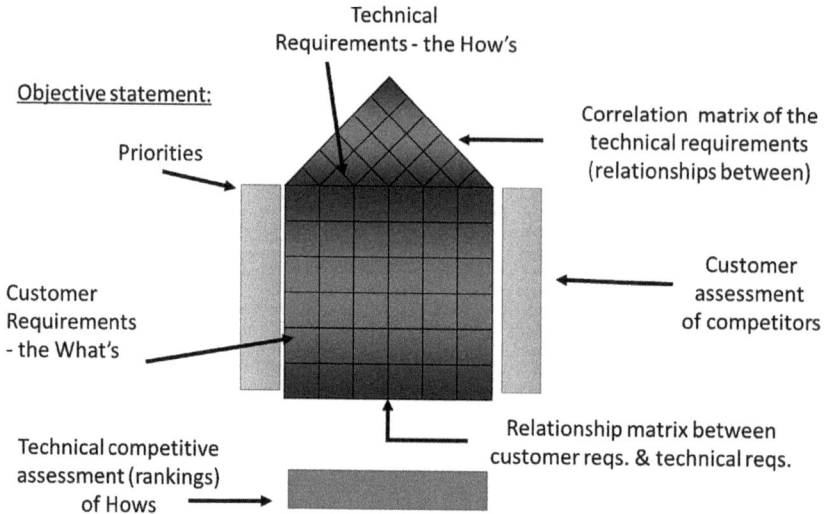

FIGURE 9.24 House of quality format and application.

Trade-off Study Steps (SEBoK 8.0):

1) Determine the system elements to be assessed for trade-offs
2) Determine the system element candidate solutions to analyze
3) Determine assessment selection criteria from the performance and other non-functional requirements and define the expected results
4) Schedule the analyses and resources to perform the analyses
5) Compare the system element candidate solutions using the criteria and provide an assessment score
6) Provide the results and make the trade-off decisions.

9.5.2 TRADE-OFF STUDIES FOR FOOD SYSTEMS

Trade-off Studies for Team 1 – Connecting Organizations

Areas of trade-offs identified by the team include data collection of orders and surveys having two methods. Paper survey has many disadvantages. Replacing paper is a trade-off point. Another trade-off would be the Forum may cause the leak of personal information and have no timeliness of communication. Therefore, it can trade off to a high effective private communication.

Trade off 1: Decrease the amount of paper surveys and increase the electronic versions.

Analysis: Handing out and collection of paper surveys face many risks: community members stained, wrote illegibly, or even lost the paper. Another issue is the wasted time sending volunteers out to collect the forms and obtain little to no response. Therefore, by replacing some paper forms with more online submissions would save time and possibly money.

Justification: Community members are made aware of the website from the ads placed around the community. The website will provide the electronic version of the form and a convenient submission method. The paper version can still be handed out with the website access information on it. Therefore, people who do not have access to computers still can partake in the paper form.

Trade off 2: Taking individual farmer food drop-offs throughout the week and accumulating them all into a single day market.

Analysis: When customers and suppliers(farmers) have meetings individually, they can only buy and sell food to a small group of people. It makes customers have only a few available choices that day. This point-to-point sale mode may limit the supplier–customer connection. In the long run, people may be more interested in going on a known set time and date instead of at random times

Project Name: Connection Organizations

Technical Requirements

Customer Requirements	Importance	1. The website shall have platform that is easy to navigate and self intuitive	2. Website shall contain drop down lists for user to select the desired section of website to go to	3. Website shal have a means for community members and volunteers to input data into the website	4. Website shall have means of means of data collection internal to the website	5. Website shall have algorithm to file sort and organize data in database	6. Website shall have a means of cyber security to protect user information	7. Website shall have enough server space for expansion of the website and prevention of website shutdowns	8. Website shall allows for a minimum of 300 persons on the server at any given time	9. Website shall have a forum or other electronic means of communication in website to allow members, farmers, advertisers, and	10. Website shall have contact information and a means to contact Citywide directly	11. The website shall have a counter that continually monitors the number of visitors to the webpage	12. Website Shall have a quick response to customer of when their order is placed	13. Website shall send out an automatic reply to customer email of online purchase order	14. Website shall have a section where customers can fill out a service survey provide feedback for the program	15. Website shall have sections with high server traffic flow where farmers products can be advertised	16. Website shall have sections with high server traffic flow where community partners can be recognized and can advertise	17. Website shall include a drop down menu where a user can easily select their language of choice and the page will translate	18. Website Shall have a dashboard display list of next pick-up times and locations for a minimum of a week prior	19. Website shall monitor and tally number of online orders placed	20. Website shall monitor and track number of orders full filled	21. Citywide shall on database or spreadsheet software offline to collect and store survey data	22. Citywide shall have means to conduct statistical studies for prioritization of foods and customer requests	23. Citywide volunteers shall be equipped with necessary technology to conduct their work inside and outside office	24. Farmers shall own means of accepting payment via cash, check and/or electronic transfer
Interface to interact with the website	9	9	9	3						3	3		3	3	3			9	3						
Website shall be easy to navigate	9	9	9	3						1								9	9				3		
Data shall be accumulated in a single database	5				9	9	3	9												3	9	3	3	1	
Data shall be accumulated in a single database	5				9	9	3	9		1										3	9	3	3	1	
Data shall be accumulated in a single database	5				9	9	3	9		1										3	9	3	3	1	
Website shall be safe for users	8			3			9	3			3			1										1	1
Website shall allow for many users at once	6			3	3			9	9			3	3	3	1	9	9			9	3				
Website shall allow for many users at once	6			3	3			9	9	3		1	3	3	1	9	9			9	3				
shall have means of communication for all persons involved	8	3	1	3					3	9	9		9	3	9								9		3
Shall have means to contact Citywide	6									9	9		1		9								9		1
CityWide shall be able to view website traffic	4	5						3	3	3		9				9	9			9	9				
notify when order is placed	4			1		3	3		1			1	9	9											
Website alert for order placed	4					3	3						9	9											
Means to obtain customer input	8	3	9	9	9	9	3	3	1	3	9	3			9										
Means for farmers to advertise	6		1	9			1	1		9		1				9									
Community partners to advertise	6		1	9			1	1		9		1					9								
Language option for website	7	9	9															9							
Notify pick-up location	7	3											9						9						
track # of orders requested	8				9	1			1			3	3	3						9	3	3	3	3	
track # of orders filled	8				9	1						3								3	9	3	3	3	
means to prioritize food	8			3	9	9									3							9	9	9	
statistical means to prioritize food	7			3											3							9	9	9	
Volunteers have internet access anytime	9			9																				3	
Farmers have way to complete transaction	7	1					3																		9
Navigation of website is easy	9	9	9	3		3		1			3							3	3						
Website shall include administrative moderator to ensure forum civility	4	1				1	1			9							1								
Website alerts members when a pick up time is delayed	9									1	1		3	1					1						
Citywide shall provide weather updates and cancelation notices	9										1			1											
Provide information to farmers for food orders	8									1			1	1					1	1	1				
Provide farmers with information of data from surveys	6														3										
Provide means to fill out orders and surveys without a computer	7																					9	3	1	
Provide a means to store paper documented information	6																					9			
Discussion with farmers when they are available for pick-up days	9									1									3						
Meeting locational shall be within community orders are placed for	6																		3						
Pick up locations shall vary from on community to another	3																								
Website is secure from leak or hacking of member information	6	9					9																		9
Farmers and Citywide responsible for cleanup of pick-up location after event	3									1	1								1						
Space not be overcrowded and provide accessible parking	6																								
Event area shall be handicap accessible	6																								
custermers shall feel safe and comfartable at market or event	9																		3						
Shall not have major disruption of traffic flow	6																								
Absolute weight		460	398	475	459	350	256	324	164	399	297	148	335	217	300	198	202	252	290	293	311	345	402	240	155
Relative weight		3	8	1	4	10	25	15	37	7	20	39	14	32	19	34	33	26	22	21	17	12	6	28	38

FIGURE 9.25 QFD house of quality 1 part 1, team 1 – connecting organizations.

throughout the week. Additionally, customers may want to evaluate the difference in quality of the product.

Justification: Both the farmer and customer need to have more exposure to others to have more choices. It may make suppliers provide better production because they are under the pressure of competition. Moreover, they can have a larger market to let more customers know their production and draw in potentially larger crowds. On the other hand, if a customer buys a variety of food from

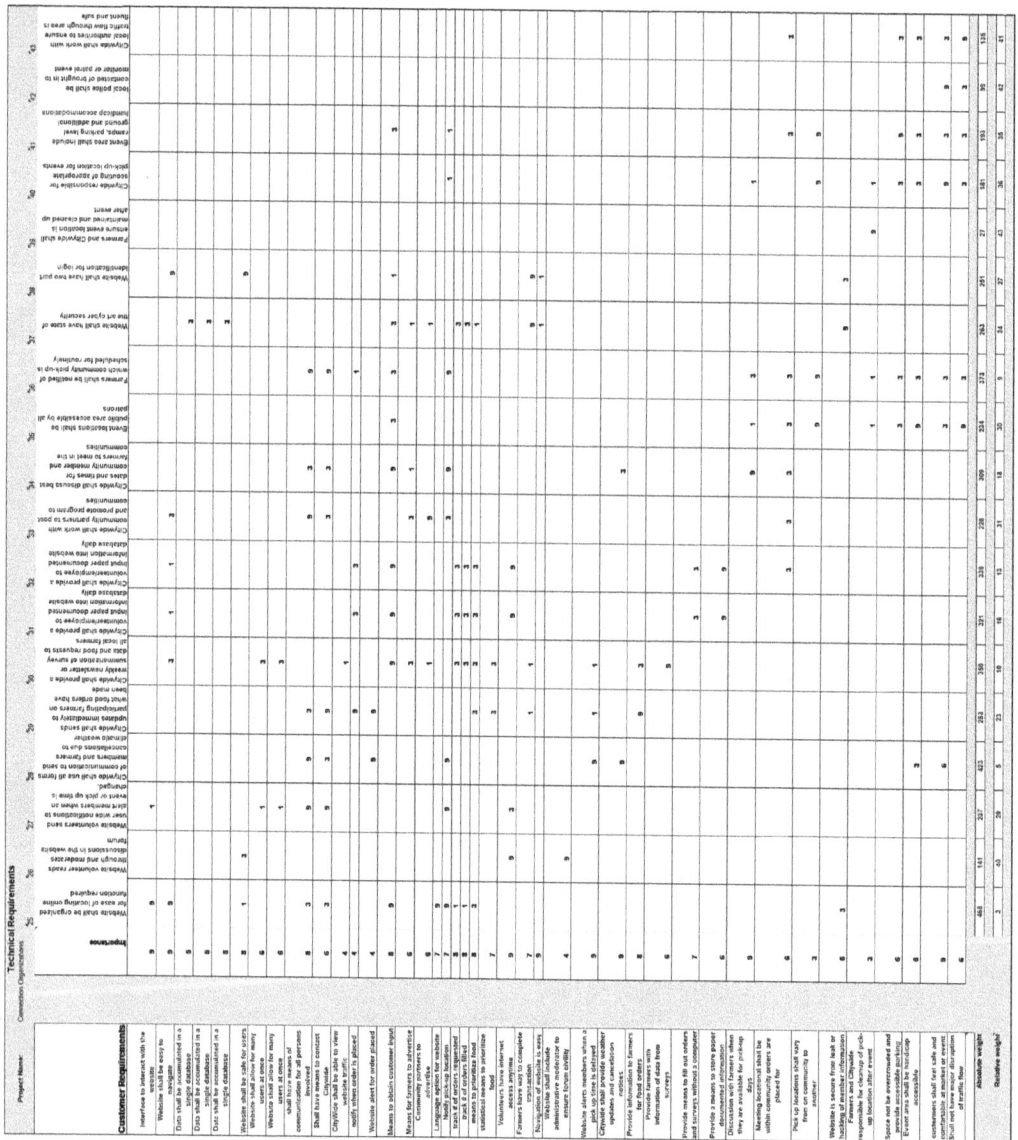

FIGURE 9.26 QFD house of quality 1 part 2, team 1 – connecting organizations.

different farmers, a single day market can save their time instead of going to multiple pickup locations. In a single day market, all the farmers who have the order comes and sells their food to people who place the order and may bring extra for someone interested in their food. They can save time and transportation costs. A larger market is also open to them. Customers can obtain their order and meet more suppliers so that they have more choices.

Trade-off Studies for Team 2 – Creating Food Policy

Public meetings can be used to identify and then involve stakeholders to build a picture of the different possible management strategies and the important criteria attached to them. These multiple criteria can then be listed out within each scenario, and the impact of different scenario criteria on key groups can be worked out. The final stages of the process involve stakeholders in agreeing which management scenario is their preferred option, and what the implications of that are on different stakeholder groups. Skills, Equipment, Time, Costs are some of the resources and requirements of Trade-off analysis (Longo, 2016).

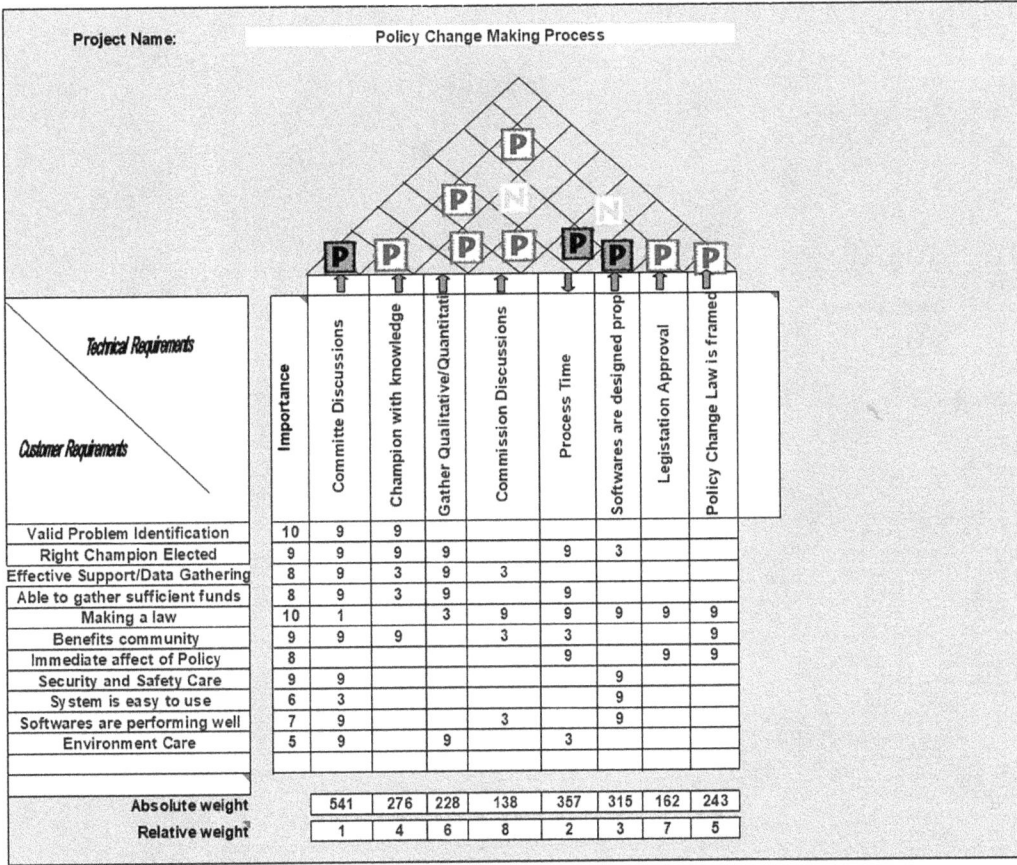

| Project Name: | Policy Change Making Process | | | | | | | |

Customer Requirements \ Technical Requirements	Importance	Committe Discussions	Champion with knowledge	Gather Qualitative/Quantitati	Commission Discussions	Process Time	Softwares are designed prop	Legistation Approval	Policy Change Law is framed
Valid Problem Identification	10	9	9						
Right Champion Elected	9	9	9	9			9	3	
Effective Support/Data Gathering	8	9	3	9	3				
Able to gather sufficient funds	8	9	3	9			9		
Making a law	10	1		3	9	9	9	9	9
Benefits community	9	9	9		3	3			9
Immediate affect of Policy	8					9		9	9
Security and Safety Care	9	9					9		
System is easy to use	6	3					9		
Softwares are performing well	7	9			3		9		
Environment Care	5	9		9		3			
Absolute weight		541	276	228	138	357	315	162	243
Relative weight		1	4	6	8	2	3	7	5

FIGURE 9.27 QFD team 2 – creating food policy.

Policy Making process also involves trade-off analysis. It is applicable and used in problem identification which is the primary step of the policy making procedure. During this step, required analysis on the policies is done in numerous ways like gathering opinions from community members, stakeholders, conducting surveys, interviewing various policy affected groups. Data collected is further analyzed and interpreted in a logical manner in order to identify the problem to be focused and worked upon further.

Trade-off Studies for Team 3 – Kids Growing Food

The trade-off analysis tool allowed the team to see the major risks that would be involved with students harvesting their own microgreens even though it would require more work for the students so that they better earned their compensation.

In the microgreen growing process, the fully-grown microgreens must be harvested by being cut with scissors near the bottom of the shaft and cleaned with hydrogen peroxide. In order for students to truly earn the money they are being given to grow microgreens, to the system development team, it made sense for the students to harvest the microgreens at home. However, after further analysis of the harvesting activities within the Processing System, the system development team put the teacher in charge of harvesting. As a result, students would not have to use scissors or the potentially harmful chemicals required to harvest the microgreens. Also, the teacher, school, or microgreen program would not have to be liable if anything were to go wrong. All in all the teacher is required to do more work, but it is worth it to keep safety a top priority for the microgreen program.

Project Name: Microgreen Growing Program

Customer Requirements	Importance	Transaction summary software	Qualified teacher	Microgreen seeds	Microgreen tray with lid	Microgreen medium	Watering apparatus	Standardardized Microgreen Quality Evaluation Chart	Processing supplies	Transportation means	Compost	Recyclable bags	Fun training deliverables and activities	Non-toxic medium	Starting Funds	Excel Workbook
Proof of purchase	7	9														
Proof of sale	7	9														
Microgreen training	9		9	3	3	3	3		3				9			
Grow high quality microgreens	10		3	9	9	9	9	3	9				1	3	9	
Transfer of microgreens	9		3						3	9						
Harvest microgreens	10		3	3	3	3	3		9	3						
Clean microgreens	10		3		3				9					3		
Bag microgreens	9		3						9			3				
Recycle waste	3										9	3				
Environmentally friendly materials	3			1							3	9				
Engage children	6		3					3					9			
Safety	10		3											9		
Sanitary conditions	10			9				3	3				3	3		
Timely compensation	5	3	1					3							9	
Thorough payment records	4	9	3					9								9
Thorough teacher and student records	4		3					3								9
Securely stored records	7	9	1													9
Easy to understand training	8		1										9			
Easy to use Standard Quality Evaluation Chart	8		1					9								
Absolute weight		240	325	150	267	147	147	213	435	81	66	73	267	240	45	135
Relative weight		4	2	6	3	7	7	5	1	9	11	10	3	4	12	8

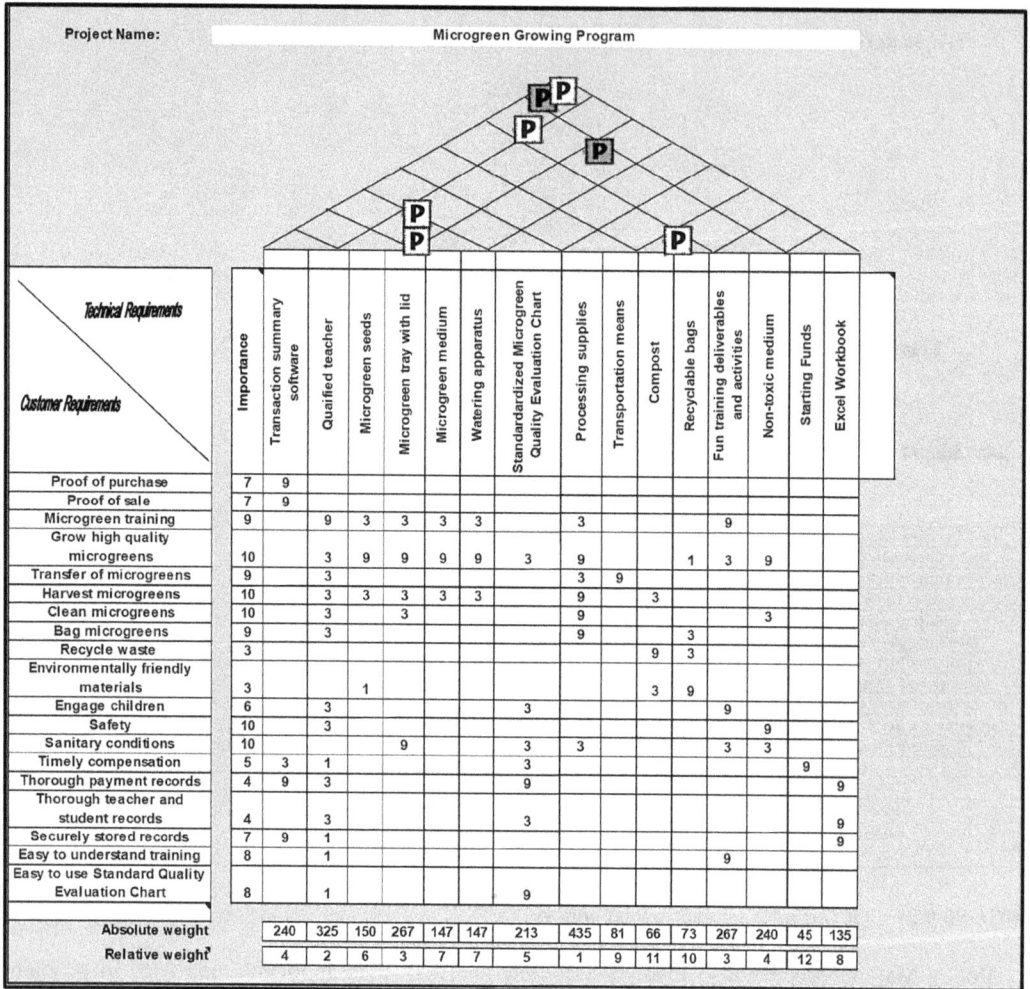

FIGURE 9.28 QFD team 3 – kids growing food.

Trade off Studies for Team 4 – Food Storage

Trade-off analysis is to compare between the characteristics of each system element to determine the solution that balances the assessment criteria. However, trade-off analysis is not needed in this project. Effectiveness analysis and cost analysis clarify it in this project. In addition, step-by-step analysis was significant in obtaining a safe path when the end results were achieved at the end of the project.

Trade-off Studies for Team 5 – Food Hub Website

One system element that has multiple solutions is the management of the website domain. The system users could manage the domain or an outside source or company could manage the website. While managing it ourselves would save cost, it would take more time and be less professional. A company devoted to managing domains will cost more but will ensure the website runs smoothly and safely. As confidential information is on our website, security is an important factor. An outside company would do a better job of managing the website's security features. For these reasons and

Quality Function Deployment - House of Quality 1

Technical Requirements

Critical To Satisfaction (from Business Plan) Customer Requirement	Importance	Food will be stored only until the desired period varying on different foods	Use of Mylar bags for long term food storage	Keep food away from harmful preservative chemicals during storage	avoid adultration of the food at the markets or storage	Helpers will maintain cleanliness	On time pick up of food from farms to not let it damage with improper storage	Delivery of food on time to markets to keep up with the demand for it	Keep a track of demand and produce of the food to make it available at all seasons	The staff is wearing hygiene gears at work	Maintaining realtime records or expiry dates	Will maintain on time pick ups from farm without delay and also mentioning of dates on food to state the expiry period	Will have skilled workers at the Storage place to operate the temperature for different food accordingly to maintain it in a fresh condition	The staff at the system will keep up with time and be punctual with work to deliver food on time	Farmers will learn methods to grow food on all seasons which is taught by homeful staff	The food is stored at desirable temperature for various foods making it available throughout the year	The transportation is maintained with safety for food, by providing a cooling system in the vehicle	The storage of food should be like avoid placing food one on top of the other.	staff should keep monitoring the storage place to avoid any failure of refrigerators during storage	Storage staff will monitor the temperature and adjust it according to the different food varieties	the staff must be taught about each and every aspect of storing for different food	Farmers or community grower should be aware of what methods of maintaining the food	The inventory should be kept up to date and thorough with details of food coming in the storage place and going out	The food collector should be provided with good information of the quantity of food to be transported	Food will be kept safely at the farm avoiding damage from the insects or others creatures.	Food should be kept safe from any fungus formation	By maintaining containers used at the storage very clean	Be preplanned with the food coming to central storage and setting space for it instead of not having space	The staff should wash hands and use of sanitizers after every activity	Food to be transported shall not exceed the limit of carriage on the vehicle to avoid overloading
Empty Row																														
Shall need the food to be less preserved and fresh	8	9		1	3					9	1		3		1	9		1	1	1	1	1						3		
Shall require food bags to be appropriate for long term	6		9			1				1	3	1													3	3			1	
Food needs to be healthy to consume	9	1		9	3	1				3					1										3	3	3		1	
Shall expect to consume only organic fresh food	8				3					1			3		3							1			1	1				
Empty Row																														
Shall demand cleanliness at the storage	9		1	9	3	9	3			3			3	1					1	1	1				3	9	9	3	3	
Shall require food to be undamaged at the farm	6		3															9			3				9	9				
Food availability for purchasing at all days	8	3	3	1	3	1	3	9	9		3	9		9	9	9	1							1						
Food is available at all seasons for consumption	8	3	9	1			1	9	9		1	3		9	9	9	1		1	3			3	1						
Shall assume the staff is maintaining hygiene methods	10		1	3	1	9				9			1		1			1			1	9			9	9	9		9	
Customers would want to consume healthy food	10	3	9	9	9	9	9			9	9	9	1	3			9		9	3	1	9	9		9	9	9		9	
Shall expect to consume only organic fresh food	7			9	9	1	9		1	9	9	9	9	9	3	3	9			9		9	9	1	9	9	3		3	
Customers would want to consume fresh food	9		1	9	3	3	9	9	9	9	9	9	9	9			1		3	3			3		3	9	9			
Food available to consume at all seasons	7	9	3	9	1	1	3	3	3	9	1	1	1	1	3	9	3	3	1	3	3	3	3	3	9	3	3	1	1	1
Food shall be transported with safe methods	9	1	3	3	3	3	9	3	1	3	1	9	3	3	3		1	9				1			3			1	3	9
Empty Row																														
Food should not be mishandelled	6	3	9	1			1	9	9	9	1		3	1	9	9	9	3	3	1	9	3	3	3	3	1	3	9	3	9
Food should be handelled with care to maintain it fresh	7	9	9	1		3	1	9	9	9	3		3	1	9	9	9	3	1	3	3	1	1	1	1	1	3	3	9	1
staff shall be skilled at handling different food	6	3	3	3	1		3	9	1	9		3		3	3	1	9	1	9	3	9	9	1	1	1	3		3	9	3
Shall expect the food grower be experienced in growing	8			3	3				9	9	9	3	9	1	9	9	1		3			9	3		9	1	1	1	3	
Customer expects the community to provide food reaching the demand	7	3					3	9	9			3		9		3	1					1								
Food shall be undamaged	10	1		9	9	9	9			9			1				9	1	3	3				1	9	9	9			
Food needs to be fresh to consume	9		3	9	9	9				3	9	9	9	9		3	9		3				9	1	3	9	3			
Expects the storage of food is clean	10				9	9													1							1	9			
Food shall always be refrigerated while storing	8		1										1				9		9	9										
Shall demand cleanliness at the storage	10				9	9													1							1	9			
Weighting		220	377	771	732	686	528	462	489	780	499	622	585	595	409	438	644	151	392	323	180	420	411	181	618	699	709	146	342	187
Ranking		24	21	2	3	6	11	15	14	1	13	8	12	10	19	16	7	27	20	23	26	17	18	29	9	5	4	28	22	25

FIGURE 9.29 QFD team 4 – food storage.

House of Quality

Project Name:

Website - Economics/Pricing

Customer Requirements	Importance	Tracking System on Distributor Vehicles	Price Drop Alerts	Shopping Interface	Clear and Minimalist Website Design	Program an IM feature	Encryption and Security Software	Contact List Available	
Live Delivery Updates	7	9	1	1	3	3	9	3	
Alert Customer of Price Drop	5	1	9	3	1	1	9	3	
Provisions against Hacking	7	3	3	9	1	3	9	3	
Produce Costs Display	9	1	9	9	9	1	1	1	
Easy-to-use Interface	8	3	9	9	9	3	3	3	
Live Communication	6	9	9	3	3	9	3	9	
Customers Involved in the System	9	9	7	4	2	2	9	9	
Business Owners	7	8	8	6	4	3	7	8	
Farmers in System	6	4	3	9	6	4	6	4	
Visibility for Level customers	8	1	7	4	7	5	8	1	
Transportation Company	7	2	5	2	9	6	7	2	
Absolute weight		190	315	270	267	176	271	134	0
Relative weight		5	1	3	4	6	2	7	

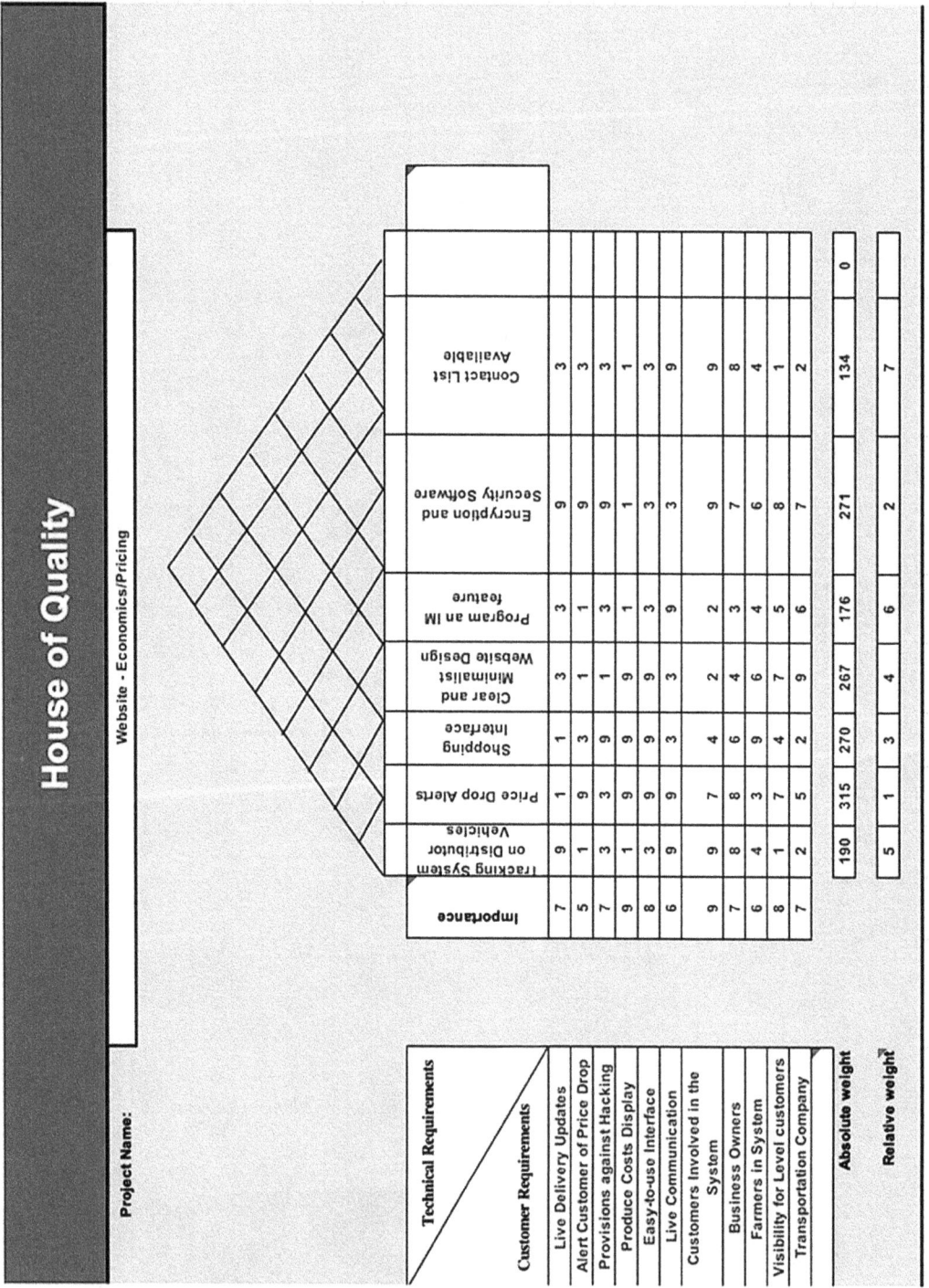

FIGURE 9.30 QFD team 5 – food hub website

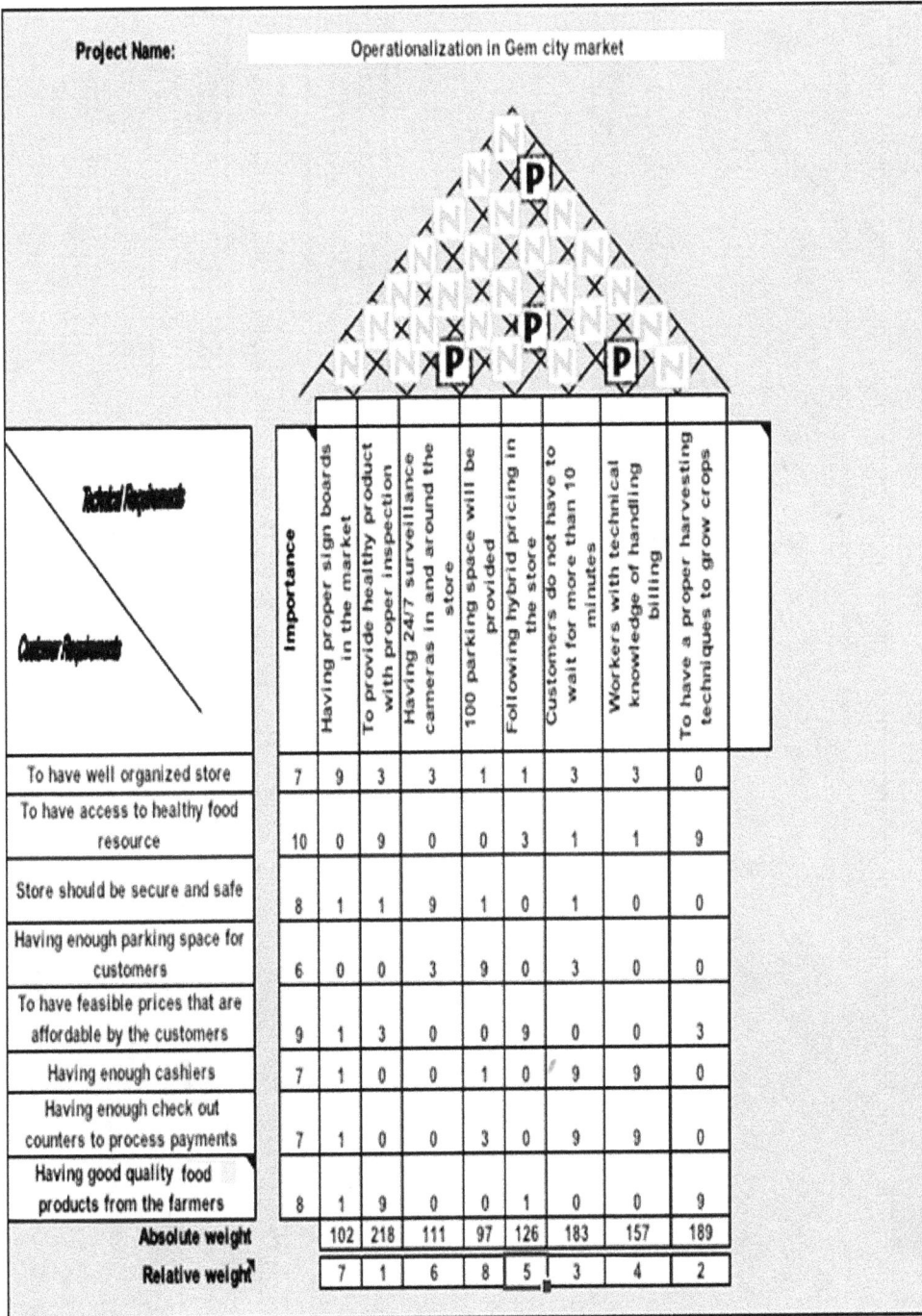

Project Name: Operationalization in Gem city market

Customer Requirements \ Technical Requirements	Importance	Having proper sign boards in the market	To provide healthy product with proper inspection	Having 24/7 surveillance cameras in and around the store	100 parking space will be provided	Following hybrid pricing in the store	Customers do not have to wait for more than 10 minutes	Workers with technical knowledge of handling billing	To have a proper harvesting techniques to grow crops
To have well organized store	7	9	3	3	1	1	3	3	0
To have access to healthy food resource	10	0	9	0	0	3	1	1	9
Store should be secure and safe	8	1	1	9	1	0	1	0	0
Having enough parking space for customers	6	0	0	3	9	0	3	0	0
To have feasible prices that are affordable by the customers	9	1	3	0	0	9	0	0	3
Having enough cashiers	7	1	0	0	1	0	9	9	0
Having enough check out counters to process payments	7	1	0	0	3	0	9	9	0
Having good quality food products from the farmers	8	1	9	0	0	1	0	0	9
Absolute weight		102	218	111	97	126	183	157	189
Relative weight		7	1	6	8	5	3	4	2

FIGURE 9.31   QFD team 6 – gem city market.

Project Name:

Table 33

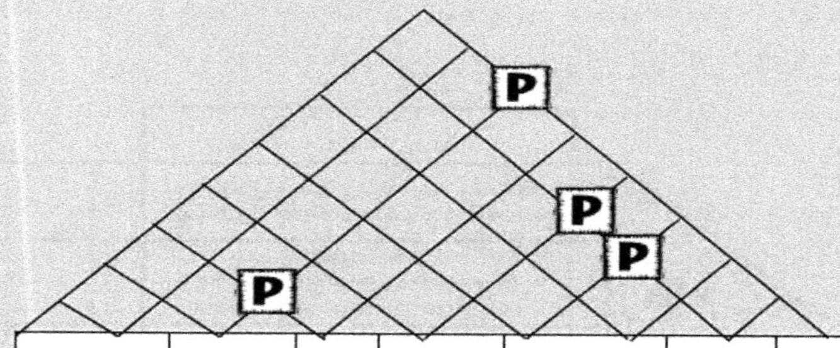

	Importance	Experianced Chief	Organic Farming Practices	Price of Food	Ingredients Reseach Team	Inventory Analyzers	Experianced Manager	Equipments and Softwares
Ingredients Quality	10	6	9	9	9		6	9
Ingredients Taste	9	9	6	6				
Ingredients Quantity	6		6	9		9	6	6
Food Cost	8		6		9	6	9	6
Variety in Menu Items	7	6					6	
Absolute weight		183	228	198	153	96	210	168
Relative weight		4	1	3	6	7	2	5

FIGURE 9.32   QFD team 7 – table 33 restaurant.

TABLE 9.6
Engineering Design Types

Engineering Discipline	Types of Designs	
	Product System	Service System (does not include buildings)
Mechanical Engineering	Mechanical and parts drawings, bill of material, product packaging, manufacturing systems and processes, production engineering.	
Electrical Engineering	Electrical and electronic components, circuits, equipment, computer hardware, networks, cybersecurity.	Computer hardware, networks, cybersecurity.
Chemical and Materials Engineering	Chemical processes, materials, and components.	
Software Engineering	Software, interfaces, databases.	Software, interfaces, databases.
Industrial Engineering	Manufacturing routings, process sheets, workspace and facilities layouts, quality plans, supply chains, production and operation plans, master schedules, work and job design, process metrics, safety plans, business continuity plans, process maps, procedures.	Process maps, procedures, work instructions, work layouts, quality plans, ergonomics, work and job design, process metrics, safety plans, business continuing plans.
Environmental Engineering	Design for operating environment, environmental impact, green design, compliance with environmental regulations (SEBoK, 8.0).	Design for operating environment, environmental impact, green design, compliance with environmental regulations (SEBoK, 8.0).
Human Factors Engineering	Human factors engineering, ergonomics, safety plans, work and job design, user interface design, usability.	Human factors engineering, ergonomics, safety plans, work and job design, user interface design, usability.

through the trade-off analysis, we have decided that the website would be owned and run by an outside company that specializes in that field.

Trade-off Studies for Team 6 – GEM City Market
Trade-off analysis is a part of the systems analysis. It is a tool that provides an assessment of anything that must be decided between certain outlined solutions, such as what type of website hosting, what kind of equipment to choose from, etc. [3]. While this tool is very helpful in certain projects where such decisions must be taken, it does not apply to our Gem City Market. This is because the solutions that the team provided are specific and concise. For example, it was specified that shopping will be conducted in the form of in-store shopping and delivery and pickup. Since both methods were specified to be used, there is no need for trade-off analysis to eliminate one of the options.

Trade-off Studies for Team 7 – Table 33 Restaurant
Trade-off analysis is a great technique to use when you have two ways to approach your system goals. It helps you compare between your methods to see which approach might be the best to reach your goals. In our system, we do not have more than one approach. Therefore, a trade-off analysis is not applicable for our system.

TABLE 9.7

Cost Analysis for Women's Healthcare Center

Type of Cost	Description and Examples	Estimated Costs
Development	Engineering, development tools (equipment and software), project management, test-benches, mock-ups and prototypes, training, etc.	$ 50,000
Product manufacturing or service realization	Raw materials and supplying, spare parts and stock assets, necessary resources to operation (water, electricity power, etc.), risks and nuances, evacuation, treatment and storage of waste or rejections produced, expenses of structure (taxes, management, purchase, documentation, quality, cleaning, regulation, controls, etc.), packing and storage, documentation required.	$ 1,500,000
Sales and after-sales	Expenses of structure (subsidiaries, stores, workshops, distribution, information acquisition, etc.), complaints and guarantees, etc.	$ 15,000
Customer utilization	Taxes, installation (customer), resources necessary to the operation of the product (water, fuel, lubricants, etc.), financial risks and nuisances, etc.	$ 600,000
Supply chain	Transportation and delivery.	$ 50,000
Maintenance	Field services, preventive maintenance, regulation controls, spare parts and stocks, cost of guarantee, etc.	$ 250,000
Disposal	Collection, dismantling, transportation, treatment, waste recycling, etc.	$ 100,000

9.5.3 COST ANALYSIS

The main life cycle costs related to the system development include (SEBok 8.0):

- Development costs: Engineering, prototyping, training, development, tools
- Product manufacturing or service realization: Production setup, manufacturing, packaging, storage, obsolescence, quality defects, rework, scrap, facilities, operations
- Sales and after-sales: Warranty, service, complaints management
- Customer utilization: Installation, taxes, costs related to product or service usage
- Supply chain: Supply chain management, transportation, delivery, logistics
- Maintenance: Field service, maintenance, spare parts
- Disposal: Collection, disposal, recycling, transportation

9.5.4 COST ANALYSIS FOR FOOD SYSTEMS

The cost analysis for the Women's Healthcare Center is shown in Table 9.7.

Cost Analysis for Team 1 – Connecting Organizations:

Table 9.8 is a table created to identify the various costs prior to and during the function of the system. For development of the website, purchase of a database and server should be considered as development costs. The database and server are required for the website platform to function and store data. Server space should be obtained to allow for 300 persons on the site at once. Additional costs include some equipment and software purchase, such as computer, statistical software, and spreadsheet software. The website is a communication platform. Therefore, it produces nothing. However, there are costs for the rent of meeting locations and events, because Citywide hosts the sale meeting between customers and farmers. To keep the website in the functional order, it is a recommendation to hire two or three network security engineers and a web admin. These individuals would monitor and provide maintenance to the website.

TABLE 9.8

Cost Analysis for Team 1 – Connecting Organizations

Type of Cost	Description and Examples	Estimated Costs
Development	Engineering, development tools (equipment and software), project management, test-benches, mock-ups and prototypes, training, etc.	Equipment: computers and appendix, office consumable material. Software: statistical software, spreadsheet software. Rent fee: server, database.
Sales and after-sales	Expenses of structure (subsidiaries, stores, workshops, distribution, information acquisition, etc.), complaints, etc.	Rent work place, meeting location permit.
Supply chain	Transportation and delivery.	No cost.
Maintenance	Field services, preventive maintenance, regulation controls, spare parts and stocks, cost of guarantee, etc.	Website maintenance: salary for security engineers and web admin.
Disposal	Collection, dismantling, transportation, treatment, waste recycling, etc.	Transportation cost for Citywide to go to event.

Cost Analysis		
Type of Cost	Description and Examples	Estimated Costs
Development	development tools (equipment and software), project management,	$100
Product manufacturing or service realization	Electricity power, or rejections produced, expenses of paper, packing and storage, documentation required.	$500
Sales and after-sales	/	
Customer utilization	/	
Supply chain	Transportation and delivery	$100
Maintenance	Commission center and Community center Maintenance	$10,000
Disposal	Transportation, waste recycling, paper waste	$200

FIGURE 9.33   Cost analysis for team 2 – creating food policy.

Cost Analysis for Team 2 – Creating Food Policy

For the change policy system, some cost may exist during the system process. Cost analysis considers the full life cycles. Some cost items are included in the cost analysis shown in Figure 9.33.

In the policy change system, equipment and software cost may exist because the commission center has its official website and scheduling software system, shown in Figure 9.33. If rejections are produced, the policy change idea has to be fixed and a new draft policy file has to be created again. Document files need paper cost. The policy change system doesn't necessarily need sales and customer utilization cost because the system is about change policy. Transportation and delivery are

TABLE 9.9
Cost Analysis for Team 3 – Kids Growing Food

	Cost Analysis	
Type of Cost	**Description and Examples**	**Estimated Costs**
Development	Engineering, development tools (equipment and software), project management, test-benches, mock-ups and prototypes, training, etc.	$27/month
Product manufacturing or service realization	Raw materials and supplying, spare parts and stock assets, necessary resources to operation (water, electricity power, etc.), risks and nuances, evacuation, treatment and storage of waste or rejections produced, expenses of structure (taxes, management, purchase, documentation, quality, cleaning, regulation, controls, etc.), packing and storage, documentation required.	Per 20 students involved: $20/month and one time cost of $150 for reusable trays
Sales and after-sales	Expenses of structure (subsidiaries, stores, workshops, distribution, information acquisition, etc.), complaints and guarantees, etc.	$0
Customer utilization	Taxes, installation (customer), resources necessary to the operation of the product (water, fuel, lubricants, etc.), financial risks and nuisances, etc.	$0
Supply chain	Transportation and delivery, gas.	$5.50/month
Maintenance	Field services, preventive maintenance, regulation controls, spare parts and stocks, cost of guarantee, etc.	$5/month
Disposal	Collection, dismantling, transportation, treatment, waste recycling, etc.	$0

needed because the champion and community members and commission center need communication and meetings with each other. For the maintenance cost, the community center and commission center need to maintain its functions, which means they need maintenance cost to keep them working well. It costs more because the community center and commission center have many small systems. Finally, disposal cost may exist because there is paper waste during the law making. And meaningless transportation could exist because some people may not be available at the specific time.

	Cost Analysis	
Type of Cost	**Description and Examples**	**Estimated Costs**
Supplies and sanitizers	Packing, labeling and cleaning staff	$5000 - $10,000
Development	Training	$5,000
Salary	Employee salary	$155,800
Supply chain	Transporation and delivery	$10,000 - $15,000
Maintenance	Refrigerator maintenance	$5,000
Disposal	Collection, dismantling, waste recylcing, etc.	$2,000 - $5,000

FIGURE 9.34 Cost analysis, team 4 – food storage.

Cost Analysis		
Type of Cost	**Description and Examples**	**Estimated Costs**
Website Development	IT Employees, Software, Web Design Costs	$1,000,000
Website Service and Servers	Server Storage, Servers, Encryption software	$5,000,000
Advertising	Website Advertisements, Commercials	$2,000,000
Distributor(s)	Transportation, Delivery, Special Handling	$3,000,000
Maintenance	Security, Updates, Bug Fixes, and Server Maintenance	$2,000,000
Quality Assurance	Produce Testing, Produce Inspection, Health and Safety Standards, FDA Approval	$3,000,000

FIGURE 9.35 Cost analysis, team 5 – food hub website.

Cost Analysis for Team 3 – Kids Growing Food

The cost analysis tool provided the team the opportunity to think through the process of our system in order to identify where costs would occur and research the amount of the costs. The team identified that the majority of the system's cost comes from the one-time purchase of trays and lids, which is unavoidable, but they can be reused which only requires a maintenance cost. Other system costs come from point of sale software, microgreen growing supplies, such as dirt, seeds, and hydrogen peroxide, and gas costs, shown in Table 9.9.

Cost Analysis for Team 4 – Food Storage

The costs are analyzed in our system (central storage location) and listed with more description, examples, and estimated costs, shown in Figure 9.34.

Cost Analysis for Team 5 – Food Hub Website

Considering the full life cycle of the system, the cost analysis considers every element of the system to determine a budget of the project along its entire existence. Since our system is a website, the costs shown in Figure 9.35 are rough estimates and could be improved from professionals with more experience in the software system field. The cost analysis is very important when developing a system in order to plan for the future cost of implementing the system.

Cost Analysis for Team 6 – GEM City Market

For our project, the tool utilizes seven types of costs which are Development, Product manufacturing or service realization, Sales and after-sales, Customer utilization, Supply chain, Maintenance, and Disposal. Each type of cost is linked with its description and examples to explain it. Using those descriptions and examples, costs were estimated for each type of cost. The development of Gem City Market, its physical construction, and initial operations would be costly, other factors include training, the website, and the software that may be used and its prototypes. As such, the estimated price below takes those factors into account. As for Product manufacturing or service realization, the necessary resources needed for operation such as water and electricity were a big factor for this type of cost. This type also considers the product defects as well as expenses of structure, packing, and storage. These factors culminated in our reflected estimated cost. For the sales and after-sales cost which contains lifetime expenses of subsidiaries, stores, and workshops, an estimated cost was placed below. The Customer utilization estimated cost takes into account taxes which is the main factor in this cost as well as resources for product operation. The estimated Supply chain cost of

Cost Analysis		
Type of Cost	**Description and Examples**	**Estimated Costs**
Development	Engineering, development tools (equipment and software), project management, test-benches, mock-ups and prototypes, training, etc.	$1,000,000
Product manufacturing or service realization	Raw materials and supplying, spare parts and stock assets, necessary resources to operation (water, electricity power, etc.), risks and nuances, evacuation, treatment and storage of waste or rejections produced, expenses of structure (taxes, management, purchase, documentation, quality, cleaning, regulation, controls, etc.), packing and storage, documentation required.	$700,000
Sales and after-sales	Expenses of structure (subsidiaries, stores, workshops, distribution, information acquisition, etc.), complaints and guarantees, etc.	$100,000
Customer utilization	Taxes, installation (customer), resources necessary to the operation of the product (water, fuel, lubricants, etc.), financial risks and nuisances, etc.	$250,000
Supply chain	Transportation and delivery	$400,000
Maintenance	Field services, preventive maintenance, regulation controls, spare parts and stocks, cost of guarantee, etc.	$50,000
Disposal	Collection, dismantling, transportation, treatment, waste recycling, etc.	$150,000

FIGURE 9.36   Cost analysis, team 6 – gem city market.

transportation and delivery for the store would be high as the transportation of products is one of the integral operations for the store. There should also be a budget allocated for Maintenance operations such as preventive maintenance, regulation controls, and field services such as cleaning. Finally, the Disposal type of cost for the store would take into account the collection, dismantling, transportation, and waste recycling for products and equipment in the store. Shown in Figure 9.36 is the tool used to include all the types of costs, their descriptions, and analysis as well as the estimated cost.

Cost Analysis for Team 7 – Table 33 Restaurant

A full description for cost analysis is provided in Figure 9.37 for Team 7. Development costs $500 per person, which include costs that will be spent on the data collection, training, and testing ingredients. As for the supply chain cost, which is the transportation and delivery in our system, it costs $200 monthly. Furthermore, all ingredients needed to fulfill the menu would cost a total estimated cost of $1000 weekly. Lastly, the equipment required for the system will cost $100 per device.

Cost Analysis		
Type of Cost	**Description and Examples**	**Estimated Costs**
Development	Data collection softwares, such as Excel and training. Testing ingredients training.	$500/person
Supply chain	Transportation and delivery	$200/month
Ingredients	All ingredients needed to fulfill the menu	$1000/week
Equipments	Equipments to test ingredients, such as Greentest and other hardwares for the inventory analyzers	$100/device

FIGURE 9.37   Cost analysis, team 7 – table 33 restaurant.

9.5.5   Technical Risk Analysis with Failure Mode and Effect Analysis

A Failure Mode and Effect Analysis (FMEA) is a systemized group of activities intended to recognize and evaluate the potential failure of a product or process, identify actions that could eliminate, or reduce the likelihood of the potential failure occurring and document the entire process.

The FMEA process includes the following steps:

1. Document process, define functions
2. Identify potential failure modes
3. List effects of each failure mode and causes
4. Quantify effects: severity, occurrence, detection
5. Define controls
6. Calculate risk and loss
7. Prioritize failure modes
8. Take action
9. Assess results

An example of a Failure Mode and Effect Analysis for the Women's Center is shown in Figure 9.38.

The Risk Priority Number (RPN) Number is calculated by multiplying the Severity times the Occurrence times the Detection. The Severity is estimated for the failure and given a numerical rating on a scale of 1 (low severity) to 10 (high severity). The Occurrence is given a numerical rating on a scale of 1 (low probability of occurrence) to 10 (high probability of occurrence). The detection scale is reversed, where a numerical rating is given on a scale of 1 (failure is easily detected) to 10 (failure is difficult to detect).

A Pareto diagram can be created based on the RPN values to identify the potential failures with the highest RPN values. Recommendations should be developed for the highest value RPN failures to ensure that they are incorporated into the system.

Only Team 2, Creating food policy, developed a FMEAs for the food system risks, shown in Figure 9.39.

9.5.5.1   Food System Failure Mode and Effect Analysis (FMEA)

Failure Mode and Effects Analysis for Team 2 – Creating Food Policy

For the Failure Modes and Effects Analysis on the system, the main failure is when the policy change idea is rejected.

FAILURE MODE AND EFFECT ANALYSIS (PROCESS FMEA)

PROCESS: PREPARED BY: Date:

DEPARTMENT: DATE:

Process Step	Potential Failure Mode	Potential Effects of Failure	Severity	Potential Causes of Failure	Occurrence	Current process controls	Detection	Risk Priority Number	Recommended Action	Owner / Estimated Completion Date
Schedule patient	Warm transfer to physician's office fails	Patient has to call back	3	Phone system drops patient	2	None	2	12		
		Patient decides not to call back	6	No process to warm transfer	9	None	10	540		
		Complains to administration	3	Technology not capable	8	None	10	240		
	No appointment available within patient's preference times and dates	Patient not able to schedule an appointment and goes elsewhere	9	Lack of resources' availability	5	Pull reports	2	90		
Register Patient	VIP patient isn't identified when scheduling	Patient doesn't receive VIP treatment	8	Technology needs manual intervention	6	None	9	432		
	Missing script / order for service	Cannot perform service, need to re-schedule	9	Patient doesn't bring script	4	Request for script at registration	2	72		
			8	Patient may have wrong script	2	Review of script by registration	4	64		
	Benefit information not valid	Need to call patient	8	Wrong benefit information given by patient	4	Verify with insurance company	1	32		
		May need to cancel appointment	9	Wrong benefit information recorded by scheduling	2	Verify with insurance company	1	18		
Authorize service	Do not receive authorization from insurance company, not valid	Patient needs to self pay	7	Insurance company benefits not allowed	2	Verify with insurance company	1	14		
		Patient does not receive service	9	Insurance company benefits not allowed	2	Verify with insurance company	1	18		
Provide spiritual care	Resources not available	Patient doesn't receive spiritual care	6	Resource availability	4	None	4	96		
Perform imaging service	Patient uncomfortable	Patient embarrassed	8	Lack of training	4	None	9	288		
	Patient has to return for additional images	Images not reviewed	9	Imaging technician skills	2	None	4	72		
	Delays in receiving service	Do not receive service	8	Resource availability	7	None	3	168		
Provide results	Patient doesn't receive results timely	Patient dissatisfied	7	Resource availability	10	None	5	350		
	Patient doesn't receive result	Patient dissatisfied	10	Process breakdown	1	None	5	50		

FIGURE 9.38 Failure mode and effect analysis for women's healthcare center.

Process Step	Potential Failure Mode	Potential Effects of Failure	Severity	Potential Causes of Failure	Occurrence	Current process controls	Detection	Risk Priority Number	Recommended Action	Owner / Estimated Completion Date

FAILURE MODE AND EFFECT ANALYSIS (PROCESS FM EA)

PROCESS: — Present policy idea to commission center — Date:
DEPARTMENT: — Commission center

Process Step	Potential Failure Mode	Potential Effects of Failure	Severity	Potential Causes of Failure	Occurrence	Current process controls	Detection	Risk Priority Number	Recommended Action	Owner / Estimated Completion Date
Present policy change idea to commission center	Inability to make policy change idea resonable	Wasting time	8	Lack communication	1	None	3	24	Keep communication	
		Wasting labor	8	Lack communication	1	None	3	24	Keep communication	
								0		

FIGURE 9.39 Failure mode and effect analysis, team 2 – creating food policy.

The FMEA documentation helps define the failure that may occur during system operation. The main failure is the rejection of policy ideas, the FMEA gives the cause of the failure and effects of the failure and the recommend action to deal with the potential failure which is useful.

9.5.6 EFFECTIVENESS ANALYSIS USING LEAN ANALYSIS FOR A MANUFACTURING OR SERVICE SYSTEM

9.5.6.1 Value Analysis

The Lean Analysis consists of performing a Value Analysis (VA) and Waste Analysis. The VA is a technique to distinguish the activities in the process as being value added, of limited value or non-value added. The VA can be performed for manufacturing or service processes that are part of the system.

The definitions follow:

- Value-added activities: Activities that the customer would pay for that add value for the customer.
- Limited value-added activities: Activities that the customer would not want to pay for but are required because of regulatory, financial reporting, or documentation purposes or because of the way the process is currently designed.
- Non-value-added activities: Activities that the customer would not want to pay for or don't add value for the customer.

To calculate the percent of value-added activities:

100% X (Number of value-added activities/
Number of total activities (value added + limited value + non-valued added))

You can calculate the percent of value-added time as:

100% X (Total time spent in value-added activities/

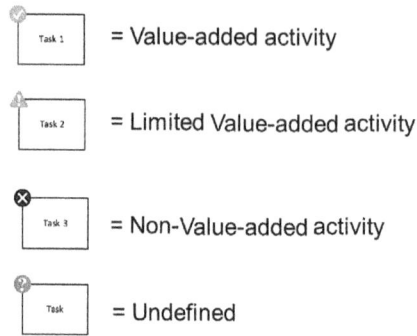

FIGURE 9.40   Symbols on process architecture map (pam) for value analysis.

Total time for process (value added time + limited value time + non-value-added time)

Typically, percent value added time is about 1% to 5% (Non-value-added time = 95% to 99%)
Steps for the VA:

1) Identify the activities (from the process maps).
2) Identify each activity as value-added, limited-value added, or non-value added.
3) Calculate either the percent of value-added activities or the percent of value-added time as a baseline.
4) Attempt to eliminate, combine, or change each activity that is non-value or limited value added.

For the Process Architecture Map (PAM) used to develop process scenarios in the requirements and architecture phase, the following symbols are used while entering the activity information to identify the type of activity as value added, limited value added, non-value added, or not defined yet, as shown in Figure 9.40.

Consider the following focus areas for optimizing the process: (Scholtes, Joiner, Streibel, 2003)

- Labor intensive processes – can they be reduced, eliminated, combined?
- Delays – can they be eliminated?
- Review or approval cycles – are all reviews and approvals necessary and value-added?
- Decisions – are they necessary?
- Rework – why is it necessary?
- Paperwork – is all of the documentation necessary? Try to limit the number of ways you handle and store it
- Duplications – across the same or multiple organizations.
- Omissions – what is slipping through the cracks causing customer dissatisfaction?
- Activities that require accessing multiple information systems
- Travel – look at the layout requiring the travel
- Storing and retrieving information – is it necessary, do we need that many copies?
- Inspections – are they necessary?
- Is the sequence of activities, or flow logical?
- Is standardization, training and documentation lacking?
- Look at inputs to processes, are they all necessary?
- Look at data and information used, is it all necessary?

- What are the outputs of processes, are they all necessary?
- Can you combine tasks?
- Is responsible person at too high or low of a level?

9.5.6.2   Waste Analysis

Also look for areas of waste. Waste Analysis is a Lean tool that identifies waste into eight different categories to help brainstorm and eliminate different types of wastes. The eight wastes are all considered non-value-added activities and should be reduced or eliminated when possible. Waste is defined as anything that adds cost to the product without adding value. The eight wastes are:

- Transportation: Moving people, equipment, materials, and tools.
- Overproduction: Producing more product or material than is necessary to satisfy the customers' orders, or faster than is needed.
- Motion: Unnecessary motion, usually at a micro or workplace level
- Defects: Any errors in not making the product or delivering the service correctly the first time.
- Delay: Wait or delay for equipment or people.
- Inventory: Storing product or materials.
- Processing: Effort which adds no value to a product or service. Incorporating requirements not requested by the customer.
- People: Not using people's skills, mental, creative and physical abilities.

An example of the Value and Waste Analysis for the Women's Healthcare Center is shown in Figure 9.41.

Optimize the processes by incorporating the improvements identified in the VA.

Other optimization tools are (beyond the scope of this text):

- Simulation: simulate changes in the process flow
- Design of Experiments: to identify factors that contribute to the most variability in the processes
- Linear regression modeling
- Lean Six Sigma
- Statistical analysis and process capability analysis
- Statistical process control
- Work sampling
- Job design

9.5.7   Food System Lean Analyses

Lean Analysis for Team 1 – Connecting Organizations

Team 1 performed a detailed lean analysis, consisting of the VA and waste analysis. They did this for each page of their process maps. Two of the lean analysis reports from the PAM is shown in Figures 9.42 through 9.43 for Team 1.

Lean Analysis for Team 2 – Creating food policy

The Lean Analysis for Team 2 is shown in Figure 9.44.

For the process map, the lean analysis chart can be used to identify which activities are Value-added, Limited-Value Added, or Non-Value-added. For the change policy system, the percent of value-added activities is $\frac{7}{9}*100\% = 77.78\%$.

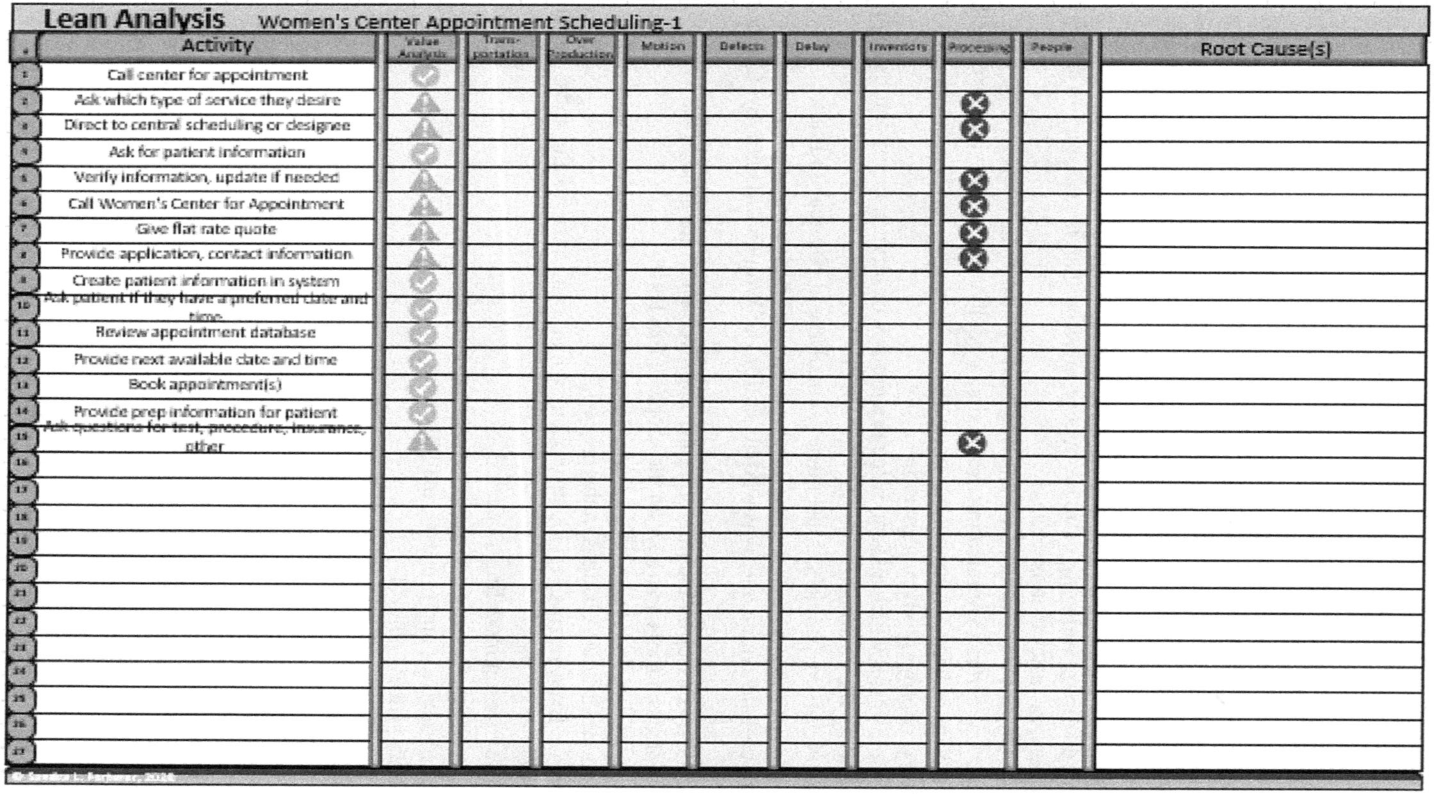

Lean Analysis — Women's Center Appointment Scheduling-1

#	Activity	Value Analysis	Trans-portation	Over-production	Motion	Defects	Delay	Inventory	Processing	People	Root Cause(s)
1	Call center for appointment	◁							⊗⊗		
2	Ask which type of service they desire	◁◁									
3	Direct to central scheduling or designee	◁									
4	Ask for patient information	◁							⊗		
5	Verify information, update if needed	◁							⊗⊗		
6	Call Women's Center for Appointment	◁◁							⊗		
7	Give flat rate quote	◁									
8	Provide application, contact information	◁◁									
9	Create patient information in system	◇									
10	Ask patient if they have a preferred date and time	◇									
11	Review appointment database	◇									
12	Provide next available date and time	◇◇									
13	Book appointment(s)	◇◇									
14	Provide prep information for patient	◁							⊗		
15	Ask questions for test, procedure, insurance, other										

FIGURE 9.41   Value and waste analysis for the women's healthcare center.

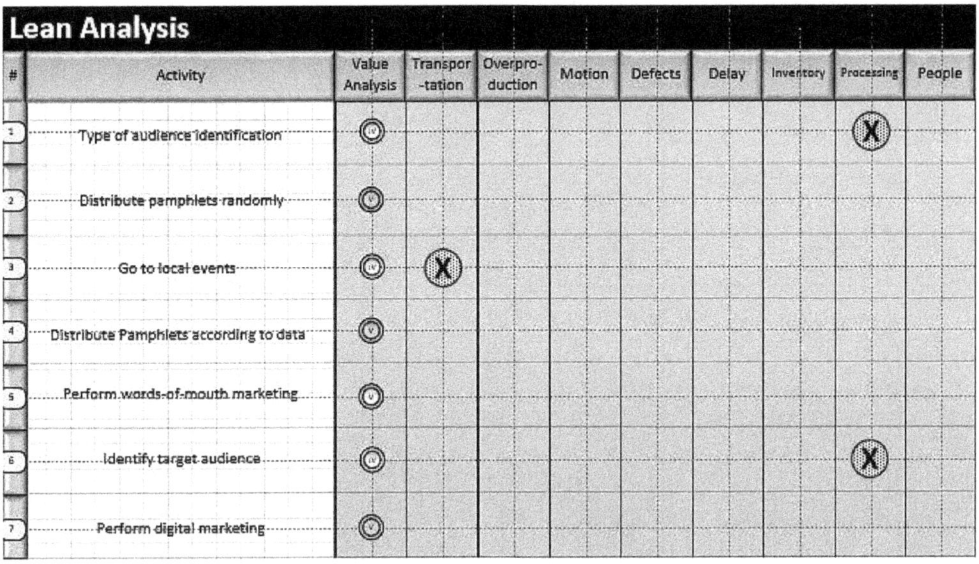

FIGURE 9.42   Page 1 of lean analysis for team 1 – connecting organizations.

For some systems, something or some process could add cost to the product without adding value. Eliminating waste is a good way to improve in profit, system performance, and customer service. Getting rid of waste is a focus of the lean analysis.

For the policy change system, defect waste could happen because a policy change idea could get rejected by the commission. If the idea is rejected, that means we have to fix the policy change idea to fulfill the customer requirement. Overproduction waste could happen because the policy change idea could be beyond the ability that the Government could do. If the policy change idea requires a large cost that the Government could afford, overproduction waste could happen. Delay Waste could happen in the policy change system when the champion is waiting for the appointment from the commission center. That means the champion couldn't give the policy change idea to the commission center immediately.

Lean Analysis for Team 3 – Kids Growing Food
The effectiveness analysis tool allowed the team to see whether resources and energy in the system were going to waste by analyzing tasks in the Process Architecture Map (PAM) in terms of value. As a result, the team found that the majority of our tasks added value and the only wastes were people and delays. The lean analysis is shown in Figures 9.45 and 9.46.

Lean Analysis for Team 4 – Food storage
There are some key aspects such as temporary storage location, transportation, central storage location, and distribution that affect the effectiveness of system outcomes. These aspects impact the central storage location and should be counted in designing the system. For example, the farmers store the food in a temporary storage location in a safe environment. It is important to mark all food packed for further easy identification and expiration dates on packaging. The lean analysis is shown in Figures 9.47 through 9.49.

Lean Analysis for Team 5 – Food Hub Website
The lean analysis for team 5 is shown in Figures 9.50 through 9.52

Advertisement

Lean Analysis

#	Activity	Value Analysis	Transpor-tation	Overpro-duction	Motion	Defects	Delay	Inventory	Processing	People
1	Ask data from website	◎							X	
2	Ask again	●					X			
3	Analyze the received data	◎								
4	Use software to analyze further	◎								
5	Interpret results	◎								
6	Use software to prioritize data	◎								
7	Report Output of analyzed data	◎								

FIGURE 9.43 Page 2 of lean analysis for team 1 – connecting organizations.

Lean Analysis DRAFT

#	Activity	Value Analysis	Transpor-tation	Overpro-duction	Motion	Defects	Delay	Inventory	Processing	People	Root Cause
1	Form Action Committee	◎									
2	Write Impact Reports	◎									
3	Form City Government Committee on Issue	◎									
4	Draft Legislation	◎									
5	Wait for Draft Legislation	●					X				
6	Wait for meeting	●					X				
7	Elect Champion	◎									
8	Identify Problem	◎									
9	Review impact reports	◎									

FIGURE 9.44 Lean analysis for team 2 – creating food policy.

Lean Analysis for Team 6 – GEM City Market

The effectiveness analysis comprises the activities from the Process Architecture Map (PAM) and maps them into the lean analysis activities and marks them into different scales like value, transportation, processing, inventory. Here, the activity of the customer walking into the store and looking at the food product is a valuable activity and it's indicated with a VA symbol, similarly moving into the different aisles is a limited value activity so it's identified as a limited VA. The analysis has been performed for both the in-store shopping and online shopping of the Gem City Market. However, there

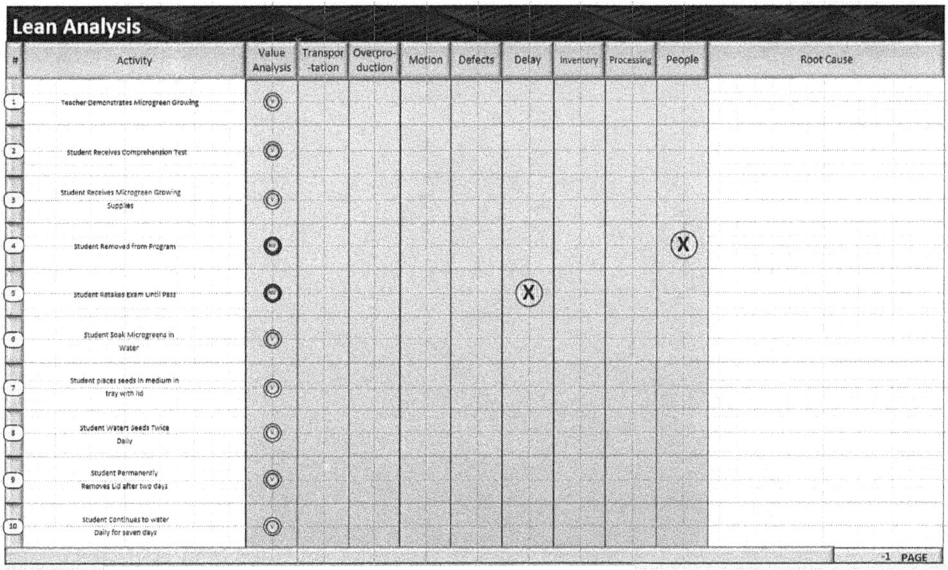

FIGURE 9.45 Page 1 of lean analysis for team 3 – kids growing food.

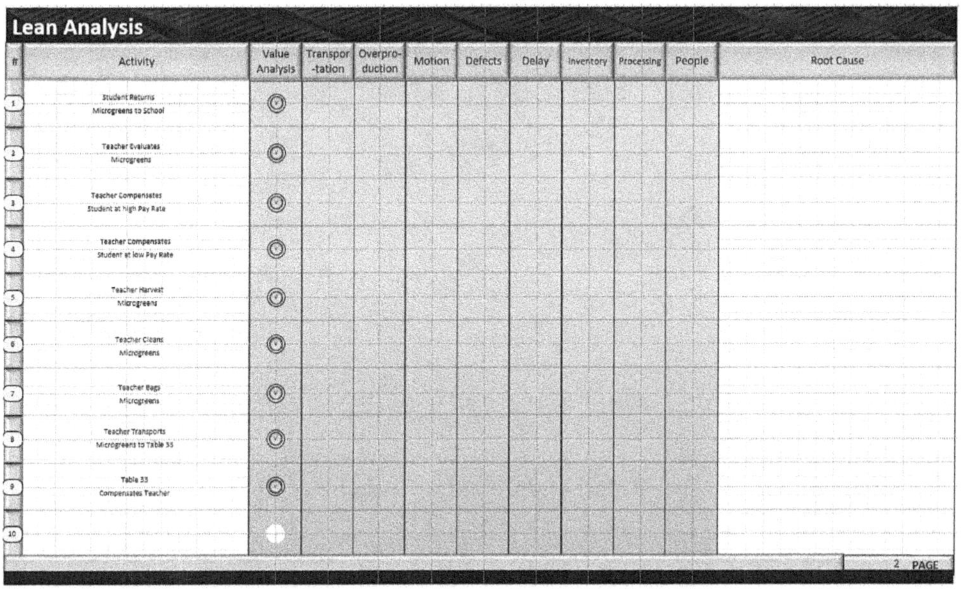

FIGURE 9.46 Page 2 of lean analysis for team 3 – kids growing food.

is no waste analysis performed because the system doesn't have any waste or delay in the analysis, it a very straight forward system where two kinds of shopping systems are involved like online and in-store shopping. The lean analysis is shown in Figures 9.53 through 9.55.

Lean Analysis for Team 7 – Table 33 Restaurant

The Lean Analysis for Team 7 is shown in Figure 9.56.

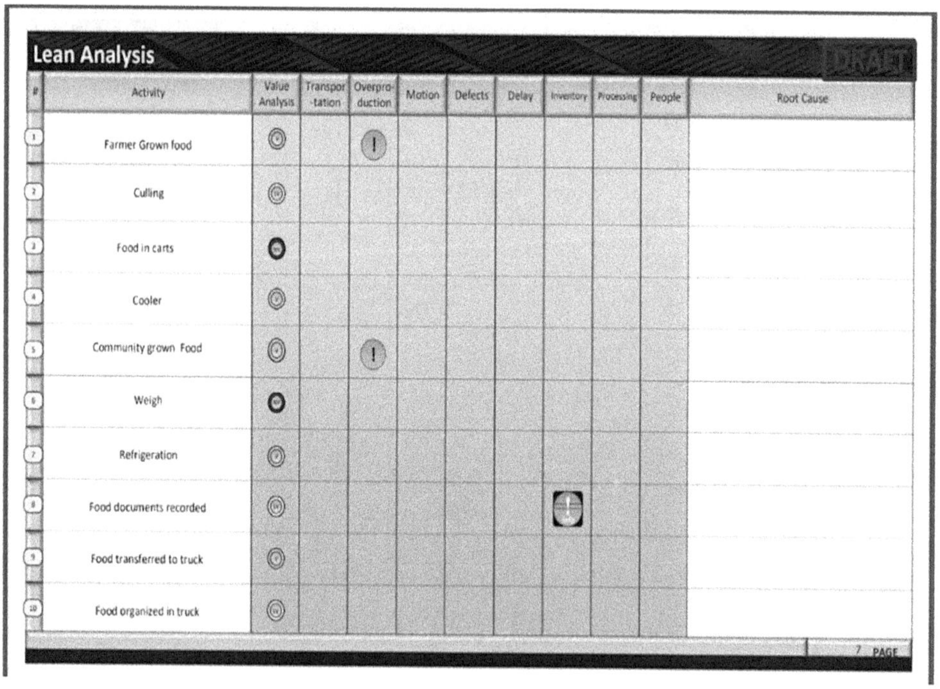

FIGURE 9.47   Page 1 of lean analysis for team 4 – food storage.

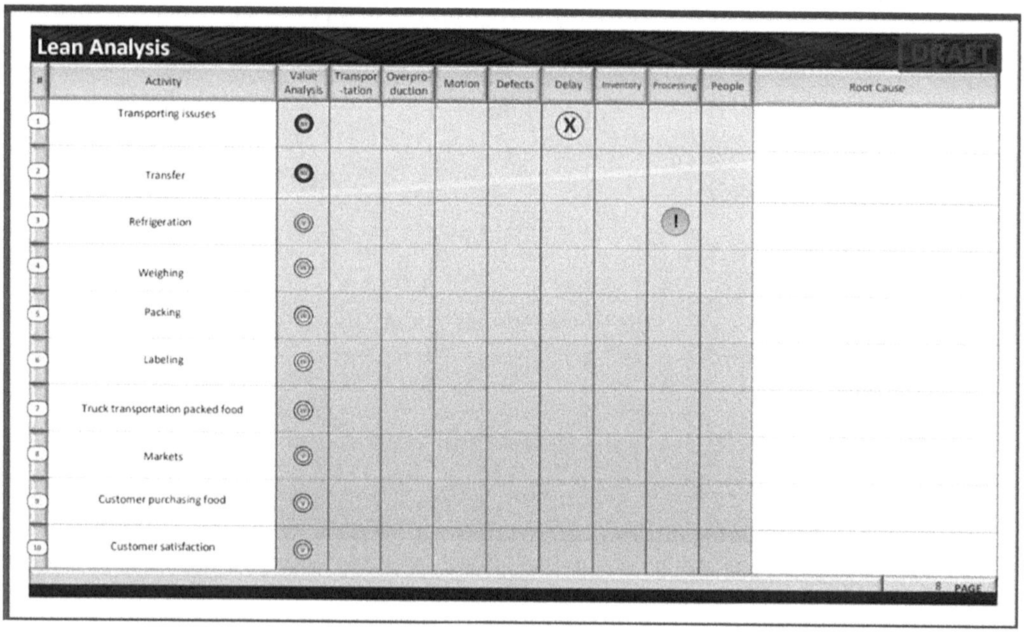

FIGURE 9.48   Page 2 of lean analysis for team 4 – food storage.

9.5.8 N-SQUARED DIAGRAM TO IDENTIFY INTERFACE POINTS

According to the SEBoK 8.0, an interface is defined as:

- A shared boundary between two functional units, defined by various characteristics pertaining to the functions, physical exchanges, and other characteristics

Lean Analysis

#	Activity	Value Analysis	Transpor-tation	Overpro-duction	Motion	Defects	Delay	Inventory	Processing	People	Root Cause
1	Mobile bikes	◎									
2	Redline region	◎									
3	Community food	◎									
4	.	○									
5	.	○									
6	.	○									
7	.	○									
8	.	○									
9	.	○									
10	.	○									

9 PAGE

FIGURE 9.49 Page 3 of lean analysis for team 4 – food storage.

Lean Analysis

#	Activity	Value Analysis	Transpor-tation	Overpro-duction	Motion	Defects	Delay	Inventory	Processing	People	Root Cause
1	Check Website for Produce	◎									
2	Contact Farmer	◎									
3	Place Order On Website	◎									
4	Await Local Availability	◉					Ⓧ				
5	Import from Small Outside Farmer	◎								Ⓧ	
6	Order as soon as Available	◎									
7	Farmer receives order confirmation e-mail	◎								Ⓧ	
8	Order notification delivered to distributor	◎									
9	Organize other Parcel service	◎					Ⓧ				
10	Distributor receives order from Farmer	◎	Ⓧ								

7 PAGE

FIGURE 9.50 Page 1 of lean analysis for team 5 – food hub website.

- The connections of two or more systems or system components for the purpose of exchanging data, materials, forces, or energy from one to the other

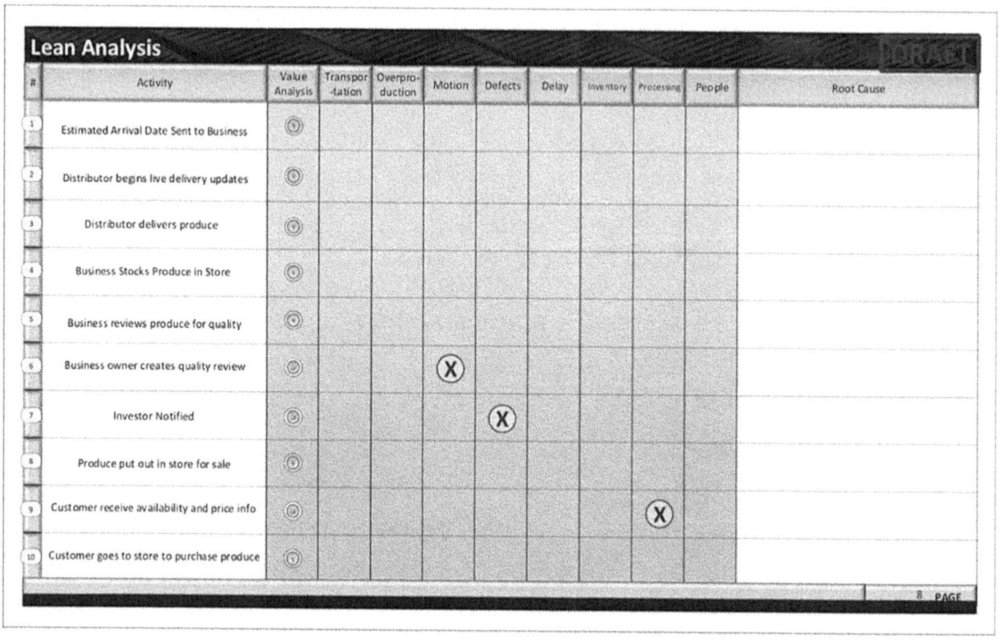

Lean Analysis

#	Activity	Value Analysis	Transpor-tation	Overpro-duction	Motion	Defects	Delay	Inventory	Processing	People	Root Cause
1	Estimated Arrival Date Sent to Business	◎									
2	Distributor begins live delivery updates	◎									
3	Distributor delivers produce	◎									
4	Business Stocks Produce in Store	◎									
5	Business reviews produce for quality	◎									
6	Business owner creates quality review	◎			Ⓧ						
7	Investor Notified	◎				Ⓧ					
8	Produce put out in store for sale	◎									
9	Customer receive availability and price info	◎							Ⓧ		
10	Customer goes to store to purchase produce	◎									

8 PAGE

FIGURE 9.51 Page 2 of lean analysis for team 5 – food hub website.

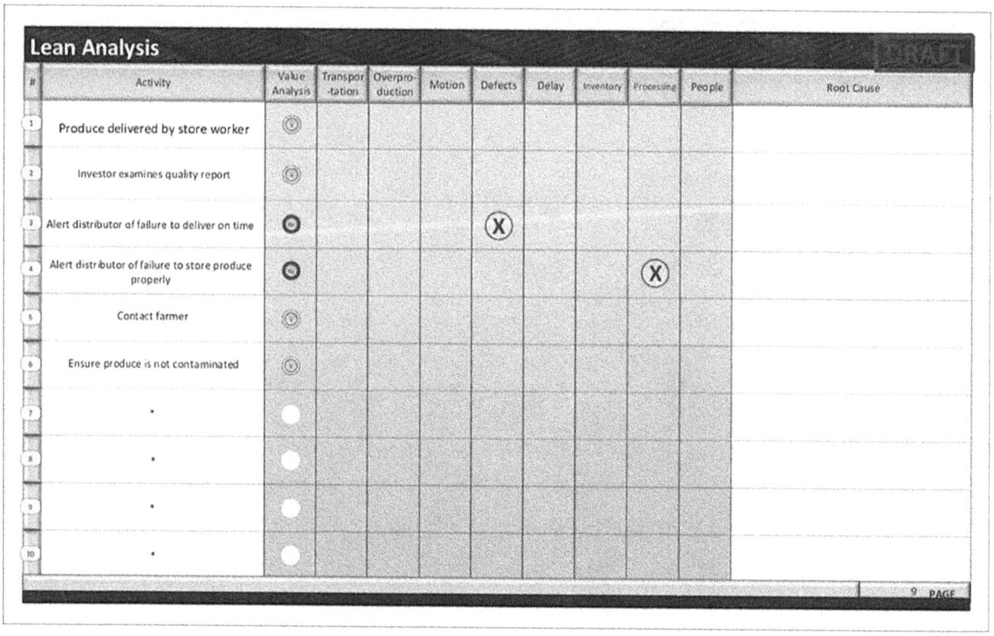

Lean Analysis

#	Activity	Value Analysis	Transpor-tation	Overpro-duction	Motion	Defects	Delay	Inventory	Processing	People	Root Cause
1	Produce delivered by store worker	◎									
2	Investor examines quality report	◎									
3	Alert distributor of failure to deliver on time	◉				Ⓧ					
4	Alert distributor of failure to store produce properly	◉							Ⓧ		
5	Contact farmer	◎									
6	Ensure produce is not contaminated	◎									
7	.	○									
8	.	○									
9	.	○									
10	.	○									

9 PAGE

FIGURE 9.52 Page 3 of lean analysis for team 5 – food hub website.

Managing interfaces is a critical activity of the systems engineer. Interfaces can exist between sub-systems and between systems elements. These interfaces can be connected either physically or functionally.

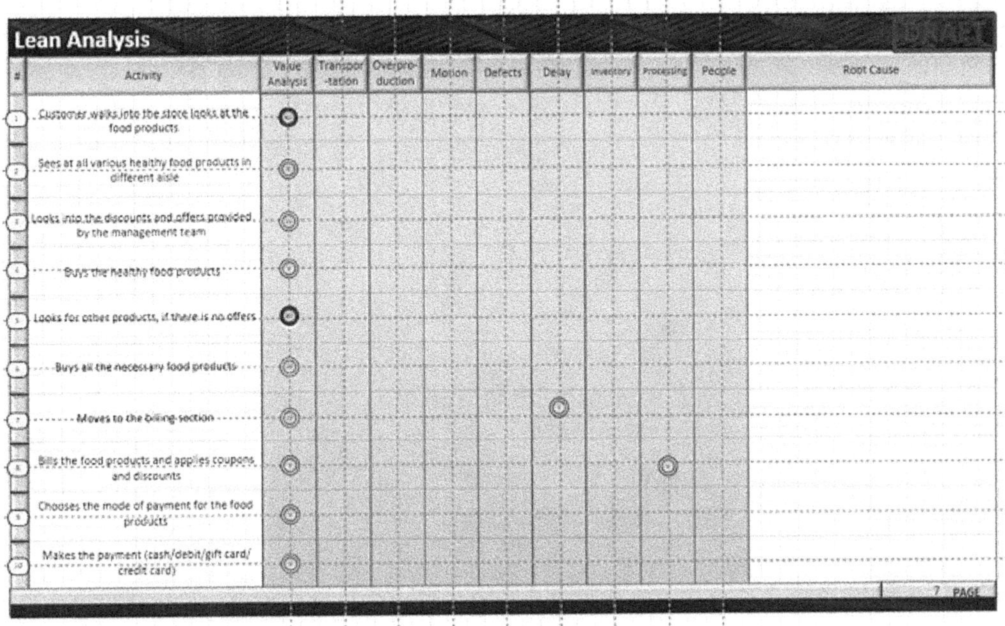

Lean Analysis

#	Activity	Value Analysis	Transportation	Overproduction	Motion	Defects	Delay	Inventory	Processing	People	Root Cause
1	Customer walks into the store looks at the food products	◎									
2	Sees at all various healthy food products in different aisle	◎									
3	Looks into the discounts and offers provided by the management team	◎									
4	Buys the healthy food products	◎									
5	Looks for other products, if there is no offers	◎									
6	Buys all the necessary food products	◎									
7	Moves to the billing section	◎					◎				
8	Bills the food products and applies coupons and discounts	◎							◎		
9	Chooses the mode of payment for the food products	◎									
10	Makes the payment (cash/debit/gift card/credit card)	◎									

7 PAGE

FIGURE 9.53 Page 1 of lean analysis for team 6 – gem city market.

Lean Analysis

#	Activity	Value Analysis	Transportation	Overproduction	Motion	Defects	Delay	Inventory	Processing	People	Root Cause
1	Gets the food products	◎									
2	Gets the receipt either as a digital copy or printed and buys the products	◎									
3	Customer looks at the online website for ordering food products	◎									
4	Looks at the various sections for ordering healthy food options and sees the different sections in the website	◎									
5	Looks for discount in website	◎									
6	Chooses the food products and adds them in the cart	◎									
7	Goes to the cart to view the food products	◎									
8	Adds discounts and coupons to the cart	◎									
9	Checks-out and chooses the desired payment methodology	◎							◎		
10	Makes the payment and confirms the order	◎									

8 PAGE

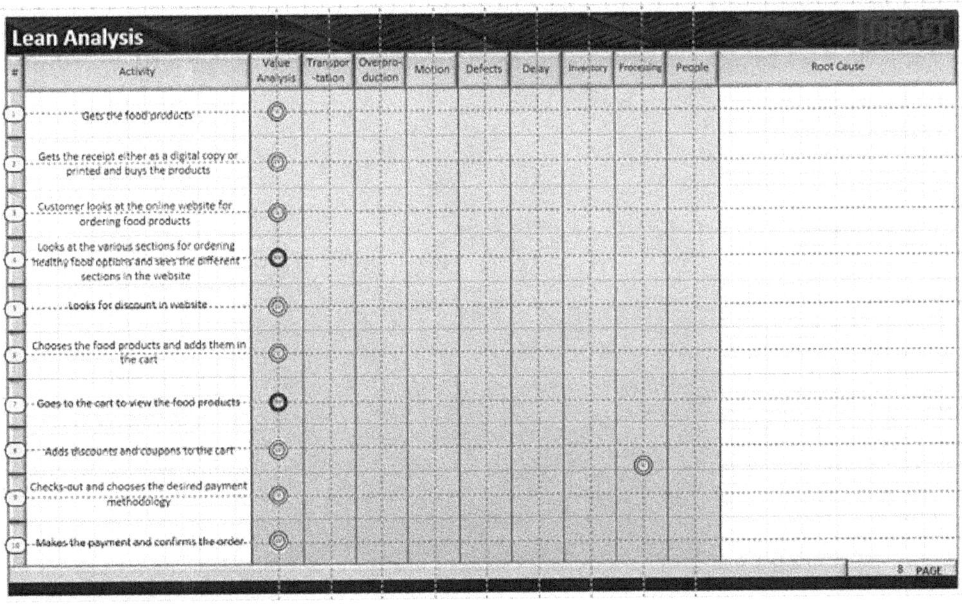

FIGURE 9.54 Page 2 of lean analysis for team 6 – gem city market.

Physical interfaces can be structural, electrical, thermal, temporal, and electromagnetic. An Interface Control Document (ICD) is used to document the major interfaces in the system. The ICD describes how to manage the interfaces. As part of the configuration management plan, which is discussed in the Systems Engineering Planning chapter, when a controlled item changes, it is important to review the interfaces to see if they are impacted by the change.

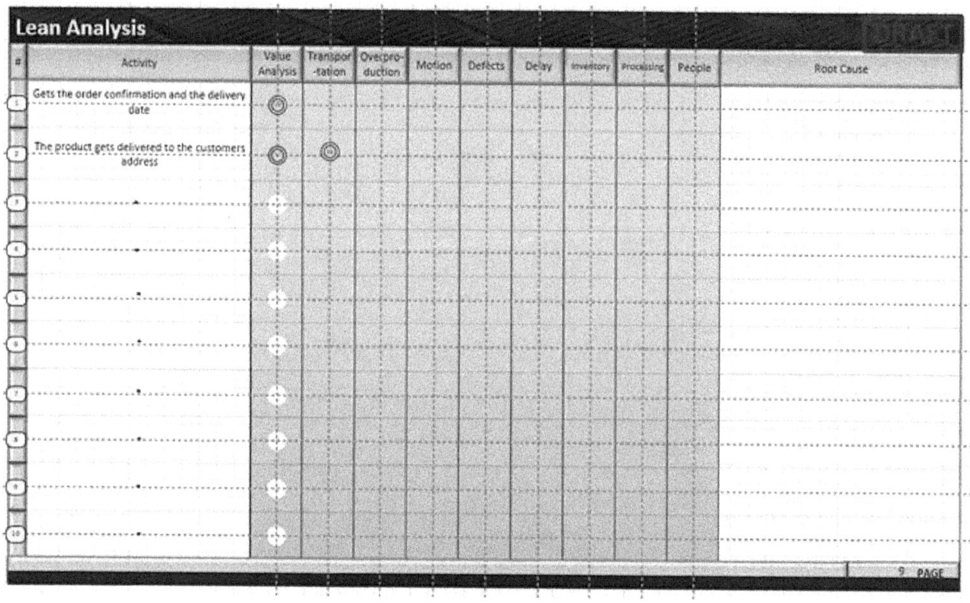

FIGURE 9.55 Page 3 of lean analysis for team 6 – gem city market.

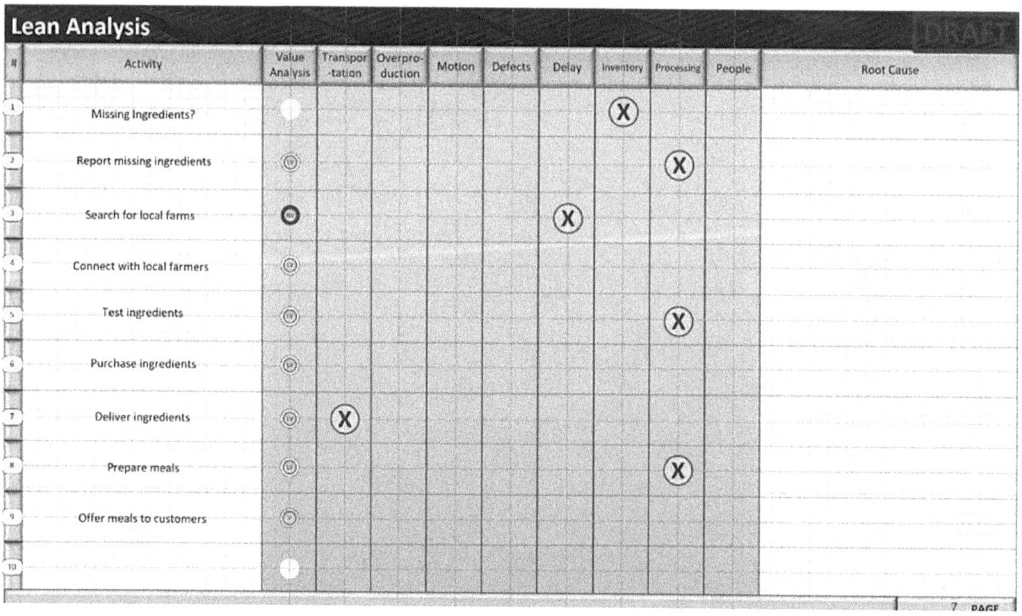

FIGURE 9.56 Lean analysis for team 7 – table 33 restaurant.

An interesting example of an ICD can be found on the Centers for Medicare and Medicaid (https://www.cms.gov/search/cms?keys=ICD+template)

The N-squared Diagram is a tool that shows the interfaces between sub-systems and elements of each sub-system. However, it doesn't describe how the interfaces connect, which would be described in more narrative detail in the ICD document. This N-squared diagram is useful to identify the interfaces as well as to ensure that the interfaces are well-tested in the verification and validation phases. Figure 9.57 shows the N-squared diagram for the Women's Healthcare Center.

Schedule Service X	X									
	Authorize Service X	X								
X		Process self-referral								
	X		Register Patient X	X						
			X	Process VIP patient						
					Perform service X	X	X	X		
					X	Connect to cancer center				
							Perform surgery			
					X	X	X	Provide results		X
					X				Provide spiritual care	
								X		Request records / film
Scheduling and Registration					Services				Spiritual Care	Records

FIGURE 9.57 N-squared diagram for the women's healthcare center.

X							
	X					X	
X							X
	X					X	X
Orders Placed							
	Survey Data						
X		Advertising	X				X
		X	Message Board		X	X	
				Website Server	X		
			X		Forum Moderator	X	
					X	F.A.Q.	
	X	X		X			Website Admin
		X	X	X			
	X						
					X		X
	X		X				
			X				
			X			X	
		X	X				
			X			X	
			X			X	
Website							

FIGURE 9.58 Snapshot of N-squared diagram, team 1 – connecting organizations.

Community Members Identify a Problem	X						
	Community Discussion	X					
		Elect a Champion	X				
	X		Champion gathers Support/Data	X			
			X	Champion Meets Commissioner	X		
				X	Commission Discussion	X	
						Passing a Legislation	X
X	X						Making a Law
Problem Recognition			Presenting Idea			Policy Framing	

FIGURE 9.59 N-squared diagram, team 2 – creating food policy.

C1			x	x								
	C2		x	x				x	x	x		
		C3	x		x	x	x				x	x
			C4	x								
				C5	x	x	x	x	x	x	x	x
					C6	x	x	x	x	x	x	x
						C7	x					
		x	x				C8					x
								C9	x	x	x	x
									C10	x	x	x
										C11	x	x
											C12	x
			x			x						C13
Food and Growing Instruction	Food Production	Food Processing									Food Distribution	

FIGURE 9.60 N-squared diagram, team 3 – kids growing food.

First matrix (collection services):

food collector	x				x	x	x	x
	food records	arrangement process			x			
			truck with cooler trail					
		x		GPS				
			x	driver	x		x	x
					collection plan			
	x			x		communication		x
x								collecting request

collection services   transportation system   homefull

Second matrix (packing food services):

Farmer Grower	x						x	x
	Community Grown food						x	
		culing food						
		food's cart	x					
		x	cooler	x				
				weighing food				
		x	x	x	x	suitable packing	dating	
x								farm helper

Gathering food services   culling system   Gathering food system   packing food services

FIGURE 9.61   N-squared diagram 1, team 4 – food storage.

distributor	x	x	x	x
	community goods	x	x	x
x	x	redline region		
x			transported plan	x
x				truck with cooler trail

Delivery System

FIGURE 9.62   N-squared diagram 2, team 4 – food storage.

N-Squared Diagram

System 1: Website	Plant Produce	x				x		
		Harvest				x		
	x		Invest	x			x	x
				Market	x			
					Profit			
		x		x		Delivery	x	
						x	Tracking	
					x			Quality Assurance

Farming	Entrepreneurs	Distribution

FIGURE 9.63   N-squared diagram, team 5 – food hub website.

Improving according to demand			x	
x	Hybrid pricing		x	x
		Delivery and pickup	x	x
x	x	x	Marketing	x
x	x	x		Store orgnzation
Database	Management			Workers

FIGURE 9.64   N-squared diagram, team 6 – gem city market.

N-Squared Diagram

System 1:

Inventory Management Systems	x		x					x
	Research System		x			x		x
		Kitchen		x	x			x
			Manager	x	x	x	x	
				Chief				x
					Waitress			
						Local Farms	x	x
							Farmers	x
								Ingredients

| Table 33 Facilities | | | Table 33 Faculties | | | Suppliers and Supplies | | |

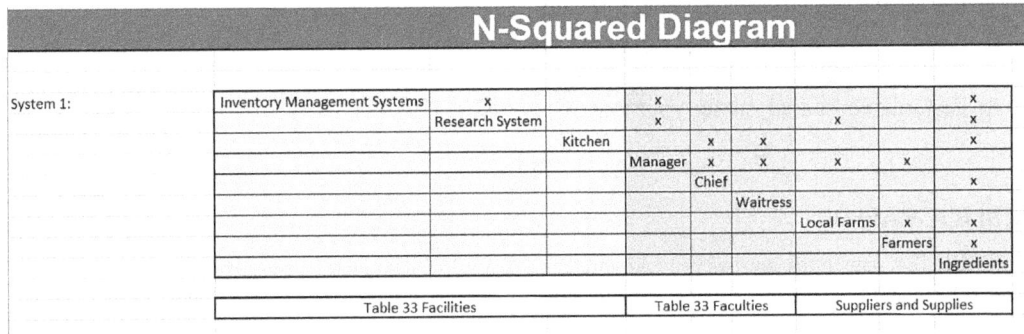

FIGURE 9.65   N-squared diagram, team 7 – table 33 restaurant.

To create the N-squared diagram you can follow the steps (Grady, 1994):

1) List the system elements or sub-systems across the diagonal of the matrix.
2) Assess whether there is an interface between each pair of elements or sub-systems.
3) Complete the chart in two directions:

- From left-to-right and down indicates an input from the first element to the second
- From right-to-left and up indicates an output from the second element to the first.

9.5.8.1   Food System N-squared Diagrams

The food system teams N-squared diagrams are shown in Figures 9.58 through 9.65.

9.6   SUMMARY

The purpose of Phase 3, Detailed Design is to provide detailed design specifications for the system, its elements, and processes. Once the detailed design is complete, the team is ready to move into Phase 4, where system implementation, integration, testing, and verification is performed.

9.7   ACTIVE LEARNING EXERCISES

1) Build a physical architecture hierarchy model for your system
2) Select one of the three physical architecture model diagrams and draw a physical architecture model for your system:
   - Physical Block Diagram (PBD)
   - SysML Block Definition Diagram (BDD)
   - SysML Internal Block Diagram (IBD)
3) Generate a list of your system elements
4) For your system, develop a QFD – House of Quality 1 matrix

5) For your system, develop House of Quality matrices 2, 3, and 4
6) Perform a Lean Analysis including the VA and the Waste Analysis
7) Perform a trade-off analysis for your system
8) Perform a cost analysis for your system

BIBLIOGRAPHY

CMS ICD Template: https://www.cms.gov/search/cms?keys=ICD+template. Accessed 1/7/2021

Evans, J. R., and Lindsay, W. M. *Managing for Quality and Performance Excellence*, Eighth Edition, South-Western Cengage Learning, Mason, Ohio, 2011, 2008.

Jeff Grady's book *System Integration* (Grady 1994).

Longo, P. "Food justice and sustainability: a new revolution." *Agriculture and Agricultural Science Procedia* 8 (2016): 31–36.

SEBoK Editorial Board, *The Guide to the Systems Engineering Body of Knowledge (SEBoK)*, vol. 2.2, ed. R. J. Cloutier (Hoboken, NJ: Trustees of the Stevens Institute of Technology, 2020), accessed May 25, 2020, www.sebokwiki.org. BKCASE is managed and maintained by the Stevens Institute of Technology Systems Engineering Research Center, the International Council on Systems Engineering, and the Institute of Electrical and Electronics Engineers Computer Society.

Scholtes, P., Joiner, B., Streibel, B., *The Team Handbook*, 3rd Edition, Oriel Inc., Madision, WI, 2003.

10 Implementation, Integration, Test and Verification, System Verification, and Validation within Phases 4 and 5

In this chapter, we perform the implementation of the system elements, where the team realizes the system elements, meaning that the system elements are physically created or implemented. This does not, however, mean that the system elements or sub-systems are integrated together or even made operational, that is done in Phase 6, Operations and Maintenance. This chapter covers Phase 4, Implementation, Integration, Test, and Verification, and Phase 5, System Verification and Validation.

10.1 PURPOSE

In this chapter, we perform the test planning and development of test cases, running the test cases, and documentation of test results. Since the term *testing* has a lower level, more detailed definition, as if you test a physical piece of equipment, the terms *verification* and *validation* are used. Verification is verifying that the system meets the requirements, and validation is validating that the system achieves its intent from a user or customer perspective. In software development, these terms align to system and integration testing (verification) and user acceptance testing (validation). The verification activities test the internal and external workings of the systems and the validation tests the customer facing parts of the system.

10.2 ACTIVITIES

The activities performed and the tools applied in Phases 4 and 5 are shown in Table 10.1. We cover the "Perform system implementation" and "Perform system integration" activities and the associated tools in Phase 4. We cover the "Perform system verification" and "Perform system validation" activities and related tools in Phase 5. The system verification and validation model is shown in Figure 10.1.

10.3 PHASE 4: IMPLEMENTATION, INTEGRATION, TEST, AND VERIFICATION

10.3.1 PERFORM SYSTEM IMPLEMENTATION

The system implementation or realization activities are performed to create and test the system as specified by the system design. Remember from the Vee model graphics shown in Chapter 2, these activities are represented at the bottom and middle of the Vee model. Within the implementation activities, we start up the right side of the Vee model, as we build the system elements, integrate the elements into the sub-systems, and if appropriate, integrate the sub-systems together. There are verification and validation activities throughout these activities. According to the SEBoK (8.0), the system realization activities include:

DOI: 10.1201/9781003081258-12

TABLE 10.1

Activities and Tools in the Detailed Design Phase

Vee Phase	Activities	Tools	Principles
Phase 4: Implementation, Integration, Test, and Verification	1. Perform system implementation 2. Perform system integration	• Integration constraints • Implementation strategy • System elements supplied • Initial operator training • Verification criteria • Verification test cases and results • N-squared diagram	• System elements • Modularity • Interactions • Networks • Relationships • Behavior
Phase 5: System Verification and Validation	1. Perform system verification 2. Perform system validation	• Verification and Validation criteria • Verification and Validation test cases and results	• Synthesis

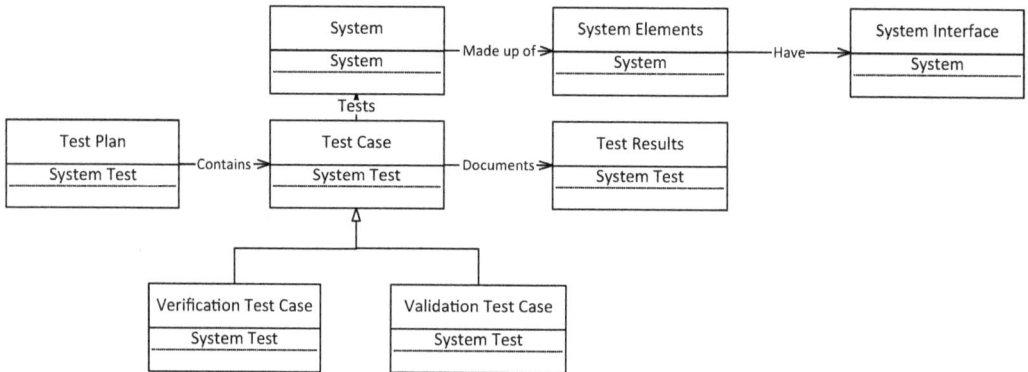

FIGURE 10.1   System verification and validation model.

- System implementation: Includes the activities that are required to build the system
- System integration: Includes the activities to integrate the system elements, sub-systems, and system.
- System verification: Includes those activities that ensures that the system aligns with the system requirements and defined architecture from Phase 2 – Requirements and Architecture.
- System validation: Includes those activities to ensure that the system meets the customers or stakeholders needs.

Figure 10.2 shows how the main artifacts from each phase trace and connect to each other. The system architecture provides the foundation infrastructure for the system. The customer requirements support the system architecture, while the system requirements are derived from the customer requirements. The system design is based upon the system requirements and the system elements are the physical instantiation of the system design. The system elements are integrated during the system implementation activities to form the sub-systems and system. The system is verified through the system integration activities and are validated by the user representatives in the system validation activities.

System Implementation:

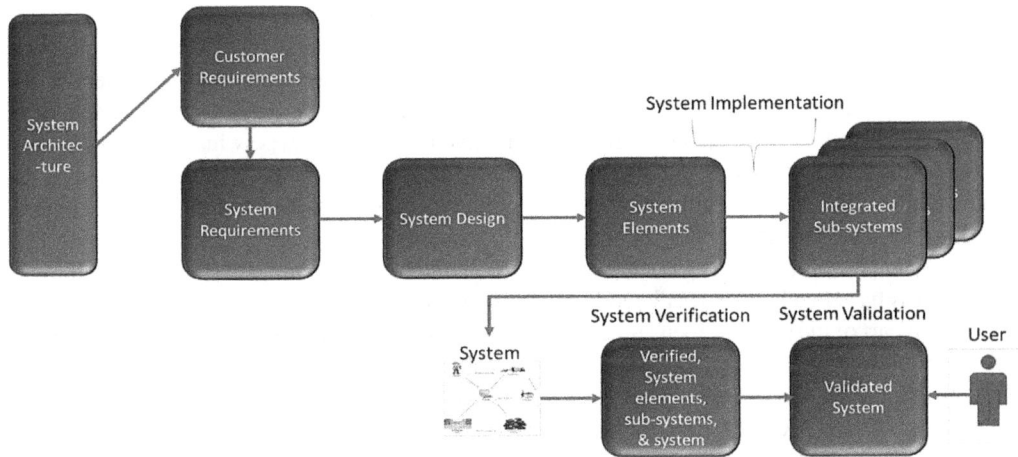

FIGURE 10.2 Traceability of artifacts by phase.

The system activities include the following:

- Define the implementation strategy.
- Realize the system: fabricate hardware elements; develop software; develop training and train initial operators; develop processes, procedures, work instructions, appropriate policies, and quality plans.
- Package, store, and provide the implemented system, once verified and validated.

The implementation artifacts include (SEBoK 8.0):

- An implemented system
- Implementation tools
- Implementation procedures
- An implementation plan or strategy
- Verification reports
- Issues and trouble reports
- Change requests if needed for design

10.3.1.1 Food System Implementation Strategies

Implementation Strategy for Team 1 Connecting Organizations

The implementation strategy is to finish or develop the website, firstly based on the website's charac-teristics or possible problems and to develop a new plan according to the development and changes of the situation. Next, the scale of implementation needs to be defined. It includes the main body of implementation, the sub-system in the website. The main body of implementation is making sure the elements operate or move in the right way. The elements are made up of four sub-systems. The form system is made up of the survey and food order form. Both electronic version and paper version are the methods of collecting the data. Therefore, they compose the initial website system, which is the basic information resource of building and running this website The database, which is used in daily running of the website, to place orders, process frequently asked questions, input survey data and provide advertising. These elements appear in the web page or run in the server. Therefore, software relating to the running of the mentioned elements, website server, and message board should also be included in this sub-system. The web admin and forum moderator, who maintain the daily operations of the website, have the main roles. The role sub-system implements the outside activity performed by the website. The community organizer, farmer, and volunteers participate and maintain the sales meeting. The community members are the main stakeholders and users of the website. The program sub-system consist of policies or documents that make up the framework of the website, as well as two kinds of software, statistical and spreadsheets, which are used in data analysis and prioritization. When the scale is ensured, all the information will be recorded in the documentation "The scale of implementation." If the scale needs to be modified, corresponding procedures need to refer to "Modifying work of implementation scale."

The second part of the strategy is estimating the potential risk in the implementation. All the risk in the process has already been predicted in the "risk assessment" part.

The third part of the strategy focuses on the outside activities, such as sales meetings, and picking up food on the weekend. Community organizer and volunteers should preserve the order and deploy the items in location according to the "Organizational behavior of activities specification."

The fourth part of the strategy is the testing strategy. When the running of the website is tested, all the behavior should be based on "Standard of testing."

The fifth part is the training strategy. Not only volunteers and other officers but also users need to be trained. The success of the website implementation depends on the degree of training. Training needs refer to the "Training strategy."

The sixth strategy is suitable for communication with Citywide and other organizations. When the offline activities need to be held, many organizations need to be communicated. These communications should be based on "Specification of communication."

The implementation strategy states the specification in every stage, analyzing the risk and potential problems in operation.

Implementation Strategy for Team 2 Creating Food Policy
Test rehearsals should be done before the system is executed. To make the system function well, rehearsals including selecting a champion, meetings, and gathering support should be performed to reduce time when the actual process is executed. The good way to implement the system is to test every process of the system. If any problem shows up during the rehearsal, it should be fixed through discussion and cooperation.

When electing the champion, the results should be clearly shown in front of all community members who attend the selection. The result should also be written as a note and saved as a document file and can be accessed through the internet. The security cameras are available and make sure that the selection is fair and recorded. Calendars are available online which include meetings' time and appointment. Once the draft policy idea is created, every member of the community should be able to access it online. Community members can give comments after they read the policy idea and if there is any problem, meetings could be set, and the policy idea can be fixed until everyone is satisfied and is confirmed by customers. The policy change idea is the most important process in the whole system, it needs data collection and support and detailed discussion, and the formal policy change idea should pass the law through the commission. Once the idea is rejected, more meetings are set to come up with the solution and fixed policy change idea. The policy change file is well protected and community members are welcome to go with the champion on the way to the commission center. Security cameras are available in the commission center and software like official website and scheduling program should function properly. Once one process is implemented, the next process should be ready to be implemented. But each process must be implemented well and correctly.

Implementation Strategy for Team 3 Kids Growing Food
The purpose of the implementation strategy tool is to determine the requirements needed to get the system up and running, including fabrication and coding procedures, tools and equipment to be used, means and criteria for auditing the configuration of resulting elements, to the detailed design documentation. The implementation strategy tool allowed the team to identify that the system would require only tools and equipment to get the system running.

A number of activities must occur to implement the microgreen growing system and for it to run as smoothly as possible. First, the teacher must confirm donation of a fair amount of supplies to be able to serve the number of interested participants. In order to determine the number of interested participants, the teacher can take a simple in-class survey or send one home with the students. The system cannot function without the supplies, therefore, anything that cannot be donated will have to be purchased by the teacher. Another implementation and system requirement is preparing the material, intellectual as well as physical, that will be presented in the microgreen training in a concise, effective manner. This includes creating colorful deliverables with directions and process steps that students can refer to when the time comes to grow microgreens themselves. The material also includes a fair, comprehensive test covering what the students learned in microgreen training and, more importantly, what they would require to successfully grow microgreens at home. Finally, for material preparation, a permission slip and Standard Quality Evaluation Chart need to be created. For quality control purposes and to ensure lack of bias, the Standard Quality Evaluation Chart should be looked at by an outside source, such as a professional in the field. The final system implementation requirement is acquiring an adequate amount of cash, depending on the interested participants, to front the first cycle of the microgreen program.

Implementation Strategy for Team 4 Food Storage
In our system, each stakeholder is fully responsible to do its part. They should open the inventory log at the temporary storage location and provide brief information about the produce. Each part of produce that is collected from farms will be examined separately, and once the produce is considered good then it is moved to the central storage location. The implementation strategy should have the requirements of the system, verification criteria, and validation criteria. The security, safety, and privacy should be considered in our central storage location when implementing the system.

Some key process elements implemented were:

- Hire professional workers
- Culling food collected
- Pack the food into bags
- Temporary storage location at farms
- Food collected and transported to central storage location
- Food is weighed, packed, and labeled
- Placed in refrigerator
- Distributed to markets

Implementation Strategy for Team 5 Food Hub
To implement the functional web-based system that is proposed, the process must start with seeking out investors and currently formed businesses that will utilize the system and its tools. After securing private business investment/interests, the help of a web design company would be necessary. This web design company would communicate with engineering and investors on what tools they think are needed in the formation of the website. The website, with Engineering's oversight, would proceed to be built. Upon completion, a trial run with a selected number of businesses, distributors, and farmers (i.e. the current working system in the Downtown Dayton Partnership) would occur. Utilizing information from the trial run, adjustments to the website will be made. The website will move forward into a general release for public use, with all investors having preferential access to the website. The website will begin implementation with occasional (every two weeks) check-ups

on website analytics and functionality. Barring no major issue, the system will remain in use with updates until retirement or website traffic is rare, in which point system retirement will be considered.

The implementation strategy is an explanation of how we intend to implement our system and what tools we need to use to do such. As explained above, our implementation strategy requires a lot of pre-work before release. Upon release the system needs little meddling to maintain.

Implementation Strategy for Team 6 GEM City Market

Implementing the planned recommended processes must begin with a strategy to effectively implement all planned solutions. This strategy must cover all the processes and system elements that are needed. To fabricate a strategy for Gem City Market, the following goals or requirements must be followed. Firstly, before our planned solutions can be implemented, the Gem City Market must be physically established and constructed. This includes parking and carts for customers. Then a highly qualified management team and workers must be employed at this point. The management that is employed will be responsible for the store organization, types of products to be stored, including proper signs, training the staff, and making sure the store is safe by taking security measures such as 24/7 camera surveillance. Different modes of payment must be made available to customers such as payment by credit cards or by cash. By this point, the various proposed systems can be implemented. Marketing will be conducted on the products and discounts will be made available to customers. Different methods will also be applied to make sure the product prices are feasible and affordable to the customers. These methods include Hybrid pricing which has been previously explained thoroughly in the report. Using sales statistics and customer surveys enable the management to "Improve according to demand" which has also been detailed previously. The last proposed method would be delivery and pickup which includes the usage of a website and a complimentary mobile application which will make shopping for customers easier. These proposed methods will give Gem City Market a competitive edge over other supermarkets while also providing healthy, attractive, and affordable end products for customers.

Implementation Strategy for Team 7 Table 33 Restaurant

System Implementation is a great way in showing how do elements interface with each other, where interfaces could be found in the N-squared Diagram. The Research System is responsible for analyzing the inventory after receiving a request from the Inventory Managements Systems. Inventory Managements Systems should first report a request to the manager and at the end the research team would search for local farms for a specific type of ingredient. The kitchen is where ingredients are being prepared using the chef's skills and then delivered to the customer by the server. Ingredients are being grown and harvested in local farms by farmers, who use organic practices.

10.3.2 Perform System Integration

The N-squared diagram, developed in the design phase, is used to derive the integration points and interfaces between system elements and sub-systems. The system integration includes taking delivery of the implemented system elements and performing the verification and validation activities while assembling. The goal of the system integration is to ensure that the system elements function properly together and that they satisfy the system design. The N-squared diagram is used to develop a list of integration constraints that should be tested.

In the performance of the system integration activities, we are moving up the right side of the Vee model, by integrating the system as a whole and verifying and validating the entire working system.

The system integration activities are (SEBoK):

- Develop the integration plan
- Determine how the integration will occur
- Take delivery of each implemented element
- Assemble the implemented elements

- Perform the verification testing
- Document the integration test results, issue report, and change requests as appropriate

The integration artifacts include (SEBoK):

- an integrated system
- assembly tools
- assembly procedures
- integration plans
- N-square diagram
- integration verification results
- issue or trouble reports
- change requests for design

The N-squared diagram for the Women's Healthcare Center is shown if Figure 10.3.

10.3.2.1 Food System Integration Constraints

Following are the food system integration constraints for each team.

Integration Constraints for Team 1 Connecting Organizations

Integration means each part of the website can work organically and harmoniously with each other to achieve overall benefits and achieve overall optimization. When the elements integrate to the system, restrictions of time, technology, and management should be considered.

Interface N-squared Diagram

Schedule Service	X	X								
	Authorize Service	X	X							
X		Process self-referral								
	X		Register Patient	X	X					
			X	Process VIP patient						
					Perform service	X	X	X	X	
					X	Connect to cancer center				
							Perform surgery			
					X	X	X	Provide results		X
					X				Provide spiritual care	
									X	Request records / film

FIGURE 10.3 N-squared diagram for women's healthcare center.

Integration Constraints:

Time Constraints: Many durations of processes are restricted. For example, the process of data transfer to customers' requirements and orders should not be slow. Otherwise, customers will lose interest in the website. Due to the timeliness of digital information, time is the main factor of restriction in the integration. A schedule which rules completion of every sub-system is needed. If one sub-system delays, the relating solutions and compromising should be prepared.

Technology Constraints: Technology constraints appear in the integration of program and website. The program needs the technology to transfer to appear in the website. For instance, the quantity of placed orders in the program may be huge. However, the processing capacity of the website is limited. Technology constraints can be decreased by hiring professional programmers or updating the server.

Management Constraints: Resulting from the different knowledge levels and educational backgrounds, management constraints should be noticed when the role system integrates with the website and the program. During the operation of the software, the differences become obvious. If the management constraints cannot be solved, the integrations will be tough. The distance between advanced and backward workers will increase. The whole integration will fail.

Integration Constraints for Team 2 Creating Food Policy
Integration Constraints

There are many constraints that exist in the system integration. Because the policy change system requires all the steps to be executed well, each element is playing an important role in the whole system. The goal of system integration is to ensure that each system element functions well as a whole and satisfies the requirements of the system. For example, if the system can't define the problem, then the system can't make any policy change because the problem is missing. And if the champion can't be selected, there are no people who can speak for the whole community members and cannot bring the draft policy change idea to the commission. The policy change system is like a bridge, if any part of the bridge is broken, then people can't go through that bridge. Finally, if the draft policy change idea is rejected, that will make the whole system go back to the beginning, the rejection makes the policy change idea a failure to satisfy the requirements of the system and it needs to be fixed through the discussion among community members and customers. So, the goal is to make sure that every step of the system needs to be carefully discussed and executed.

Integration Constraints for Team 3 Kids Growing Food
The integration constraints tool prompted the team to assess the N-Squared Diagram for not only functions that would influence one another, but how they would influence one another. The team identified two sets of functions.

Functions that the team foresees having difficulty coming together to provide a smooth overall process consists of, first, microgreen growing and transporting the microgreens back to school. The fully-grown microgreens are transported back to school by the students. Microgreen trays are bulky by themselves, with 8 inches by 12 inches by 2 inches dimensions consuming a total volume of 192 square inches. The average high school student gets back and forth to school using the bus, walking, or with their own vehicle. It is safe to assume that the majority of at-risk high school students are taking the bus or walking. This presents a risk of students dropping their fully-grown microgreen trays and, thereby, losing their product to earn money with and, possibly, patience with the program. Another potential integration difficulty with these functions is the students remembering the fully-grown microgreens to transport back to school. It is a time sensitive process, therefore, students neglecting to bring in the microgreens on time can have detrimental effects on the program. Since microgreens are not a normal school supply transported daily, it may be difficult for students to remember to take them, especially when they already have a number of other things to deal with. Additional functions that may run into integration difficulties are evaluating the microgreens using the Standard Quality Evaluation Chart and compensating the student. Features of microgreens of different qualities can still be very similar and different people may view the same microgreen crop in different ways, such as the teacher and the student. The student, for obvious reasons, may feel as though their microgreens match the higher standard quality and they deserve more compensation. As a result, flow between the two functions may be difficult and the process may have to be reversed and re-evaluated in some situations.

Integration Constraints for Team 4 Food Storage

The purpose of the integration constraints tool is to identify which system functions will hinder the system working properly as a whole or lack satisfying the design properties or characteristics of the system when they are integrated together. As a result, special care can be taken when developing the integration plan and performing the integration to allow these functions to come together. Functions have difficulty coming together to provide a smooth process consisting of, first, temporary storage and transporting the food from farm to the Central storage location. The stored food at farms is transported to central storage by Community food collectors. The central storage location should be very well monitored, and the food should be undamaged. Further it should be packed in a suitable packaging and delivered to the market location or redline region. There are some failures of the system that could happen between the interfaces. The interfaces are shown in the N-Squared Diagram which is mentioned previously.

Integration Constraints for Team 5 Food Hub Website

Concerning integration constraints there was concern when it came to integrating tracking with distributors trucks and personnel. The tracking system would allow users, namely business owners, to track their produce deliveries from farm to store. The concerns involved halting delivery processes to install proper tracking equipment and conflict of interest with the distributor. If the distributor services more than the businesses within the network, this could be a conflict of interest. The distributor may deliver to competing areas or businesses. For this reason, limited visibility on tracking should be considered in that tracking data is only made available to the business that ordered the produce and display "busy" to other businesses seeking to use that particular distributor. The integration constraints are set forth to address concerns in integrating certain system elements, namely, will the implementation of certain elements affect the ability for another to operate? For our system, much of the process is pretty seamless; however, we had concerns with the tracking system we mentioned in the design elements and thought it necessary to address.

Integration Constraints for Team 6 GEM City Market

In almost all systems, there are elements of the system that integrate with one another, such is the case for our project as well. The goal of system integration is to ensure that individual system elements function as a whole satisfying the characteristics of the system. The N-Squared diagram tool enables us to view the systems or elements that interface with each other for purposes of analysis. When certain elements interface with one another, there is a chance that some negative aspects or problems may occur. Such problems are data lost when they are moved from one server to another. The Integration constraints discuss these potential problems for our project using the N-Squared diagram to determine interfaces. As Improving according to demand and Hybrid pricing inputs into Marketing, the negative constraints of this would be that some products would be advertised and promoted less. Hybrid pricing would also affect the Store organization in such a way that might negatively place certain products in aisles less traveled to by customers. However, the benefits of effectively implementing Hybrid pricing, which wouldimprove with increased demand would offset any negative aspects that may occur.

Connection	Problem	Solution
IMS & Research Team	Failure in reporting all missing ingredients	Make sure to work carefully using excel to report all missing ingredients
Manager & Research Team	Manager doesn't supply research team with enough equipment or proper training	Make sure to provide all analyzing and testing equipment and proper training
Local Farms & Manager	Manager unable to find good local farmers	Search for local farmers continuously
Farmers & Local Farms	Local farmers use non-organic growing practices	Look at farmers certificate in growing and harvesting ingredients before purchasing
Ingredients & Research Team	Research team is not well trained to verify ingredients quality	Provide proper training to make sure that people who test ingredients are experienced

FIGURE 10.4   Integration constraints for team 7 table 33 restaurant.

Integration Constraints for Team 7 Table 33 Restaurant

Integration constraints can be very useful in trying to avoid future conflicts. By studying this connection from the N-square diagram, we were able to create Figure 10.4 which is the constraints or problems that might occur in the future and ways to avoid them.

10.4   PHASE 5: SYSTEM VERIFICATION AND VALIDATION

10.4.1   PERFORM SYSTEM VERIFICATION

System verification:

System verification activities include (SEBoK):

- Develop verification test plans and cases
- Schedule verification activities
- Setup test environment
- Execute verification test cases against the requirements, architecture, and design
- Document the test results and reports

System verification artifacts include (SEBoK):

- verification test plans
- verification test cases
- verification procedures
- verification results reports
- verification tools
- list of verified elements
- issues and trouble reports
- change requests to the design

If detailed use cases were developed in the requirements phase, these can be used to develop the verification test cases. This helps to ensure that the test cases are traceable to the requirements and that the requirements are verified. An example test case is shown in Figure 10.5 from the Women's Healthcare Center.

TASK #	TASK/STEPS	EXPECTED RESULT	PASS/FAIL	ACTUAL RESULT/DESCRIPTION	NOTES
Scenario 1: Schedule appointment					
1	Patient calls Central Scheduling and requests an imaging appointment	Central Scheduling answers phone and greets patient			
2	Ask for name, social security number from patient	Obtain patient information			
3	If Medical information file exists, verify demographic and update if necessary; if doesn't exist, create it	Information verified or entered			
4	Ask patient which services would like to schedule for, could be multiple appointments	Patient provides list of services			
5	Ask patient if they have a preferred date and time of day	Patient provides preferred dates and times			
6	Review central schedule database for preferred date and time	Obtain appointment times			
7	Provide next available date and time	Obtain appointment times			
8	Confirm with patient, if not repeat steps 6 through 7	Appointment time confirmed (or start at task 6)			
9	Book appointments	Book appointment in system			
10	Provide prep information for patient	Provide prep information for patient			
11	Ask questions on test, procedure, insurance, other information	Patient asks questions			
12	Reconfirm date, time, remind patient to bring photo id, copay/deductible and script and provide any other instructions	Reconfirm appointment			

FIGURE 10.5   Example system verification test case for the women's healthcare system.

TABLE 10.2
System Verification List of Test Cases for Team 1 Connecting Organizations

Test Name	Test Case Number	Test Case Description
Receive form	Receive form 004-1	Collect data to better understand food needs of the customer
Analyze and prioritize data	Analyze and prioritize data 004-2	Evaluate survey information to prioritize food needs and challenges
Distribute data to farmer	Distribute data to farmer 004-3	Provide farmer with what foods are ordered and requested foods by customers
Obtain farmer feedback	obtain farmer feedback 004-4	Determine food availability and whether additional resources are needed
Advertise foods/events	advertise food/events 004-5	Provide information on website of available foods, contact information, and hosted community events

TABLE 10.3
System Verification List of Test Cases for Team 2 Creating Food Policy

Test Name	Test Case Number	Test Case Description
Problem Identification	001 Problem Identification	The process for identifying a community problem that needs a new or changed policy and regulation
Elect a champion	002 Elect a champion	The process for electing a champion for the problem
Gather supporting data	003 Gather supporting data	To identify and collect supporting data for the policy and/or regulation
Present Policy change Idea to Commission	004 Present policy change idea to commission	The process for presenting the policy change idea to the commission
Commission Discussion	005 Commission Discussion	Commission discusses the potential policy and/or regulation change
Pass legislation	006 Pass legislation	The process for passing legislation

10.4.1.1   Food System – System Verification List of Test Cases

The system verification list of test cases for each team are shown in Tables 10.2 through 10.8

10.4.1.2   Food System – System Verification Sample Test Cases

The sample system verification test cases for each team are shown in Tables 10.9 through 10.11 and Figures 10.6 through 10.9.

10.4.2   Perform System Validation

The system validation is also called operation or user acceptance testing. The system validation testing validates that the system is what the customer wants and that it meets the customer requirements. The activities can be planned and executed throughout the Vee model life cycle. Some of the elements that are tested in the validation phase include:

TABLE 10.4

System Verification List of Test Cases for Team 3 Kids Growing Food

Test Name	Test Case Number	Test Case Description
Food and growing instruction	001 Food and growing instruction	The process for developing food and growing instructions
Food production	002 Food production	Producing or growing the food
Food processing	003 Food processing	Cleaning and packaging the food
Food distribution	004 Food distribution	Distributing the food for sale

TABLE 10.5

System Verification List of Test Cases for Team 4 Food Storage

Test Name	Test Case Number	Test Case Description
Temporary storage at farm until collection	001 Temporary storage of food	Storing the food temporarily on the farm
Collection of food for storage	002 Food collection	Collection of food from farm to central storage location
Food storage	003 Food storage	Food storage at central location
Food distribution	004 Food distribution	Food distribution by bikes

TABLE 10.6

System Verification List of Test Cases for Team 5 Food Hub

Test Name	Test Case Number	Test Case Description
Harvest crops/produce	001 Harvest crops/produce	Harvest crops or produce on farm
Market produce	002 Market produce	Market produce availability across farms
Real time costs	003 Real time costs	Develop real time costs
Forecast food availability	004 Forecast food availability	Forecast food availability for all farms
Communicate availability	005 Communicate availability	Communicate food availability to the market
Real time management	006 Real time management	Perform time management in real time

- Documents: Such as Process Architecture Maps, design specifications, drawings, etc.
- Customer and system requirements
- Design artifacts
- System elements
- Processes, functions, and activities

There are several different techniques that can be applied for verification and validation test purposes, including (Sugarman, 2016):

Inspection: A visual review of the system element to ensure compliance with requirements.
Analysis: Use of modeling and simulation analysis to ensure compliance with requirements.

TABLE 10.7
System Verification List of Test Cases for Team 6 GEM City Market

Test Name	Test Case Number	Test Case Description
Improving according to demand	001 Improving according to demand	Improve products in store based upon customer feedback
Hybrid pricing	002 Hybrid pricing	Implement hybrid pricing
Make food products for customer	003 Make food products for customers	Make food products for customers in the market
Delivery and pickup service	004 Delivery and pickup service	Implement delivery and pickup service for customers

TABLE 10.8
System Verification List of Test Cases for Team 7 Table 33 Restaurant

Test Name	Test Case Number	Test Case Description
Report missing ingredients	001 Report missing ingredients	Report missing ingredients for purchase
Find local farms	002 Find local farms	Find local farms that want to sell products
Ingredient transportation	003 Ingredient transportation	Transport ingredients from farm to market
Move ingredients from inventory to kitchen	004 Move ingredients from inventory to kitchen	Process for moving ingredients from inventory to kitchen

TABLE 10.9
System Verification Sample Test Cases for Team 1 Connecting Organizations

Task #	Task/Steps	Expected Result	Pass/ Fail	Actual Result/ Description	Notes
		Scenario 1: Obtain Form			
1	Create Newsletters	Newsletter to appeal to customers			
2	Distribute Newsletters	Letters sent to homes			
3	Contact people going home to home	Successful sign ups going door to door			
4	Discuss purpose of form	Customer understanding of form and program			

Demonstration: Observing the use of the system
Test: Explicit measure of performance data under controlled conditions.
The system validation activities are (SEBoK):

- Develop a validation strategy
- Execute validation test cases
- Analyze validation results to ensure compliance to customer requirements
- Document test case results

TABLE 10.10
System Verification Sample Test Cases for Team 2 Creating Food Policy

Task 1	Task/Steps	Expected Result	Pass/ Fail	Actual Result/ Description	Notes
		Scenario 1: Problem Identification			
1	Check if community members face any problem	Community Members face problem			
2	Categorize the challenges faced	Challenges are classified			
3	Community Members identify a problem	Problem is identified			

TABLE 10.11
System Verification Sample Test Cases for Team 3 Kids Growing Food

Test Cases

Task #	Task/Steps	Expected Result	Pass/ Fail	Actual Result/ Description	Notes
		Scenario 1: Food and Growing Instructions			
1	Microgreen supplies (tray, medium, seeds) are delivered to school by sponsor donor.	Microgreen trays, medium, and seed packet counts are taken to classroom and verified for predetermined number of participants.			
2	Teacher spends week demonstrating proper steps and care for growing microgreens (amount of student involvement dependent on supply amount and interest).	Students follow teacher's demonstration by growing their own microgreens in groups, thereby, gaining hands on experience in addition to detailed microgreen growing knowledge.			
3	Teacher distributes test that students must pass with a minimum score to ensure understanding and retention of training knowledge.	The majority of students pass the test verifying adequate understanding of microgreen growing to participate in the program.			
4	Teacher seeks and achieves permission from student's guardian.	Teacher receives signed permission slips back from students.			

System Validation Artifacts:

- Validation strategy
- Validation plan
- Validation procedures
- Validation test results reports

TASK	TASK/STEPS	EXPECTED RESULT	PASS/FAIL	ACTUAL	NOTES
	Scenario 2: Collection of food from farm to central storage location				
1	Farmers stored food at temporary storage after culling, and they prepare to make the food ready to go	the food is in good condition and is packaged and prepared to be			
2	Farmers call HomeFull community staff (collectors) and request a time to gather their food	HomeFull community staff respond farmers call and approve			
3	HomeFull collectors manage time with other schedules and provide the confirmation to the farmers on pick up	the farmer receive the confirmation regarding to pick up time.			
4	Collectors check the packaged food to make sure it is safe to consume by their authorization.	the packaged food is checked and safe to consume			
5	HomeFull (collectors) with the help of farm workers move the sorted food into the truck.	the packaged food is organized and stored inside the truck			
6	After a neat arrangement of food, the truck is started and moved to a second pick up location	the truck and collectors go on time to second location.			
7	The truck goes to the second farm to collect the food.	the truck and collectors arrived on time at second location			
8	Collectors check the packaged food to make sure it is safe to consume by their authorization.	the packaged food is checked and safe to consume			
9	HomeFull (collectors) with the help of farm workers move the sorted food into the truck.	the packaged food is organized and stored inside the truck			
10	Depending on the numbers of pick up locations, after all, pick up, it moves to the central storage location	the collectors are done with whole requests and go to Central			
11	The truck arrives at the central storage location parking.	the truck arrived and parked in suitable location at central			
12	The food is checked if there is any damage or loss and record the inventory	the packaged food is in good condition and it is fresh, and the			
13	The food moved from truck to central storage location	the packaged food is stored in suitable place in central storage.			

FIGURE 10.6 System verification sample test cases for team 4 food storage.

TASK #	TASK/STEPS	EXPECTED RESULT	PASS/FAIL	ACTUAL RESULT/DESCRIPTION	NOTES
		Scenario 1: Harvest Crops/Produce			
1	Business Owner uses website to request produce that is marketed as "available" or "coming soon"	Request for produce			
2	Farmer confirms produce is available and provides an estimated harvest time	Expected Delivery Date			
3	Business Owner contacts a Distributor to organize pick-up and delivery of produce from farmer	Pick-up/Delivery Date			
4	Distributor coordinates with Farmer proper pick-up procedures	Successful Pick-up			
5	Distributor delivers the produce to the appropriate business	Successful Delivery			
6	Business Owners place produce out for sale	Sell Produce			

FIGURE 10.7   System Verification Sample Test Cases for Team 5 Food Hub

TASK #	TASK/STEPS	EXPECTED RESULT	PASS/FAIL	ACTUAL RESULT/DESCRIPTION	NOTES
		Scenario 1: Improving according to demand			
1	Customers Check The available products	Customers finds a wide range of products			
2	Customers gives their feedback to the store in store	Customer fills a feedback form after checkout			
3	Customers gives their feedback to the store online	Customer fills a feedback Online			
4	Management checks the Feedbacks	Feedbacks being checked by the management team			
5	Management improves the store after checking the majority of the feedbacks	Improving the store according to the feedbacks			

FIGURE 10.8   System verification sample test cases s for team 6 gem city market.

Test Cases				
TASK #	TASK/STEPS	EXPECTED RESULT	PASS/FAIL	ACTUAL RESULT/DESCRIPTION
		Scenario 1: Report missing ingredients		
1	Check the restaurant inventory	Manager knows the quantity of ingredients in inventory		
2	Write a report of missing ingredients	The exact quantity needed of missing ingredients		
3	Send the report to manager	Prepare the budget for missing ingredients		

FIGURE 10.9   System verification sample test cases for team 7 table 33 restaurant.

TABLE 10.12
List of System Verification and Validation Test Case for Women's Healthcare Center

Test Case Names	Test case order
Scheduling and Registration	1.0
Schedule Service	1.1
Authorize Service	1.2
Process Self-Referral	1.3
Register Patient	1.4
Process VIP patient	1.5
Provide Spiritual Care	3.0
Request records/films	4.0
Services	2.0
Perform Service	2.1
Perform Surgery	2.2
Connect to Cancer Center	2.3
Provide Results	2.4

TABLE 10.13
List of Validation Test Cases for Team 1 Connecting Organizations

Number	Name	Description
1	Obtain Form	Customer understanding of form and program
2	Complete Form	Completed order form
3	Submit/Return Form	Documents are stored and maintained safely
4	Kick off website	Community members informed about expanded food program
5	Meet the Farmer	Produce or desired goods from the Farmer available to the customer
6	Pick up and Pay for Ordered Food	Customer leaves with correct order
7	Receive Form	Data ready to be analyzed
8	Analyze and Prioritize Data	Data is analyzed and prioritized correctly
9	Distribute Data to Farmers	Farmers receive to info
10	Obtain Farmer Feedback	Farmer is in website database
11	Advertisement	Information digitally delivered to people

- Validation tools
- List of validated elements
- Issue, non-conformance, and trouble reports
- Change requests on requirements, products, services, and enterprises, as appropriate

TABLE 10.14

List of Validation Test Cases for Team 2 Creating Food Policy

Test Name	Test Case Number	Test Case Description
Problem Identification	001 Problem Identification	The process for identifying a community problem that needs a new or changed policy and regulation
Elect a champion	002 Elect a champion	The process for electing a champion for the problem
Gather supporting data	003 Gather supporting data	To identify and collect supporting data for the policy and/or regulation
Present Policy change Idea to Commission	004 Present policy change idea to commission	The process for presenting the policy change idea to the commission
Commission Discussion	005 Commission Discussion	Commission discusses the potential policy and/or regulation change
Pass legislation	006 Pass legislation	The process for passing legislation

TABLE 10.15

List of Validation Test Cases for Team 3 Kids Growing Food

Scenario	Title	Pertinent Step Number(s)
Scenario 1	Food and Growing Instruction	1–4
Scenario 2	Food Production	1
Scenario 3	Food Processing	1,2
Scenario 4	Food Distribution	1–2

An example of a list of validation test cases is shown in Table 10.12 for the Women's Healthcare Center.

10.4.2.1   Food System – System Validation List of Test Cases

The list of system validation test cases for each team is shown in Tables 10.13 through 10.17 and Figure 10.10 and Table 10.18.

10.4.2.2   Food System Validation Sample Test Cases

The food system validation sample test cases are shown in Figure 10.11 and Tables 10.19 through 10.21 and Figure 10.12 through 10.14.

For service systems, many of the test cases will be the same for both verification and validation since the system consists mainly of processes for providing services and/or software, although verification test cases can be focused also on moving data between modules, applications, databases, and sub-systems.

TABLE 10.16
List of Validation Test Cases for Team 4 Food Storage

Test Cases	Step Number #
Scenario 1: Temporary storage at farm until collection	None
Scenario 2: Collection of food from farm to central storage location	6,10
Scenario 3: Food storage at Central Storage Location	5,6,8,10
Scenario 4: Food distribution by electric bikes	1–8

TABLE 10.17
List of Validation Test Cases for Team 5 Food Hub

Test Case Names	Test Case Order
Food Production	1
Website Order Placed for Produce	1.1
Produce Harvested	1.2
Distributor Contacted	1.3
Food Distribution	2
Food Picked-up	2.1
Food Delivered	2.2
Food Sales	3
Food Inspected	3.1
Food Sold	3.3
Produce Available	3.2

TASK #	TASK/STEPS	EXPECTED RESULT	PASS/FAIL	ACTUAL	NOTES
1	Store Organization	Customers finds an organized easy to navigate store			
2	Availability of carts and parking	All the customers finds parking spaces and carts			
3	in store shopping experience	Workers help the customers to find what they need			
4	Checkout process	Customer finds all payment moods to choose from			
5	Delivery and Pickup application	Easy to use and clear			
6	Delivery Time	The workers delivers the products on time			
7	Pickup Time	the worker makes sure that everything is correct and its ready for pickup			
8	Improve according to demand	The management team makes sure the store is improving to the satisfy the customers			

FIGURE 10.10 List of validation test cases for team 6 gem city market.

TABLE 10.18
List of Validation Test Cases for Team 7 Table 33 Restaurant

Test Name	Test Case Number	Test Case Description
Report missing ingredients	001 Report missing ingredients	Report missing ingredients for purchase
Find local farms	002 Find local farms	Find local farms that want to sell products
Ingredient transportation	003 Ingredient transportation	Transport ingredients from farm to market
Move ingredients from inventory to kitchen	004 Move ingredients from inventory to kitchen	Process for moving ingredients from inventory to kitchen
Meals preparation	005 Meals preparation	Prepare meals with locally sourced ingredients

TABLE 10.19
Validation Test Cases for Team 2 Creating Food Policy

Task #	Task/Steps	Expected Result	Pass/Fail	Actual Result	Notes
		Scenario 3: Gather Support/Data			
1	Data collection sources	Ensure the data collection process includes multiple ways in order to ensure the authenticity of the data presented			
2	Awareness in Community	Create awareness in the community, so that the right participation can be ensure during any surveys conducted to gather data			
3	Community support	Be sure to collect the enough support from Community			
4	Facts based data	At least some part of the data collected should be based on facts			

TABLE 10.20
Validation Test Cases for Team 3 Kids Growing Food

Scenario	Title	Pertinent Step Number(s)
Scenario 1	Food and Growing Instruction	1–4
Scenario 2	Food Production	1
Scenario 3	Food Processing	1,2
Scenario 4	Food Distribution	1–2

TABLE 10.21
Validation Test Cases for Team 4 Food Storage

Test cases	Step Number #
Scenario 1: Temporary storage at farm until collection	None
Scenario 2: Collection of food from farm to central storage location	6,10
Scenario 3: Food storage at Central Storage Location	5,6,8,10
Scenario 4: Food distribution by electric bikes	1–8

TASK #	TASK/STEPS	EXPECTED RESULT	PASS/FAIL	ACTUAL RESULT/DESCRIPTION	NOTES
		Scenario 2: Complete Form			
1	Ask for the questions	Feedback for each question			
2	Write them down	Completed Survey			
3	Ask for conditional question/suggestion	Community concerned			
4	Interest in placing order	Answer of yes or no			
5	Customer fill out the order form	Completed order form			
6	Pitch info. for website order form	Increased awareness			
7	Tell when and where stand/ food pick up point	phone call reminder			

FIGURE 10.11 Validation test cases for team 1 connecting organizations.

TASK #	TASK/STEPS	EXPECTED RESULT	PASS/FAIL	ACTUAL	NOTES
1	Business Owner seeks produce on website	Finds produce			
2	Locates produce and places order	Order is placed			
3	Order notification is delivered to farmer	Farmer is notified			
4	Farmer harvests/gathers produce	Product is readied			
5	Distributor is contacted	Distributor prepares for order			
6	Produces is readied at pick-up location	Produce is picked up			

FIGURE 10.12 Validation test cases for team 5 food hub.

TASK #	TASK/STEPS	EXPECTED RESULT	PASS/FAIL	ACTUAL	NOTES
1	Store Organization	Customers finds an organized easy to navigate store			
2	Availability of carts and parking	All the customers finds parking spaces and carts			
3	in store shopping experience	Workers help the customers to find what they need			
4	Checkout process	Customer finds all payment moods to choose from			
5	Delivery and Pickup application	Easy to use and clear			
6	Delivery Time	The workers delivers the products on time			
7	Pickup Time	the worker makes sure that everything is correct and its ready for pickup			
8	Improve according to demand	The management team makes sure the store is improving to the satisfy the customers			

FIGURE 10.13 Validation test cases for team 6 gem city market.

TASK #	TASK/STEPS	EXPECTED RESULT	PASS/FAIL	ACTUAL RESULT/DESCRIPTION
		Scenario 5: Meals preparation		
1	Hand menu to customers	Customer chooses between the meals in the menu		
2	Choose meals	Customer ready to order		
3	Take order	Take order to chief		
4	Prepare Ingredients	Processing speed		
5	Wash Ingredients	Clean and healthy ingredients		
6	Cook meals	Customer satisfaction		
7	Take meals to customers	Customer satisfaction, money, and order completion		

FIGURE 10.14 Validation test cases for team 7 table 33 restaurant.

10.5   SUMMARY

In this chapter, we discussed the verification and validation of the systems, which consists of Phases 4 and 5 of the Vee model. Although the testing activities can be planned and executed throughout the Vee Life Cycle. The verification and validation phases are critical to ensure that the system is meeting the requirements and design through verification and the customer requirements in the validation phase. We'll now move into the last phase of the Vee model, Phase 6 Operations and Maintenance.

10.6   ACTIVE LEARNING EXERCISES

1) Develop a verification and validation test plan
2) Develop a list of verification test cases
3) Develop a verification test case
4) Develop a list of validation test cases
5) Develop a validation test case that is different from the verification test case above
6) Develop a list of integration constraints

BIBLIOGRAPHY

SEBoK Editorial Board, *The Guide to the Systems Engineering Body of Knowledge (SEBoK)*, vol. 2.2, ed. R. J. Cloutier (Hoboken, NJ: Trustees of the Stevens Institute of Technology, 2020), accessed May 25, 2020, www.sebokwiki.org. BKCASE is managed and maintained by the Stevens Institute of Technology Systems Engineering Research Center, the International Council on Systems Engineering, and the Institute of Electrical and Electronics Engineers Computer Society.

Sugarman, R., *Management of Engineering Systems Class Notes*, University of Dayton, Dayton, Ohio, 2016.

11 Operation and Maintenance Phase 6

In this chapter, we discuss placing the system into operation, maintaining, and eventually dispose the system. The phase is called the Operation and Maintenance phase. It is the last phase in the Vee Life Cycle model.

11.1 PURPOSE

In this chapter, we perform activities related to placing the system into operation with its users completing the verification and validation activities from Phases 4 and 5. There are many tools that facilitate the operation and maintenance activities, including training plans and materials, certification plans and materials, operation manuals, performance reports, maintenance and service plans, Failure Mode and Effect Analysis (FMEA) for assessing maintenance risks, and Disposal and Retirement Plans.

11.2 ACTIVITIES

The activities performed and the tools applied in the Operation and Maintenance phase are shown in Table 11.1. We cover the "Deploy System," "Perform training," "Perform certification," "Perform risk assessment and planning for maintenance," and Perform disposal and retirement" activities, and the associated Operation and Maintenance tools. The Operations and Maintenance model is shown in Figure 11.1.

11.2.1 Deploy System

The system is deployed, or made operational, as the first step of this phase. The systems engineer should ensure that the system is operationally acceptable. The responsibility and accountability, from the perspective of the effective, efficient, and safe operation of the system is now on the owners. Once the system is deployed and installed, including the system elements, sub-system, system, processes, procedures, policies, etc., it should be integrated and tested to ensure that the system works properly as installed or deployed. There should be formal hand-off processes between the systems engineering team to logistics, operations, maintenance, and support (SEBoK 8.0).

The activities for operationalizing the system are (SEBoK 8.0):

- Develop a deployment or transition strategy, including installation, checkout, integration, and testing. This strategy should also include development of operational criteria, responsibility, and procedures for system installation.
- Perform system installation, including connecting interfaces and integration to other systems such as electrical, computer, security, and software systems. Installation planning should consider the factors impacted in the environment, human factors, and safety procedures.
- On-site verification and validation tests may need to be performed to ensure that the system is operational.

DOI: 10.1201/9781003081258-13

TABLE 11.1

Activities and Tools in the Operation and Maintenance Phase

Vee Phase	Activities	Tools	Principles
Phase 6: Operation and Maintenance	1. Deploy System 2. Perform training 3. Perform certification 4. Perform risk assessment 5. Perform improvement and maintenance planning 6. Perform disposal and retirement activities	• Training plan and materials • Certification plan and materials • Operations Manuals • Performance reports • Improvement and Maintenance plans; FMEA • Disposal and retirement plan	• Control behavior and feedback • Encapsulation (hide internal workings of system) • Stability and change

- Systems performance and related data should be collected to ensure:
  - Operational performance
  - System effectiveness
  - System efficiency
  - Failure reporting and trending
  - Training effectiveness and need for training and development for operational and support users.

11.2.2 PERFORM TRAINING AND PERFORM CERTIFICATION

A training plan identifies the resources, activities, and materials that are developed and deployed to ensure that everyone is trained properly to operate the system. The training plan for the Women's Healthcare Center is shown in Figure 11.2. The fields on the training plan are:

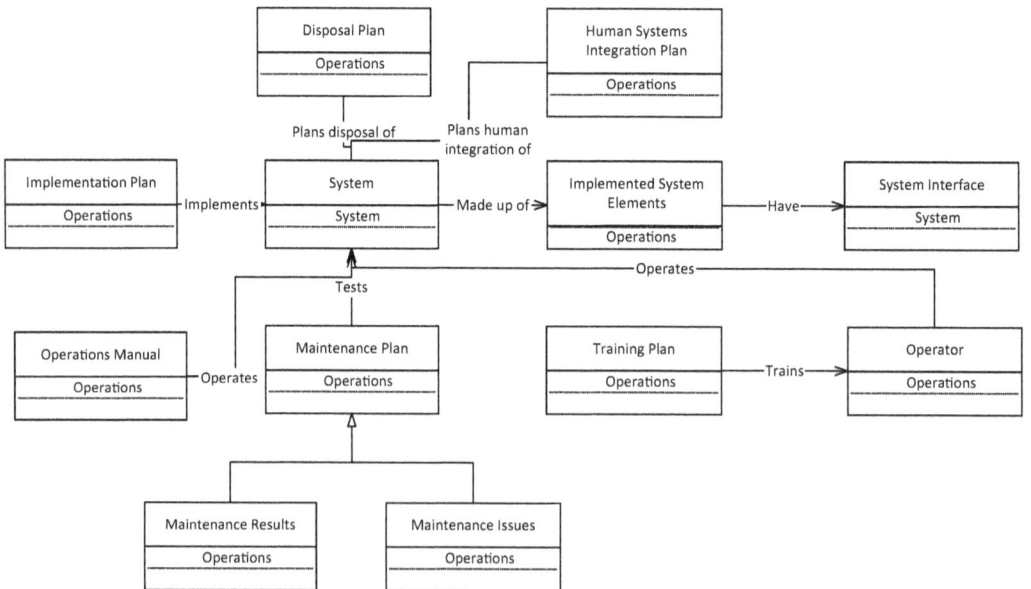

FIGURE 11.1 System operations model.

Training Plan	
Develop a training plan of how the process changes will be rolled out to the stakeholders.	
Training Objectives and Outcomes:	Train on process improvement changes
Training Audience and Resources:	ED Medical Staff
Instructional Strategies	
Exercise and case studies:	None
Assessments and exams:	Director Assessment
Homework assignments:	None
Team and individual work:	Individual
Project work:	None
Presentations:	
Training mode:	*Live Classroom *Instructor Led *Train the Trainer *Other - Weekly update poster
Training schedule and syllabus:	Just-in-time training
Training location:	ED Conference room

FIGURE 11.2   Training plan.

- Training objectives and outcomes: The purpose of the training and the expected outcomes for the training participants.
- Training audience and resources: The proposed audience to be trained and any resources required whether they are subject matter experts, training equipment, software, etc.
- Instructional strategies: Strategies for how the training are provided includes;
  - Exercise and case studies: whether active learning or other type of learning exercises that mimic use of the system or case studies that provide a similar learning experience related to the training objectives will be used.
  - Assessments and exams: a definition of how the learning will be assessed, through exams or other assessment methods, such as case studies, project-based learning, demonstration on use of the system, certification, etc.
  - Homework assignments: whether outside of class work will be assigned to enhance learning.
  - Project work: whether the participants will work on projects or other experiential learning.
  - Presentations: whether presentations, lectures, etc. will be provided.
  - Training mode: the modality for how the participants will be training
    - training will be in a classroom live,
    - training will be instructor led,
    - application of train the trainer – where a subject matter expert trains others in the organization to train additional system operators.
  - Training schedule and syllabus: the proposed schedule and syllabus that provides course objectives, training information, assessments used, instructor information, class policies, grading criteria, etc.
- Training location: the location of where the training will be held.

Certification typically includes a formalized method of ensuring that the system operator has the appropriate skills to operate the system for their role. Certification can include training, required skills, and qualifications. The certification is usually reassessed periodically to ensure continual skills are maintained. Certification can also be part of a person's job performance and periodic evaluation.

11.2.2.1   Food Systems Training Approach and Plans

The training approach and plans for the food system teams are as follows.

Training approach and plans for Team 1 connecting organizations

Training plans are completed for the project to ensure all members are trained properly in different elements of the system. Training plans for this system are primarily for Citywide employees and volunteers. This is because they will be maintaining the website functions and the bulk of the data collection and analysis. There is a total of five training plans created for the website system. They include:

1. Train volunteers to use website
2. Train volunteers to communicate with customers and fill out form
3. Statistical software training
4. Ethics training
5. Website Admin training

Volunteers receive trainings 1–4 because they complete most of the front-end tasks for the system. This includes talking with the customers, obtaining their information, demonstrating use of the online website, and statistical analysis of data. The Website Admin training is conducted yearly to ensure the competence of information of the Website Admin. It is important to ensure that the person is aware of their duties, knows how the website functions, and can fix problems with the website if they occur. Ethics training is not only provided to the Citywide employees and volunteers but the farmers and community partners as well. The training is given annually. It is important to provide this training, because Citywide wants to ensure that all members of the program are aware of their standards and prevent ethics issues from occurring. It is important to involve farmers and community partners as well because, they are interfacing with the customers. The training plans can be found below in Figure 11.3 through 11.7.

Training approach and plans for Team 2 creating food policy

Training for this system will be quite simple, as that was the goal of this project. Those involved in a policy change system would be briefed on the basic structure of the policy change diagram. This would be to make sure they understand all the channels they would be working through. Participants would also be trained on record keeping procedures to make sure all proper documentation was in place and accounted for. Other training would include proper research techniques in order to create better economic and community impact reports, general computer training, and door-to-door petitioning strategies.

Down the road, when it becomes apparent that there is potential for a policy to change, participants are briefed on how to work with the government in order to provide a good experience for all parties involved.

Training Plan	
Develop a training plan of how the process changes will be rolled out to the stakeholders.	
Training Objectives and Outcomes:	Train Volunteers to use the website
Training Audience and Resources:	CityWide Volunteers
Instructional Strategies	
Exercise and case studies:	None
Assessments and exams:	Web Admin Assessment
Homework assignments:	None
Team and individual work:	Individual
Project work:	None
Presentations:	Website Overview
Training mode:	*Instructor Led *Train the Trainer
Training schedule and syllabus:	Just-in-time training
Training location:	Citywide Office

FIGURE 11.3   Train volunteers to use website.

Training Plan 2	
Develop a training plan of how the process changes will be rolled out to the stakeholders.	
Training Objectives and Outcomes:	Train Volunteers to comunicate and fill out form
Training Audience and Resources:	CityWide Volunteers
Instructional Strategies	
Exercise and case studies:	None
Assessments and exams:	Fill out example Form
Homework assignments:	None
Team and individual work:	Individual
Project work:	None
Presentations:	None
Training mode:	*Instructor Led *Role Play *Train the Trainer
Training schedule and syllabus:	Just-in-time training
Training location:	Citywide Office

FIGURE 11.4   Train volunteers to communicate and fill out form.

Training Plan 3	
Develop a training plan of how the process changes will be rolled out to the stakeholders.	
Training Objectives and Outcomes:	Statistical Software Training
Training Audience and Resources:	CityWide Volunteers
Instructional Strategies	
Exercise and case studies:	None
Assessments and exams:	Crunch Data sample sucessfully
Homework assignments:	Example Problems
Team and individual work:	Individual
Project work:	None
Presentations:	None
Training mode:	*Instructor Led *Example Data Sets
Training schedule and syllabus:	Just-in-time training
Training location:	Citywide Office

FIGURE 11.5   Statistical software training.

Training Plan 4	
Develop a training plan of how the process changes will be rolled out to the stakeholders.	
Training Objectives and Outcomes:	Website Admin Training
Training Audience and Resources:	CityWide Employee
Instructional Strategies	
Exercise and case studies:	Case study on website security
Assessments and exams:	Admin Assessment
Homework assignments:	None
Team and individual work:	Team
Project work:	None
Presentations:	Presentation of website security, coding, database, server
Training mode:	*Live Classroom *Instructor Led *Train the Trainer
Training schedule and syllabus:	Annual training
Training location:	Citywide Office

FIGURE 11.6   Website admin training.

Training Plan 5	
Develop a training plan of how the process changes will be rolled out to the stakeholders.	
Training Objectives and Outcomes:	Ethics
Training Audience and Resources:	CityWide Volunteers, Employees, Farmers, community partners
Instructional Strategies	
Exercise and case studies:	Exercise of different sitiuations, Case Studies on past scandals
Assessments and exams:	None
Homework assignments:	None
Team and individual work:	Team
Project work:	None
Presentations:	Overview of Citywide Ethics Standards
Training mode:	*Live Classroom *Instructor Led
Training schedule and syllabus:	Annual Training
Training location:	Citywide Office

FIGURE 11.7   Ethics training.

Training approach and plans for Team 3 kids growing food
The training plan is useful to provide the initial training for all operators in order to equip them with the skill and knowledge to operate the system and provide a continuation training to ensure currency of knowledge. The training objectives and outcomes are to train students how to grow microgreens alone. The trained audience and resources are students. There are four instructional strategies for the training plan. The first strategy is assessments and exams which is the microgreen program comprehension assessment. The second is team and individual work which is team growing microgreens during training. The third is training schedule and syllabus which must be at school time. The fourth is training location which must be in school. The training plan is shown in Table 11.2.

Training approach and plans for Team 4 food storage
A training plan is an important part of any program or activity to ensure that the users have the knowledge to perform their tasks. One of the ways to create a training approach is by developing instructional materials, which start by examining the training plan and available resources. Depending on the learning objectives and length of the training program, training materials may provide workbooks. In our system, the most important training activities that occur are the training of Homeful staff for doing tasks and delivering food. The training plan for Team 4 is shown in Figure 11.8.

Training approach and plans for Team 5 food hub website
In any situation, training for all operators of a system is an integral part of implementing a new system. Our website will require training for all users (business owners, entrepreneurs, farmers, and distributors) in order for the system to work as designed. To train our users, we would have a training session in a classroom setting with PowerPoint/hands on computer training on how to use the tools of the website. We would also create an online training module for those users who cannot attend the in-person training and for continuous availability of help for the users. The training plan for Team 5 is shown in Figure 11.9.

TABLE 11.2
Training Plan for Team 3 Kids Growing Food

Training Plan	
Training Objectives and Outcomes:	The objectives include teaching students about the nutritional benefits of eating microgreens, the microgreen plant in general, and, most importantly, how to grow microgreens at home. Training outcomes include students who know and have retained enough information about microgreens to be able to successfully grow them on their own at home.
Training Audience and Resources:	Students, all seasons greens business
Instructional Strategies	
Exercise and case studies:	None
Assessments and exams:	Microgreen Program Comprehension Assessment
Homework assignments:	None
Team and individual work:	Team growing microgreens during training
Project work:	None
Presentations:	None
Training mode:	Demonstration along with a lecture followed by hands on work
Training schedule and syllabus:	One week long during the beginning portion of class time. Receive Comprehension Assessment and permission slip at conclusion of training.
Training location:	In school

Training Plan	
Develop a training plan of how the process changes will be rolled out to the stakeholders.	
Training Objectives and Outcomes:	Train the farmers on growing the organic food, train community truck operators, train the community helpers at central storage location
Training Audience and Resources:	Homefull community, farmers, community helpers
Instructional Strategies	
Exercise and case studies:	methods on improving their respective skills
Assessments and exams:	assessed by their level of improvement in completing the tasks
Homework assignments:	None
Team and individual work:	Work as team along with the other co-workers
Project work:	None
Presentations:	None
Training mode:	Provide an teaching session for the farmers to understand the concepts of organic food growing, train the drivers with the knowledge of each food transportation and storing temperatures and condition to apply while transporting the food, Orgainze a community experienced advisor give teaching session for the community helpers on undestanding the temperature settings for storing food and other conditions for each particular food item.
Training schedule and syllabus:	At least two weeks before beginning tasks
Training location:	Homefull Community center or Farm locations

FIGURE 11.8 Training plan for team 4 food storage.

Training Plan	
Develop a training plan of how the process changes will be rolled out to the stakeholders.	
Training Objectives and Outcomes:	Train all stakeholders how to effectively use Website
Training Audience and Resources:	Business Owners, Entrepreneurs, Farmers, and Distributors
Instructional Strategies	
Exercise and case studies:	None
Assessments and exams:	None
Homework assignments:	None
Team and individual work:	Complete Training Course
Project work:	None
Presentations:	Targeted Presentations on Website Tutorial
Training mode:	Live Classroom, Online Tutorial
Training schedule and syllabus:	Planned Training Presentations @ Public Location, Online Tutorial Offer
Training location:	Dayton Convention Center

FIGURE 11.9 Training plan for team 5 food hub website.

Training approach and plans for Team 6 GEM City Market
The training plan is used to train the staff workers on how to give a world class customer service. A standard protocol file describing the different work procedures and rules will be created for the market to train the staff with respect to every aspect of the store. The training would be for 15 days in the market itself and the training would be about how to bill and checkout, stock maintenance, stack maintenance, organizing the store segregating types of food items, floor cleaning and maintenance, pickup and delivery services, marketing about the new offers or items introduced into the store, disposal of trash and expired items, works to be taken care of while opening and closing the market. The training plan for Team 6 is shown in Figure 11.10.

Training approach and plans for Team 7 Table 33 Restaurant
To make sure that all operators are equipped with the proper skills, a training plan has been designed for the Table 33 system. This plan can help to ensure the currency of knowledge where the job done by the system's operators can be monitored and evaluated as well as the quality of operators. In our system, shown in Figures 11.11 and 11.12, local farms and the inventory of the restaurant are the two operators who have been chosen to train.

Training Plan	
Develop a training plan of how the process changes will be rolled out to the stakeholders.	
Training Objectives and Outcomes:	Training the staff workers in all the wings of the Store management
Training Audience and Resources:	Store Supervisors and managers
Instructional Strategies	
Exercise and case studies:	None
Assessments and exams:	None
Homework assignments:	None
Team and individual work:	Yes
Project work:	None
Presentations:	None
Training mode:	A standard protocol file will be given and followed, describing all the works, rules, regulations, requirements and the procedure of the work to be done in all the wings of the market like Billing checkouts, stock and stack maintenance,floor maintenance, pickup & delivery services, customer assistance, closing and opening process of the store etc.
Training schedule and syllabus:	15 days
Training location:	Gem City Market

FIGURE 11.10   Training plan for team 6 gem city market.

Training Plan	
Develop a training plan of how the process changes will be rolled out to the stakeholders.	
Training Objectives and Outcomes:	Train Personnel to be an expert in testing and choosing local ingredients
Training Audience and Resources:	Any person
Instructional Strategies	
Exercise and case studies:	None
Assessments and exams:	
Homework assignments:	None
Team and individual work:	Individual
Project work:	None
Presentations:	
Training mode:	* Kitchen Classes * Live Classes * Train the Trainer
Training schedule and syllabus:	* In time training and during local farm visits
Training location:	* Kitchen and Local farms

FIGURE 11.11   Training plan 1 for team 7 table 33 restaurant.

Training Plan	
Develop a training plan of how the process changes will be rolled out to the stakeholders.	
Training Objectives and Outcomes:	Train Personnel to be an expert in tracking ingredients using softwares like Excel
Training Audience and Resources:	Any person
Instructional Strategies	
Exercise and case studies:	None
Assessments and exams:	CFI Excel exam
Homework assignments:	None
Team and individual work:	Individual
Project work:	None
Presentations:	
Training mode:	* Live Classes
Training schedule and syllabus:	* In time training
Training location:	* Restaurant's Inventory

FIGURE 11.12   Training plan 2 for team 7 table 33 restaurant.

11.2.3   PERFORM RISK ASSESSMENT

A risk assessment, similar to the one provided in Phase 1, Concept of Operations, should be updated based on risk events that are possible to occur and may have a major impact on system operations. See Chapter 5 for the risk analysis tools that can be used to focus on the operations and maintenance phases.

11.2.4 PERFORM IMPROVEMENT AND MAINTENANCE PLANNING

For a service system, the maintenance activities could include:

1) Improving the processes
2) Maintaining facilities and equipment

The Improvement and Maintenance Plan for the Women's Healthcare Center is shown in Figure 11.13.

The maintenance plan activities and the owners are the fields developed to create the maintenance plan. The maintenance plan should include preventive and corrective maintenance plans as well as continuous improvement plans, especially desirable for a service or enterprise system.

The improvement and maintenance artifacts include:

- Improvement and Maintenance Plan
- Maintenance results
- Maintenance issues
- Maintenance training materials and plan
- Maintenance changes, as required
- Improvement project ideas

Improvement & Maintenance Plan Activities	
Improvement Plan	**Owners**
Identify areas for improvement via customer satisfaction surveys, complaints, collection of process metric data, improvement efforts	Director of Operational Excellence
Develop improvement ideas	Director of Operational Excellence
Pilot improvements and train	Director of Operational Excellence
Update documentation, share results	Director of Operational Excellence
Implement Changes	Director of Operational Excellence
Maintenance Plan	**Owners**
Preventive maintenance activities per schedule	Director of Facilities
Corrective maintenance activities per facilities requests	Director of Facilities
Train and update documentation and results	Director of Facilities
Preventive maintenance activities for biomedical equipment	Director of Biomedical Engineering
Corrective maintenance activities per staff request	Director of Biomedical Engineering
Train and update documentation and results	Director of Biomedical Engineering

FIGURE 11.13 Improvement and maintenance plan for women's healthcare center.

11.2.4.1 Food System Improvement and Maintenance Plans

Improvement and Maintenance Plan for Team 1 connecting organizations

Maintenance and service plans are created to ensure that the system is maintained and can continue to provide a service. For the website, it is important to maintain the site since it is a critical form of communication between Citywide, communities, and the farmers. The system includes two types of plans. One is an improvement plan to identify ways to improve the website system and program process. Another is a maintenance plan with activities to update the website, program processes, documentation, etc. Figure 11.14 provides a list of the improvement plans and maintenance plans for the Citywide website system. For both plans, Citywide and its volunteers are the primary owners of each step listed. Some improvement plan steps include collecting metrics, obtaining new ideas, piloting the ideas, updating documentation, and implementing changes. Some maintenance steps include preventative maintenance of the website server, software updates, archiving messages from the message board, and maintaining public relations.

Improvement and Maintenance Plan for Team 2 creating food policy

A maintenance plan needs to be established to let community members know who is responsible for which part of the system. The maintenance and improvement plan for Team 2 is shown in Figure 11.15.

Basically, the policy change system needs maintenance for the community center and commission center, because the center needs power to keep working and community members need communication to develop improvement ideas and share information to reduce operation time, while training the staff can save operation time.

A Failure Mode and Effects Analysis on the system is used to assess where the system may fail. The main failure is when the policy change idea is rejected. The FMEA is shown in Figure 11.16. The FMEA documentation helps define the failure that may occur during system operation. The FMEA gives the cause of the failure and effects of the failure and the recommended action to deal with the potential failure which is useful.

Improvement & Maintenance Plan Activities	
Improvement Plan	**Owners**
Identify areas for improvement via customer satisfaction surveys, complaints, collection of process metric data, improvement efforts	CityWide
Develop improvement ideas	CityWide
Pilot improvements and train	CityWide
Update documentation, share results	CityWide
Implement Changes	CityWide
Maintenance Plan	**Owners**
Preventive maintenance of website server	CityWide
Maintaince of software to keep it up-to-date	Citywide
Train and update documentation and results	CityWide
Archiving forum messages everyweek to reduce clutter	Volunteer/Moderator
Weekly check of paper form condition and volume at pickup locations	Volunteer
Maintain City property and public relationships	Volunteer & CityWide

FIGURE 11.14 Improvement and maintenance plans for website system.

Improvement & Maintenance Plan Activities	
Improvement Plan	**Owners**
Trainings of staff to save operation time	Community Staff
Develop improvement ideas	Community members
Keep communicating with community members	Community members, Champion, Customers
Update documentation, share results	Community Staff
Maintenance Plan	**Owners**
Calibration of equipment	Community staff and commission staff
Keep writing and storing maintenance documentation	Community staff and commission staff
Train and update documentation and results	Community staff and commission staff

FIGURE 11.15   Improvement and maintenance plan for team 2 creating food policy.

FAILURE MODE AND EFFECT ANALYSIS (PROCESS FMEA)										
PROCESS:		Present policy idea to commission center		Date:						
DEPARTMENT:		Commission center								
Process Step	Potential Failure Mode	Potential Effects of Failure	Severity	Potential Causes of Failure	Occurrence	Current process controls	Detection	Risk Priority Number	Recommended Action	Owner / Estimated Completion Date
Present policy change idea to commission center	Inability to make policy change idea resonable	Wasting time	8	Lack communication	1	None	3	24	Keep communication	
		Wasting labor	8	Lack communication	1	None	3	24	Keep communication	
								0		

FIGURE 11.16   FMEA for team 1 creating food policy.

Improvement and Maintenance Plan for Team 3 kids growing food

The purpose of the maintenance and improvement is to sustain the capability of a system to provide a service. The improvement plan identifies areas for improvement via student surveys, meetings or phone calls with Table 33, complaints, collection of process metric data, and the owner of this plan is the teacher. The next step in the improvement plan is to develop improvement ideas and the owners are the teacher and students. The owner of the next four steps of the improvement plan is the teacher and these four steps are the pilot improvement and train, the update documentation and share results, update documentation share results, and implement changes. In addition, for this project there are five maintenance plans which are 1) Preventive maintenance activities on microgreen growing supplies per cycle, 2) Corrective maintenance activities on microgreen growing supplies per student request, 3) Train and update documentation and results, 4) Preventive maintenance activities on microgreen harvesting equipment, 5) Corrective maintenance activities on microgreen harvesting equipment per the teacher's request. The teacher is the owner of all the five portions of the maintenance plan. The Improvement and Maintenance Plan is shown in Table 11.3.

TABLE 11.3
Improvement and Maintenance Plan for Team 3 Kids Growing Food

Improvement and Maintenance Plan Activities

Improvement Plan	Owners
Identify areas for improvement via student surveys, meetings or phone calls with Table 33, complaints, collection of process metric data	Teacher
Develop improvement ideas	Teacher, students
Conduct a system test with improvement ideas temporarily implemented	Teacher
Pilot improvements and implement into training based on test results	Teacher
Update documentation, share results	Teacher
Implement changes	Teacher
Maintenance Plan	**Owners**
1) Preventive maintenance activities on microgreen growing supplies per cycle	Teacher
2) Corrective maintenance activities on microgreen growing supplies per student request	Teacher
3) Train and update documentation and results	Teacher
4) Preventive maintenance activities on microgreen harvesting equipment	Teacher
5) Corrective maintenance activities on microgreen harvesting equipment per teacher's request	Teacher

Improvement and Maintenance Plan for Team 4 food storage
The Improvement and Maintenance Plan for Team 4 is shown in Figure 11.17.

Improvement and Maintenance Plan for Team 5 food hub website
The purpose of our Improvement and Maintenance Plan is to ensure that our website works as designed, not only at the start, but throughout the system's entire lifetime. All stakeholders of the system will have input into the improvement and maintenance of the website and the website developer will carry out the improvements and maintenance as desired. The Improvement and Maintenance Plan is shown in Figure 11.18.

Improvement and Maintenance Plan for Team 6 GEM City Market
The maintenance is an important aspect in the long run of the market. Maintaining the sales floor, cleanliness of the market, food quality will automatically maintain the reputation and growth in sales of the market. Improving according to demand is very important and hence improvement plans must be defined and implemented. Each aspect is handled by each person/team in the store and that will also make the work look simple and successful. The Improvement and Maintenance Plan is shown in Figure 11.19.

Improvement and Maintenance Plan for Team 7 Table 33 Restaurant
As a way to ensure that the capability of the system is sustainable to provide the proposed service, a maintenance plan has been created along with the responsible owner for each maintenance plan as

Improvement & Maintenance Plan Activities	
Improvement Plan	**Owners**
Identify areas for improvement via customer satisfaction surveys, complaints, collection of process metric data, improvement efforts	Homefull community & Staff
Develop improvement ideas	Homefull community & Staff
Pilot improvements and train	Homefull community & Staff
Update documentation, share results	Homefull community & Staff
Implement Changes	Homefull community & Staff
Maintenance Plan	**Owners**
Preventive maintenance activities per schedule	Central storage Staff & Homefull community
Corrective maintenance activities per facilities requests	Central storage Staff & Homefull community
Train and update documentation and results	Central storage Staff & Homefull community
Preventive maintenance activities for refrigerators equipment	Central storage Staff & Homefull community
Corrective maintenance activities per staff request	Central storage Staff & Homefull community
Train and update documentation and results	Central storage Staff & Homefull community

FIGURE 11.17 Improvement and maintenance plan for team 4 food storage.

Improvement & Maintenance Plan Activities	
Improvement Plan	**Owners**
Identify areas for improvement via customer satisfaction surveys, complaints, collection of process metric data, improvement efforts	System Investors and Entreprenuers
Develop improvement ideas	All Users
Pilot improvements and train	Web Developer
Update documentation, share results	All Users
Implement Changes	Web Developer
Maintenance Plan	**Owners**
Preventive maintenance activities per schedule	Web Developer
Corrective maintenance activities per facilities requests	Web Developer
Train and update documentation and results	All Users
Preventive maintenance activities for Health and Safety	Farmers, Business Owners, Distributors
Corrective maintenance activities per quality reports	Farmers, Business Owners, Distributors
Train and update documentation and results	Farmers, Business Owners, Distributors

FIGURE 11.18 Improvement and Maintenance Plan for Team 5 food hub website.

shown in Figure 11.20. By having the maintenance plan, system's requirements can be considered as maintenance constraints and the stakeholder requirements are sustained. Overall, the maintenance plan can ensure the improvement of the processes in the system.

The Operation and Maintenance phase should be the longest phase from a time perspective. The system should operate for a much longer time that it took to develop it.

Improvement & Maintenance Plan Activities	
Improvement Plan	**Owners**
Identify areas for improvement via customer satisfaction surveys, complaints, collection of process metric data, improvement efforts	Customer Service Manager (Customer Service department of the management)
Develop improvement ideas	Operational manager and Customer Service Manager
Pilot improvements and train	Operational manager
Update documentation, share results	Operational manager
Implement Changes	Operational manager
Maintenance Plan	**Owners**
Sales Floor Maintenance (Cleanliness)	Maintenance Supervisor
Stock and Stack maintenance	Maintenance supervisor and manager
Food Quality maintenance in the store	Maintenance management
Corrective maintenance activities by staff	Maintenance Supervisor
Train and update documentation and results	Maintenance supervisor and manager

FIGURE 11.19 Improvement and maintenance plan for team 6 gem city market.

Improvement & Maintenance Plan Activities	
Improvement Plan	**Owners**
Use Excel or any other type of data analysis softwares to help manage the restaurant's inventory	Manager
Use new techniques and devices, such as greentest, to test ingredients	Chief and Manager
Find the best way to deliver ingredients from local farms to restaurant	Manager
Schedule appropriate times to move ingredients from inventory to kitchen	Manager
Create a website or application to connect local farmers with Table 33	Manager
Maintenance Plan	**Owners**
Check ingredients inventory regularly	Inventory Management Systems
Update inventory documentation and results	Inventory Management Systems
Update website or application to insure a strong connection between local farmers and restaurant	Manager
Keep searching for local farmers from time to another	Manager

FIGURE 11.20 Improvement and maintenance plan for team 7 table 33 restaurant.

11.2.5 PERFORM DISPOSAL AND RETIREMENT

There is quite a bit of subject matter dedicated to defining and subsequently extending the service life of a system. There are many systems that have exceeded their original expected length of service, through the discipline of service life extension. Refer to references to learn more about service life extension, if interested.

The last part of the operations and maintenance phase is the disposal and retirement of the system. At the near end of the system's useful life the systems engineer should focus on how to retire and dispose the system. At some point, the system will be obsolete, unrepairable or difficult, and/or expensive to maintain. A plan to phase out and replace the system and then dispose equipment, facilities, and processes should be developed by the systems engineer. There should also be

Disposal and Retirement Activities	
Disposal and Retirement Activities	**Owners and Resources**
Evaluate disposal of medical equipment when ready to retire.	Director of Biomedical engineering Equipment vendor
Evaluate disposal of facilities materials when rehabbing and constructing new facility from old facility	Director of construction
Consider water pollution impact for disposal of materials and medical waste	Chief Operating Officer Director of Facilities
Develop plan for disposal of medical waste	Chief Operating Officer Director of Facilities

FIGURE 11.21   Disposal and retirement plan for the women's healthcare center.

consideration to deal with any hazardous or toxic materials or waste in accordance with applicable guidance, policy, regulations, and statutes. Areas should be considered related to (SEBOK 8.0):

- Air pollution and control
- Water pollution and control
- Noise pollution and control
- Radiation
- Solid waste

Even though the systems engineer who is involved in the design and deployment of the original system may not be responsible for the system when it is ready to be disposed of, the systems engineer should still plan and design for the retirement and disposal of the system.

The plan for disposal and retirement activities for the Women's Center is shown in Figure 11.21.

11.2.5.1   Food System Disposal and Retirement Plan

The food system Disposal and Retirement plans are discussed as follows.

Disposal and Retirement Plan for Team 1 connecting organizations

Disposal and Retirement Plans for the website and other elements of the program involved data collection and cleanup. The goal of the website is to continue its service to the communities for communication if possible. Thus, there is no set time as to when it is disposed. However, the website technology may become outdated in the future and require a transfer of information to another method. Thus, Citywide would be responsible for establishing the migration of data to a more advanced form of technological communication, such as a phone app. Another disposal activity to be considered is ensuring that paper forms are scanned into an electronic system and recycled properly. This would reduce the amount of information clutter Citywide would have and have the forms available in an electronic space. Another important activity is the cleanup and disposal of waste at

Disposal and Retirement Activities	
Disposal and Retirement Activities	**Owners and Resources**
Website becomes outdated and information is migrated to an App	Citywide
Ensure paper forms that have been scaned are recycled properly	Volunteer
Develop plan for clean up and desposal at markets/events	Citywide
Develop a plan of an annual purge of unneccesary data	Web Admin

FIGURE 11.22   Website disposal and retirement plans for team 1 connecting organizations.

markets/events. This is because Citywide is working with these communities to improve and expand the program. Leaving an area that they rented as a mess would give Citywide a bad reputation. This is not desired and should be why there is a plan in place to ensure cleanup is properly done. Finally, information clutter in both paper and electronic forms will eventually become a problem. Thus, the system needs a developed plan where old and unnecessary data is disposed on a yearly basis. Figure 11.22 provides a list of the plans discussed for the website system.

Disposal and Retirement Plan for Team 2 Creating Food Policy
Product or service disposal and retirement is an important part of system life management. Sometimes some parts of the system are no longer useful due to its useful life and accidents. They can't be functional forever, so disposal and retirement activities can be helpful to make the system function well. Disposal and retirement activities define the disposal parts and those parts can be replaced or cleared after evaluation.

Because the policy change system is about changing policy steps, there are no disposal activities for changing policy steps. But during the policy-making process, paper waste is the disposal. Paper waste needs to be cleared and ensure that it is harmless to the environment. And expired paper also needs to be evaluated if it needs to be cleared. The Disposal and Retirement Plan is shown in Figure 11.23.

Disposal and Retirement Plan for Team 3 kids growing food
Disposal plans are created to think about the retirement of a system. In this project, we have four disposal and retirement activities. These four activities are to evaluate disposal of microgreen trays

Disposal and Retirement Activities	
Disposal and Retirement Activities	**Owners and Resources**
Evaluate disposal of paper waste	Community staff and commission staff
Evaluate disposal of expired paper	Community staff and commission staff

FIGURE 11.23   Disposal and retirement plans for team 2 creating food policy.

TABLE 11.4

Disposal and Retirement Plan for Team 3 Kids Growing Food

Disposal and Retirement Activities	
Disposal and Retirement Activities	**Owners and Resources**
Evaluate disposal of microgreen trays and lids when ready to retire.	Owner – Teacher, Resources – School science department, recycling facility, Goodwill
Develop plan for disposal of microgreen medium after each harvest.	Owner – Teacher, Resources – Local Environmental Services
Evaluate disposal of leftover hydrogen peroxide when ready to retire.	Owner – Teacher, Resources – Local Environmental Services
Develop plan for disposal of leftover hydrogen peroxide when ready to retire.	Owner – Teacher, Resources – Local Environmental Services

and lids when they are ready to retire, develop plan for disposal of microgreen medium after each harvest, evaluate disposal of leftover hydrogen peroxide when ready to retire, and develop a plan for disposal of leftover hydrogen peroxide when ready to retire. The first disposal and retirement activity is to evaluate disposal of microgreen trays and lids when they are ready to retire and this activity has the teacher as an owner and the school science department, recycling facility, and Goodwill as resources. The teacher is the owner, and local environmental services is the resource for the other three activities. The Disposal and Retirement Plan is shown in Table 11.4.

Disposal and Retirement Plan for Team 4 food storage

There are four disposal and retirement activities that need to be researched on in this project. Each one of them has its plan and solution. The first one is to evaluate disposal of equipment of growing when it is ready to retire, with the owners being the Homeful community staff and the food collectors. The second one is to evaluate disposal of facilities of storing and make a plan to govern the disposal. The third and fourth activities are to consider the food pollution impact for disposal of materials and develop a plan for disposal of growing food waste. The Disposal and Retirement Plan is shown in Figure 11.24.

Disposal and Retirement Activities	
Disposal and Retirement Activities	**Owners and Resources**
Evaluate disposal of equipment of growing when ready to retire.	Homefull community staff, food collectors.
Evaluate disposal of facilities of storging and make a plan to govern the disposal	Homefull community staff, collectors- Local Center Storage Workers.
Consider food pollution impact for disposal of materials.	Homefull community staff - disterbutors
Develop plan for disposal of growing food waste	Homefull community staff - disterbutors - food collectors

FIGURE 11.24   Disposal and retirement plan for team 4 food storage.

Disposal and Retirement Plan for Team 5 food hub website
All systems will come to an end at some point and it is important as an engineer to plan for that point in the system's life. Our website will be an effective tool for the food system and will come to an end only if another more efficient and cost-effective system is designed. The stakeholders of the food system will also have input to the life cycle of the website in terms of its continued effectiveness and ease of use. The Disposal and Retirement Plan for Team 5 is shown in Figure 11.25.

Disposal and Retirement Plan for Team 6 GEM City Market
Disposals are a daily activity that should be taken care of for the sake of good health. Removal of trash, spoiled/rotten items, and the expired items are important tasks in every day's work. Also, checking the life of the equipment used in the market-like filling machinery, weight balances, computers, printers and their ink cartridge etc., to change after a certain useful time. The Disposal and Retirement Plan for Team 6 is shown in Figure 11.26

Disposal and Retirement Plan for Team 7 Table 33 Restaurant
For the case if the system is to be disposed and removed from the operation environment, the Disposal and Retirement Plan has been created to save time and money. Figure 11.27 illustrates the disposal plan for the Table 33 system where it can provide the required activities to be removed and the owner responsible.

Disposal and Retirement Activities	
Disposal and Retirement Activities	**Owners and Resources**
Evaluate Website effectivness, use traffic before retire	Web Developer
Consider new web alternative or system management software	Entrepreneurs
Develop website phase out plan, effective date TBA	Web Developer and Entrepreneurs

FIGURE 11.25   Disposal and retirement plan for team 5 food hub website.

Disposal and Retirement Activities	
Disposal and Retirement Activities	**Owners and Resources**
Evaluate disposal of food which crossed their expiry date.	Maintenance supervisor and staff
Evaluate disposal of rotten or spoiled food items	Maintenance supervisor and staff
Evaluate the retirement of various equipments, printers, weigh balances, filling machines and computers in the market	Store management (Technical and Maintenance)
Evaluate disposal of trash from the store	Maintenance supervisor and staff

FIGURE 11.26   Disposal and retirement plan for team 6 gem city market.

Disposal and Retirement Activities	
Disposal and Retirement Activities	**Owners and Resources**
Evaluate disposal of testing ingredients equipment when ready	Manager and restaurant owners
Evaluate disposal of facilities materials when rehabbing and constructing new facility from old facility	Manager and restaurant owners
Evaluate disposal of documentations and results related to Inventory Management Systems	Manager and restaurant owners
Develop plan for disposal of unused ingredients	Chief

FIGURE 11.27   Disposal and retirement plan for team 7 table 33 restaurant.

11.3 SUMMARY

In this chapter, we discussed placing the system into operation, the longest length of time in any phase, assuming that the system is successful and meets the customers' needs. We also discussed the disposal and retirement of the system that should be planned by the systems engineer, even if ultimately they have little control over the actual retirement and disposal activities.

11.4 ACTIVE LEARNING EXERCISES

1. Develop a training approach and plan for your system
2. Develop an Improvement and Maintenance Plan for your system
3. Develop a Disposal and Retirement Plan for your system

BIBLIOGRAPHY

SEBoK Editorial Board, *The Guide to the Systems Engineering Body of Knowledge (SEBoK)*, vol. 2.2, ed. R. J. Cloutier (Hoboken, NJ: Trustees of the Stevens Institute of Technology, 2020), accessed May 25, 2020, www.sebokwiki.org. BKCASE is managed and maintained by the Stevens Institute of Technology Systems Engineering Research Center, the International Council on Systems Engineering, and the Institute of Electrical and Electronics Engineers Computer Society.

12 Systems Engineering Planning

In this Chapter we discuss the Systems Engineering planning activities and artifacts.

12.1 PURPOSE

In this chapter, we discuss the systems engineering management planning activities and artifacts. This chapter gives an overview of systems engineering management planning based on the view of the SEBoK (Original) and the Systems Engineering Management Plan that consists of multiple plans that are important to the planning efforts throughout the Vee Life Cycle Model. This chapter is not meant to give enough detail to be an expert in this area but gives the systems engineer an understanding of the planning and activities that they may be asked to contribute to as part of their role in the systems engineering life cycle.

12.2 ACTIVITIES

The activities and tools to perform systems engineering planning throughout all the Vee Life Cycle phases are shown in Table 12.1. We cover the "Develop Systems Engineering Management Plan and associated plans" and provide examples of the plans. The Systems Engineering Management Plan Model is shown in Figure 12.1.

The plans that are part of the Systems Engineering Management Plan are listed below. Each plan is described in more detail in the following sections.

1. Assessment and Control Plan: The Assessment and Control Plan describes how the key artifacts will be assessed and controlled throughout the systems engineering life cycle.
2. Configuration Management Plan: The Configuration Management Plan is used to plan how the processes and data will be configured and changed.
3. Contractor Management Plan: The Contractor Management Plan is used to describe how the contractors will be managed in relationship to the government or system owner organization.
4. Deployment Plan: The Deployment Plan describes how the system will be deployed within the Phase 6 Operations and Maintenance.
5. Disposal and Retirement Plan: The Disposal and Retirement Plan is used to safely dispose the system elements when the system is retired.
6. Information Management Plan: The Information Management Plan is used to manage the technical information for the development, design, and deployment of the system.
7. Interface Management Plan: The Interface Management Plan is used to manage the system interfaces, to ensure appropriation verification testing.
8. Maintainability Program Plan: The Maintainability Program Plan is used to maintain the system in the operations phase.
9. Measurement Plan: The Measurement Plan is used to design and manage metrics for measuring the key performance indicators for the system.
10. Quality Management Plan: The Quality Management Plan is used to plan and implement a quality management system.
11. Risk Management Plan: The Risk Management Plan is used to identify, assess, monitor, mitigate, and control system risks.

TABLE 12.1
Activities and Tools within Systems Planning

Vee Phase	Activities	Tools	Principles
PLANNING (throughout all phases)	1. Develop Systems Engineering Management Plan and associated plans.	• Systems Engineering Management Plan • Assessment and Control Plan; Configuration Management Plan; Contractor Management Plan; Deployment Plan, Disposal, and Retirement Plan; Information Management Plan; Interface Management Plan; Maintainability Program Plan; Measurement Plan; Quality Management Plan; Risk Management Plan; Specialty Engineering Plan; System Development Plan; System Integration Plan; System Integration Plan	

12. Specialty Engineering Plan: The Specialty Engineering Plan is used to plan the work of the required specialty engineering disciplines.
13. System Development Plan: The System Development Plan is used to plan the development of the system in the Phase 3 Detailed Design.
14. System Integration Plan: The System Integration Plan is used to plan how the system will be integrated to ensure verification and validation of the system.

12.3 DEVELOP SYSTEMS ENGINEERING MANAGEMENT PLAN AND ASSOCIATED PLANS

12.3.1 ASSESSMENT AND CONTROL PLAN

The Assessment and Control Plan, shown in Figure 12.2, describes how the key artifacts are assessed and controlled throughout the systems engineering life cycle.

The Systems Engineering Assessment and Control (SEAC) process includes the following activities (SEBoK 8.0):

3.1 Plan the reviews that will be performed and document them in an Assessment and Control Plan, with an example for the Women's Healthcare Center shown in Figure 12.2. This example includes a description of the major reviews for each phase of the Vee model.

3.2 Perform appropriate reviews within the systems engineering life cycle.

3.3 Identify need for changes to artifacts assessed during the reviews, analyze issues, and incorporate resolutions.

3.4 Identify system technical risks, document, and escalate significant risks, as appropriate.

3.5 Manage risks and changes to artifacts.

3.6 Hold post-delivery assessments to identify systems engineering process and project learnings.

A typical review includes pre-work activities, holding the review, and post-work activities, as shown in Figure 12.3 (SEBoK 8.0). For the pre-work activities, the systems engineer or review facilitator identifies who should participate in the review and what their roles will be. The systems engineer will develop the agenda and establish assessment criteria. They then distribute all the materials that will be reviewed prior to the actual review, so that the participants will have adequate

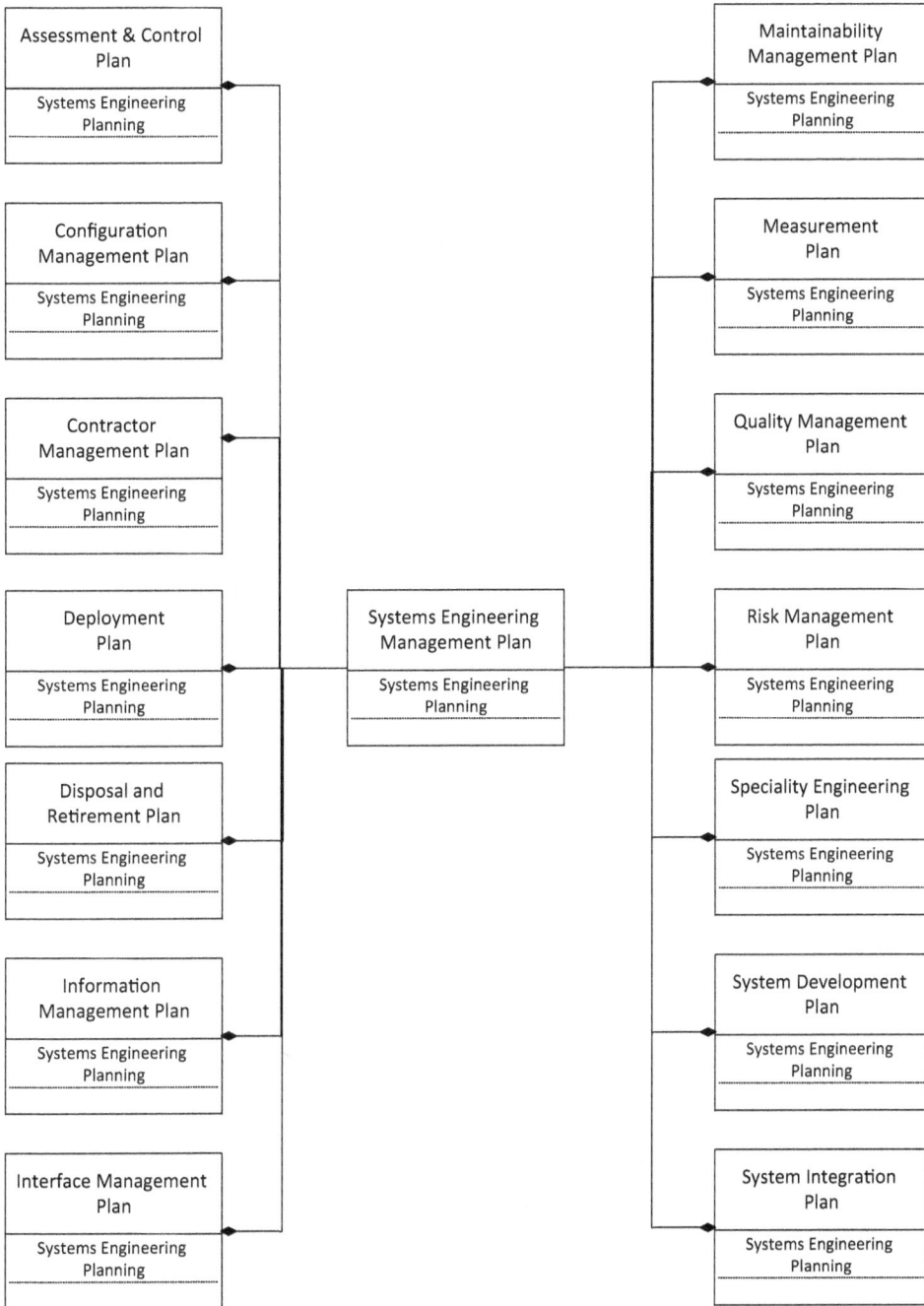

FIGURE 12.1 Systems engineering management plan model.

time to review the materials, typically at least two weeks prior to the review. The systems engineer will facilitate the review, identify issues, action items, and document results when the review is performed. The post work includes the appropriate participants resolving the action items, revising documents, and gaining approval on the changes, through the change control process facilitated by the systems engineer (SEBoK 8.0).

Assessment & Control Plan			
Review Type	**Review Description**	**Life Cycle Phase**	**Have this Review?**
Initial Technical Review	A multi-disciplined review that supports a program's initial program objective memorandum submission.	1 - Concept	Yes
System Requirements Review (SRR) or Conceptual Design Review	A multi-disciplined review to ensure that the system under review can proceed into initial systems development and that all system requirements and performance requirements derived from the initial capabilities document or draft capability development document are defined and testable, as well as being consistent with cost, schedule, risk, technology readiness, and other system constraints.	2 - Requirements & Arch.	Yes
Alternative Systems Review	A multi-disciplined review to ensure the resulting set of requirements agrees with the customers' needs and expectations.	2 - Requirements & Arch.	No
System Functional Review (SFR)	A multi-disciplined review to ensure that the system's functional baseline is established and has a reasonable expectation of satisfying the requirements of the initial capabilities document or draft capability development document within the currently allocated budget and schedule.	2 - Requirements & Arch.	Yes
Prototype Technical Review	To review the prototypes developed.	2 - Requirements & Arch.	No
Preliminary Design Review (PDR)	A technical assessment establishing the physically allocated baseline to ensure that the system under review has a reasonable expectation of being judged operationally effective and suitable.	3 - Design	Yes
Critical Design Review (CDR)	A multi-disciplined review establishing the initial product baseline to ensure that the system under review has a reasonable expectation of satisfying the requirements of the capability development document within the currently allocated budget and schedule.	3 - Design	Yes
Test Readiness Review (TRR)	A multi-disciplined review designed to ensure that the subsystem or system under review is ready to proceed into formal testing.	4 - Integration Test	Yes
Production Readiness Review (PRR)	The examination of a program to determine if the design is ready for production and if the prime contractor and major subcontractors have accomplished adequate production planning without incurring unacceptable risks that will breach thresholds of schedule, performance, cost, or other established criteria.	5 - System validation	No
Physical Configuration Audit	An examination of the actual configuration of an item being produced around the time of the full-rate production decision.	5 - System validation	No
System Verification Review (SVR)	A multi-disciplined product and process assessment to ensure the system under review can proceed into low-rate initial production and full-rate production within cost (program budget), schedule (program schedule), risk, and other system constraints.	5 - System validation	Yes
Technology Readiness Assessment	A systematic, metrics-based process that assesses the maturity of critical technology elements, such as sustainment drivers.	5 - System validation	Yes
Integrated Baseline Review	A joint assessment conducted by the government program manager and the contractor to establish the performance measurement baseline.	5 - System validation	No
Operational Test Readiness Review	A multi-disciplined product and process assessment to ensure that the system can proceed into initial operational test and evaluation with a high probability of success, and also that the system is effective and suitable for service introduction.	6 - Op Maintenance	Yes
Functional Configuration Audit	Formal examination of the as tested characteristics of a configuration item (hardware and software) with the objective of verifying that actual performance complies with design and interface requirements in the functional baseline.	6 - Op Maintenance	Yes
In-Service Review	A multi-disciplined product and process assessment that is performed to ensure that the system under review is operationally employed with well-understood and managed risk.	6 - Op Maintenance	Yes

FIGURE 12.2   Assessment and control plan example for women's healthcare center.

12.3.2   CONFIGURATION MANAGEMENT PLAN

The Configuration Management Plan is used to plan how the processes and data will be configured and changed.

Configuration management is a critical part of designing and developing a system, because it controls the information that is used to build the system, through the Vee Life Cycle Model. Two definitions of configuration management include:

Plan, review and follow up for Assessment & Control Reviews

Pre-work
- Create participants list
- Develop agenda
- Establish assessment criteria
- Distribute materials for review

Perform review
- Facilitate review
- Identify issues
- Identify action items
- Document results

Post work
- Resolve action items
- Create change controls
- Revise documents
- Gain approval to move forward

FIGURE 12.3   Typical design review activities.

Configuration Management (CM) deals with managing changes across the life cycle of a product. Its importance is, fully recognised in various disciplines such as design, engineering, production and services.

(Xu, Malisetty and Round, 2013)

Configuration Management is a management discipline that applies technical and administrative direction to the development, production and support lifecycle of a Configuration Item. The discipline is applicable to hardware, software, processed materials, services, and related technical documentation. Configuration Management is an integral part of life-cycle management.

(Burgess, Byrne and Kidd, 2003)

The configuration management activities are shown in Figure 12.4.

The configuration management activities with a description of each are as follows:

1) Identification: In this activity, the systems engineering team identifies the system elements that should be under configuration control or not and in which phase of the Vee model that it should be placed under configuration control. The phase that the element is placed under configuration control is when the artifact is baselined. The baseline is the first version of the artifact when it is approved and placed under configuration control. The types of elements that could be under configuration control can include:
   - Database
   - Document
   - Equipment
   - Facility

Configuration Management Activities

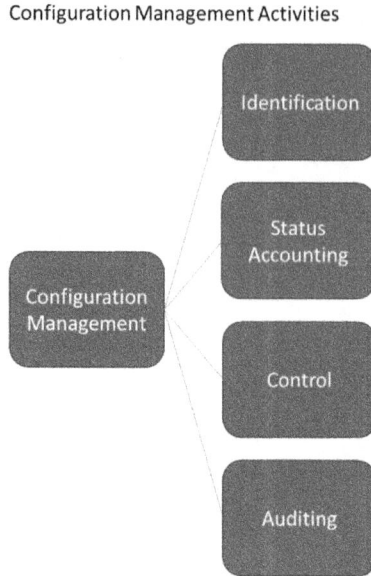

FIGURE 12.4   Configuration management activities.

- Hardware
- Materials
- Operator roles
- Organizational unit
- Procedure
- Policies
- Process
- Product
- Project
- Protocol
- Services
- Software
- Supplies
- Telecommunication
- Other

To identify whether an element should be under configuration control, you can consider whether:

- An item that is uniquely identified, such as by serial number,
and
- Would alter or affect the function of the system or sub-system if changed, modified, or adjusted in any way.

2) Status accounting: This activity includes assessing the status of each element that is identified to be under configuration control. The status usually includes whether the element has been baselined, it is not yet under configuration control, or it has been disposed or retired. The history for status accounting can include:

a. The purchase data: date purchased, price, shipping number, lot number, and other traceability data
b. Installation date
c. Removal date
d. Serial number

  e. Maintenance and repair data
  f. Version number
  g. Revision date
  h. Revision author
  i. Reason for revision or change

3) Control: This is the process of controlling any changes to elements and artifacts that are under configuration control. If the form, fit, or function of a system element that is under configuration control is to be changed, a change control process should be followed. Many organizations have a Change Control Board (CCB) that manages the change control process. Changes are typically categorized into two main classes (SEBoK 8.0):

  a. Class I: A major or signification change that affects cost, schedule, or technical performance. These normally require customer approval.
  b. Class II: A minor change that affects documentation or internal design details. It will not affect system specification requirements and does not require customer approval.

4) Auditing: In an audit there is an evaluation of the current status of the system's elements that are under configuration control, to determine their conformance to the functional and physical characteristics that are stated in the latest version of the artifacts. For documentation, a comparison between the document artifact and ensuring that the requirements are met can be part of the audit. For a physical element, the physical element can be compared to the requirements or design documents (drawings, specifications, etc.).

An example Configuration Management Plan for the Women's Healthcare Center is shown in Figure 12.5. The system elements (Figure 12.6) identify the system, the system elements, the type of elements, whether the element will be under configuration control, and the phase that it would go under configuration control. Figure 12.7 shows the configuration elements that are under configuration control, their description, and the current status of the elements with respect to configuration control.

Configuration Management Plan		
Topics	**Description**	**Activities**
Configuration Management Purpose	To control documents and technology changes	Define the purpose of configuration management.
Identification of elements to be under configuration control	Identify system elements to be under configuration control and the phase that they will go under control	Identify system elements to be under configuration control and the phase they will be baseline and go under control.
Configuration status accounting	Center Director will provide status accounting	Update status of system elements when they change.
Configuration control	Center Director will control	Perform change control process and document.
Configuration auditing	Center Director will audit	Perform audits, document results, resolve issues.

FIGURE 12.5 Configuration management plan for women's healthcare center.

System Elements				
System	System Elements	Type	Under Configuration Control?	Phase Goes Into Config. Control
Connect to Cancer Center	Patient Navigator	Operator roles	No	
Provide results	Radiologist	Operator roles	No	
Construction of Facility	Facility	Operator roles	Yes	5 - Op Maintenance
Women's Center	Services	Operator roles	Yes	5 - Op Maintenance
Women's Center	Physician	Operator roles	No	
Perform service	Technician	Operator roles	No	
Spiritual Care	Spiritual care resources	Operator roles	No	
Schedule Service	Resources	Operator roles	No	
Women's Center	Concierge	Operator roles	No	
Women's Center	Volunteers	Operator roles	No	
Process VIP patient	VIP service members	Operator roles	No	
Women's Center	Process dashboard	Software	Yes	4 - System validation
Perform service	Imaging technology	Equipment	Yes	5 - Op Maintenance
Women's Center	Process maps	Process	Yes	6 - Op Maintenance
Women's Center	Policies	Policies	Yes	6 - Op Maintenance
Women's Center	Procedures	Procedure	Yes	6 - Op Maintenance
Women's Center	Facility plans / drawings	Document	Yes	3 - Design
Women's Center	Test plans	Document	Yes	4 - Integration Test

FIGURE 12.6   System elements and identification of configuration control for women's healthcare center.

Configuration Elements		
Configuration Elements can include: physical parts, components, software code, requirements, specifications, architecture documents, etc.		
Configuration Elements	Description	Status
Facility	The newly constructed or rennovated facility	Not placed under configuration control
Services	Any new Women's Center services	Not placed under configuration control
Imaging technology	Any new imaging technology	Baselined and under configuration control
Process dashboard	The software that will display the latest process metrics for the system	Revised and under configuration control
Process maps	All process maps for the Women's Center, current & future state	Baselined and under configuration control
Policies	Any required policy changes	Baselined and under configuration control
Procedures	All procedures related to the Women's Center	Baselined and under configuration control
Facility plans / drawings	The architecture drawings and plans	Baselined and under configuration control
Test plans	Test plans for verification and validation	Revised and under configuration control

FIGURE 12.7   Configuration elements and current status for women's healthcare center.

12.3.3   CONTRACTOR MANAGEMENT PLAN

The Contractor Management Plan is used to describe how the contractors will be managed in relationship to the government or system owner organization. The Contractor Management Plan will vary based on the infrastructure already existing in the organization for contract management. Some resources and infrastructure may need to be developed. An example Contractor Management Plan is shown in Figure 12.8.

Contractor Management Plan		
Topics	**Description**	**Activities**
1) Develop request for proposals	Develop request for proposal documents	1) Identify request for proposals (RFPs) to be developed
		2) Develop appropriate request for proposals
		3) Review and revise RFPs
		4) Post RFPs
Hold bid meetings	Plan and hold bid meetings with potential contractors	1) Review questions from contractors on RFPs
		2) Plan bid meetings
		3) Hold bid meetings
3) Review request for proposals	Develop criteria and rubrics for evaluating bids and contractors. Review requests for proposals	1) Develop criteria and rubrics for bid evaluations
		2) Review RFPs
		3) Evaluation RFPS against rubrics
		4) Select contractors
		5) Award contracts
		6) Update documentation
4) Develop infraasture for managing contractors	Develop contract management infrastructure	1) Develop required contract management infrastructure
		2) Identify resources for contract management
5) Manage contractors	Manage contractors and contracts	1) Manage contractors and contracts
6) Track contractor metrics	Track contractor metrics and feedback to contractors	Track contractor metrics and feedback to contractors

FIGURE 12.8   Contractor management plan.

12.3.4   DEPLOYMENT PLAN

The Deployment Plan describes how the system will be deployed within the Phase 6 Operations and Maintenance. The Deployment Plan includes development of a deployment or transition strategy, performing system installation and on-site testing, and implementing system performance measurement and training for operational and support users. A Deployment Plan is shown in Figure 12.9.

12.3.5   DISPOSAL AND RETIREMENT PLAN

The Disposal and Retirement Plan is used to safely dispose the system elements when the system is retired. It includes evaluating the disposal and retirement approach, identification of the requirements for what needs to be disposed and, also retired. The plans should include how the team will perform the system disposal and retirement. A sample Disposal and Retirement Plan for the Women's Center is shown in Figure 12.10.

Deployment Plan		
Topics	**Description**	**Activities**
1) Develop deployment or transition stragety	Develop deployment or transition strategy to make the system operational after testing is complete.	1) Develop deployment or transition strategy 2) Develop installation activities 3) Develop integration activities 4) Develop testing approach on site
2) Perform system installation	Perform system installation including connecting interfaces and integration to other systems.	1) Perform on site installation of the system 2) Perform on site system testing 3) Document and obtain system sign-offs
3) Implement system performance measurement	Implement system performance measurement including: operationaltion performance; system effectiveness; system efficiency, failure reporting and trending; training effectiveness and need for traing and development for operational and support users.	1) Define criteri for system performance 2) Roll out system performance data collection 3) Define training plan and approach 4) Implement training and development for operational and support users.

FIGURE 12.9   Deployment plan.

Disposal and Retirement Plan	
Disposal and Retirement Activities	**Owners and Resources**
Evaluate disposal of medical equipment when ready to retire.	Director of Biomedical engineering Equipment vendor
Evaluate disposal of facilities materials when rehabbing and constructing new facility from old facility	Director of construction
Consider water pollution impact for disposal of materials and medical waste	Chief Operating Officer Director of Facilities
Develop plan for disposal of medical waste	Chief Operating Officer Director of Facilities

FIGURE 12.10   Disposal and retirement plan.

12.3.6   INFORMATION MANAGEMENT PLAN

The Information Management Plan is used to manage the technical information for the development, design, and deployment of the system. This includes dissemination of the information to designated stakeholders.

The information for a systems development project includes the technical, project, organizational, and user information. Examples of information types are shown in Table 12.2. It is important that the process related to managing information ensures that information is relevant to the stakeholder groups, timely, complete, validated, appropriate, and confidential when required. Information must also ensure that the data is properly stored, maintained, secured, and accessible to the identified stakeholders.

TABLE 12.2
Examples of Information Types

Technical	Project	Organizational	User Information
Drawings	Systems Engineering Management Plans	Policies	Users' names, information rights
Specifications	Project and program plans	Procedures	
Requirements	Work breakdown structures	Processes	
Interface documents	Gantt charts	Organizational charts	
Test plans, cases, and results			
Class information models			

The steps to develop an Information Management Plan align with the phases of the Vee model, with respect to developing information management elements (SEBoK):

1) Perform systems operational needs analysis for information needs
2) Identify system requirements and architecture for information
3) Develop component design for information
4) Implement system, verify, and validate for information
5) Operationalize system maintenance and sustainment for information

Information Management planning is very extensive, dealing with data management as well. Data management planning develops an approach for developing data requirements, collecting, validating, storing, and maintaining data used in the system. The steps for performing data management include (SeBok):

1) Identify data requirements
2) Acquire data
3) Receive, verify, and accept data
4) Store, maintain, and control data
5) Use and exchange data

An example Information Management Plan is shown in Figure 12.11.

12.3.7 INTERFACE MANAGEMENT PLAN

The Interface Management Plan is used to manage the system interfaces, to ensure appropriate verification testing. An interface is "*A shared boundary between two functional units, defined by various characteristics pertaining to the functions, physical exchanges, and other characteristics.*" An interface can also connect "*... two or more systems or system components for the purpose of exchanging data, materials, forces, or energy from one to the other.*" (SEBok)

Hardware and software interfaces exist in systems and can be described and measured in multiple ways. The physical types of interfaces include:

- Structural: related to the parts and components and how they connect
- Electrical: related to power and electricity
- Thermal: related to heat transfer
- Temporal: related to timing of activities
- Electromagnetic: related to electromagnetic connectivity

Information Management Plan		
Topics	**Description**	**Activities**
1) Perform systems operational needs analysis for information needs	Develop the system's information management needs for technical, project, organizational, and user information	1) Develop information management needs
		2) Validate information management needs
		3) Dcoument information management need
2) Identify system requirements and architecture for information	Identify system requirements and architecture for information	1) Translate information needs to information requirements
		2) Develop information architecture
		3) Validate and document requirements and architecture
3) Develop component design for information	Develop components design and specifications for information design	1) Develop component design
		2) Develop data requirements
		3) Acquire data
		4) Receive, verify and accept data
		5) Store, maintain and control data
		6) use and exchange data
4) Implement system, verify and validate for information	Verify, validate and implement system information and data	1) Implement system
		2) Verify and validate information
5) Operationalize system maintenance and sustainment for information	Center Director will audit	1) Develop operationalized information management plan
		2) Development information management plan

FIGURE 12.11   Information management plan example.

Software systems typically connect users to the systems, information systems to information systems or applications to applications, and data to users or data to systems and databases for storage and retrieval. The N-squared diagram shown in Chapters 9 and 10 enables the identification of integration or interface points within and between systems. The Interface Management Plan, shown in Figure 12.12, describes the planning of interface planning and management.

12.3.8   Improvement and Maintainability Program Plan

The Maintainability Program Plan is used to maintain the system in the operations phase. A sample improvement and Maintainability Program Plan is shown in Figure 12.13.

Interface Management Plan		
Topics	**Description**	**Activities**
1) Define interface and integration requirements	Define interface and integration requirements with application of N-Squared Diagram.	1) Define interface and integration requjirements
		2) Document interface and integration reqjuirements
		3) Develop test plans and test cases- for testing interfaces and integration
		4) Develop criteria for testing success
2) Manage interfaces	Manage interfaces and testing	1) Manage interfaces as they are interfaced and integrated
		2) Test interfaces and integration points
		3) Document test case results

FIGURE 12.12   Interface management plan.

Maintainability Program Plan	
Activities	**Owners**
Identify list of supppliers	Director of Maintenance
Develop maintainability program review schedule	Director of Maintenance
Perform maintainability program reviews	Director of Maintenance
Collect, analyze, and develop a corrective action system	Director of Maintenance
Perform maintainability modeling	Director of Maintenance
Perform maintainability allocation	Director of Maintenance
Perform maintainability allocation	Director of Maintenance
Perform FMECA analysis	Director of Maintenance
Perform maintenance task analysis (MTA)	Director of Maintenance
Evaluate level-of-repair analysis	Director of Maintenance
Identify and prepare maintainability data	Director of Maintenance
Plan and implement maintainability program	Director of Maintenance

FIGURE 12.13   Maintainability plan.

12.3.9   MEASUREMENT PLAN

The Measurement Plan is used to design and manage metrics for measuring the key performance indicators for the system. The Critical to Satisfaction (CTS) characteristics measurement definition and collection approach used with Six Sigma projects are an effective way to identify metrics to measure the product or service being developed. CTSs are the characteristics of a product, process, or service that significantly affect the output. There are categories of CTSs that can be organized as follows:

CTQ: Critical to Quality

- Customer Satisfaction/Service: This relates to how satisfied the customer is with the product, process, or service.
- Quality/Defects: Quality and defects relate to the quality of a product or the type and number of defects related to the quality of the product, process, or service. Quality of the systems engineering project or program can also be assessed in this category.

CTQ: Critical to Delivery

- Timeliness/Turnaround: Timeliness and turnaround relate to the timeliness or delivery of the product, process, or service. Timelines of the systems engineering project or program can also be measured.

Cost/Productivity

- Cost to develop and/or deliver the product, process, or service. Cost can be measured, at many different levels, by the project, program, through to the individual components, parts, process, or service, including an assessment of the resources applied to the systems engineering project or program.
- Productivity is a measure of the resources applied to perform desired activities, develop a product, process, or service. This can be applied at multiple levels within the systems engineering project and program.

Revenue/Market Growth

- The Revenue that a certain product or service brings to the organization can be assessed in this category.
- The Market Growth that the organization achieves for the product or service can be assessed in this category.

The Measurement Plan tool can be used to identify the following data elements to define the data to be collected for the metrics related to the product, process, or service being developed and delivered. The following provides definitions of each element:

CTS: The CTS characteristics of the product, process, or service are used to measure and assess the outcomes or outputs. The CTS is collected by listening to the Voice of the Customer (VOC). VOC is a term used to describe listening to the customers to understand their needs related to the product, process, or service.

Metric: A standard of measurement.

Operational definition: A clear, concise definition of how to measure the metric. The operational definition should include a definition, purpose, and description of how to measure the metric.

Data collection source: Here the data is collected manually with methods such as via a time study or a check sheet or from an established database, such as an Emergency Department Bed Board system.

Analysis mechanism: The mechanisms of how the data will be analyzed, such as using descriptive statistics or Analysis of Variance (ANOVA) techniques.

Sampling plan: The sampling plan that will be used, including the sample size, sampling type, and how to ensure that the sample is representative of the population.

Sampling instructions: A description of how the data will be collected, so that multiple analysts can collect the data in a consistent and valid manner.

A sample Measurement Plan for the Women's Center is shown in Figure 12.14.
Steps for creating a Measurement Plan:

The steps to develop the Measurement Plan are:

1) Define the CTS characteristics
2) Develop metrics
3) Identify data collection mechanism (s)
4) Identify analysis mechanism (s)
5) Develop sampling plans
6) Develop sampling instructions

We now discuss each of the steps for creating a Measurement Plan.

1) **Define the CTS characteristics**
   To define the CTS, the following steps are performed:
   1. Gather VOC data relevant to the product, process, or service. Many techniques can be used, such as surveys, interviews, focus groups, to name just a few methods.
   2. Identify relevant statements in transcripts (verbatims) of customer comments and copy them onto self-stick notes. Focus on statements that relate to why a customer would or would not buy your product/service.
   3. Use Affinity Diagrams or Tree Diagrams to sort ideas and find themes.
   4. Start with the themes or representative comments and probe for why the customer feels that way. Do follow-up with customers to clarify their statements. Be specific to identify why.
   5. Conduct further customer contact as needed to establish quantifiable targets and tolerances (specification limits) associated with the need.
   6. When you've completed the work step back and examine all the CTSs or customers' needs as a set. Fill in gaps as needed.

Measurement Plan						
Identify metrics to measure and assess improvement that relate to the CTS's from the Define Phase.						
Critical to Satisfaction (CTS)	Metric (short title)	Operational Definition (metric description)	Data Collection Source	Analysis Mechanism	Sampling Plan (size, frequency)	Process to Collect and Report
Timeliness	Procedure time	Time from when patient goes into procedure room to when they leave	Medical information system	Statistical analysis	July through September 2014	IT Director will provide reports
Timeliness	Wait time for procedure	Time from when the patient is registered to when they get into the procedure room	Medical information system	Statistical analysis	July through September 2014	IT Director will provide reports
Timeliness	Time to register patient	Time from when patient arrives in the facility to when registration is complete	Registration system	Statistical analysis	July through September 2014	IT Director will provide reports
Timeliness	Time to receive results	Time to receive results of exam, from when you finished the procedure	Imaging system	Statistical analysis	July through September 2014	IT Director will provide reports
Timeliness	Wait to get an appointment	Time to wait until get an appointment	Scheduling system	Statistical analysis	July through September 2014	IT Director will provide reports

FIGURE 12.14   Measurement plan.

2) **Develop Metrics**

To develop the metrics, you have to generate and define at least one metric for each CTS, although you could have multiple metrics that you want to measure for each CTS. Think about a title that everyone understands for the metric. Frequently, there already exists metrics that are established and measured in the organization. Ensure that they meet the goal for measurement for each CTS. Ensure that a clear and concise operational definition is documented and agreed upon with the stakeholders. The operational definition includes a documented definition, with a purpose and description of how to measure the metric. This can include the time frame of the data to be collected for measurement. A baseline for the metric should be collected before or when the system becomes operational to ensure a reference is established for measuring improvements as well as trends in the operation of the system.

An example of an operational definition for measuring the time that a vision center takes to process eyewear orders follows:

Defining the Measure – The focus is on the time that it takes for a customer to receive their eyewear order.
Purpose – We want to improve customer satisfaction by measuring the time it takes to deliver the eyewear order and then improve this time and compare it to our USL = Four days.
Clear way to measure the process
- We will measure the time from the register transaction to when the call is placed notifying the customer that the glasses are received.
- We will measure this in 10 vision centers in four states.
- We will collect data from May 1 to June 30.

3) **Identify data collection mechanism (s)**

The data collection mechanism defines how data can and will be collected. Data may need to be collected manually, such as through performing a time study of times to complete a process or collecting defects in a process with a manual check sheet. Many times, there are already established databases that can be used to download data from the system to perform the analysis that lends to using a dashboard or scorecard related to the system of interest.

4) **Identify analysis mechanism (s)**

Thinking about how the data will be analyzed sometimes clarifies how the data will be collected or should be collected. If we are collecting defect data from a manufacturing process and want to perform descriptive and inferential statistics, by collecting dimensional data from a part, we have to ensure that the data is collected to accommodate this analysis mechanism. You would want to collect the measured dimension for each part measured, instead of just tracking whether the part met the specification or not (conforming or non-conforming). If we want to go back to the original data to collect a mean and standard deviation, if we don't collect and track the data in the correct manner, we would have lost out on the ability for generating descriptive statistics and performing hypotheses tests in the future.

5) **Develop sampling plans**

In developing sampling plans, we have to identify the sample that will be collected, the sample size that ensures the statistical power of our test, the sampling method that we will use and ensure that the sample is representative of the population, representing all possible or probable conditions (operators, days, hours, weeks, seasonality, current state of the system, process, or service). There are many sample size formulas that can be used to determine a statistically valid sample size for continuous data. For survey data, typically a 20% response rate is considered "good" for surveys sent to mostly anonymous potential respondents. For a more focused convenience sample, where the researchers or analysts selected the potential respondents, a much higher response rate should be possible. For focus groups, there are no set response rates, rather focus groups are run until no new themes are generated from the focus groups. There are usually less than 10 participants in an individual focus group. Following are different types of sampling methods (Quality Council of Indiana Black Belt Primer, n.d.):

- Simple Random Sample: Each unit has an equal chance of being sampled.
- Stratified Sample: The N items are divided into subpopulations or strata and then a simple random sample is taken from each stratum. This method is used to decrease the sample size and the cost of sampling. Ex. Sample from each location.
- Systematic Sample: N items are placed into k groups. The first item is chosen at random; the rest of the sample selects every kth item.
- Cluster Sample: N items are divided into clusters and used for wide geographic regions, although it is not as efficient as other sampling methods.

6) **Develop sampling instructions**

To ensure that everyone who collects the data is collecting it in a consistent manner, there should be clearly documented sampling instructions. The instructions should include:

- Who: Who will collect the data?
- Where: Where will the data be collected?
- When: When will the data be collected? Consider the elements of a representative sample discussed above.
- How: How will the data be collected?

12.3.10   QUALITY MANAGEMENT PLAN

The Quality Management Plan is used to plan and implement a quality management system. We describe a Quality Management Plan based on the Baldrige National Quality Award Criteria for Performance Excellence. A Quality Management System is defined as a way of how an organization can meet the requirements of its customers and stakeholders for whom the system is being developed (ISO, 2015).

The criteria for Performance Excellence includes the following (Baldrige, 2019) elements:

1. Leadership
2. Strategy
3. Customers
4. Measurement, Analysis, and Knowledge Management
5. Workforce
6. Process or Operations
7. Results

The Malcolm Baldrige National Quality Award's (MBNQA) Criteria for Performance Excellence elements are shown in Figure 12.15.

Leadership:

The description of senior leadership, how they lead, how they foster achievement of the mission are part of the leadership criteria. Leadership should define how the organization creates a sustainable organization and how the leadership elements relate to the workforce engagement criteria. The leadership should describe how leadership will take action to accomplish the organizational objectives as well as the organizational governance structure. The Leadership criteria helps to describe how leadership ensures ethical behavior and what to do when ethics are breached. Leadership helps to describe how the organization ensures societal responsibilities and connection to the key communities.

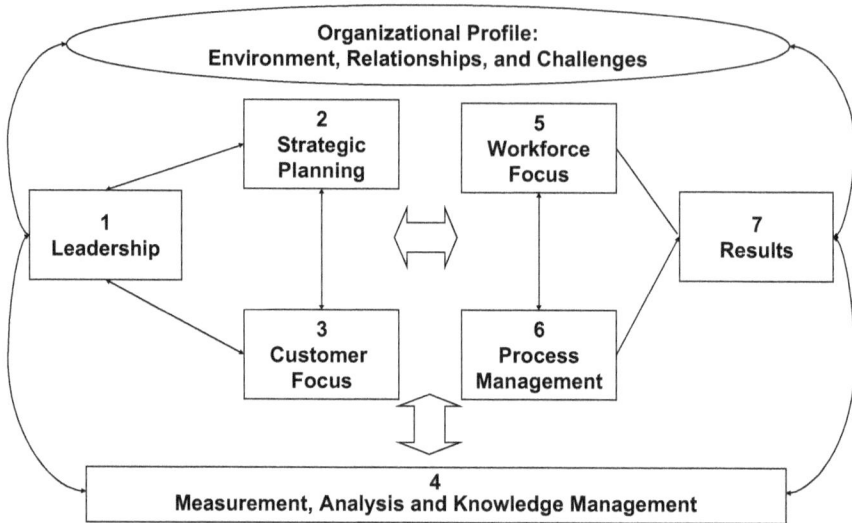

FIGURE 12.15 Baldrige criteria for performance excellence.

Strategic Planning:
The Strategy criteria is used to extract the information for identifying the strategic planning process, the deployment processes, and the performance indicators related to the strategic plan. The organization can develop a SWOT analysis to understand key external factors that impact the strategic objectives.

Customer Focus:
The next QMS element is the customer focus element that captures how to collect and manage VOC data and information. The VOC information would include understanding and managing customers' expectations and requirements. The leadership defines how to create and sustain the culture that focuses on customer and stakeholder satisfaction. In this element the organization would also describe how to measure and improve patient satisfaction and patient dissatisfaction as well as manage and recover from customers' complaints.

Measurement, Analysis, and Knowledge Management
The next QMS element is the Measurement, Analysis, and Knowledge Management. The organization describes the metrics that are used to measure, analyze, and improve organizational improvement. The metrics are developed to manage processes, customer metrics, service line metrics, leadership metrics, financial, market, and workforce metrics. Leadership describes how to manage information, data, and knowledge for the organization, through definition of the leadership policies as well as describe the policies for managing information resources and technology.

Workforce Management:
The next QMS element is Workforce Management. The organization describes the culture and how it encourages workforce engagement and how the organization assesses workforce engagement. The Workforce Management element documentation would also describe how the organization builds an effective and supportive workforce environment and creates a diverse work environment.

Quality Management Plan		
Topics	**Description**	**Activities**
1) Define how each criteria for the quality management plan will be applied	Define the quality management plan approach	1) Develop leadership approach
		2) Develop strategy approach
		3) Develop customer focus approach
		4) Develop Measurement, Analysis, and Knowledge Management approach
		5) Develop workforce management approach
		6) Devlop Process or Operations approach
		7) Develop Results approach
2) Implement the quality management plan	Implement the quality management plan	1) Develop selection criteria for design elements
		2) Perform trade off analysis
		3) Perform effectiveness analysis
		4) Perform cost analysis
		5) Perform technical risk analysis
		6) Develop justification report

FIGURE 12.16   Quality management plan.

Process or Operations Management:
The next QMS element is Process or Operations Management. The organization describes how the organization designs and innovates their processes. The organization describes the key and support processes and ensures that the patients' requirements are captured and incorporated into the business processes. This element helps to describe how the processes will be controlled and improved and which process metrics will be used to understand the costs, efficiency, and effectiveness of the work processes. This element will also help the organization to identify their process improvement strategies.

Results:
The last QMS element is the Results. This element describes the different types of metrics and the results of the metrics. The result categories include:

- Process outcomes
- Customer-focused outcomes
- Workforce-focused outcomes
- Leadership and governance outcomes
- Financial and market outcomes

A Quality Management Plan is shown in Figure 12.16.

Risk Management Plan		
Topics	Description	Activities
1) Risk planning	Define interface and integration requjirements with appliation of N-Squared Diagram.	1) Develop risk planning strategy
		2) Develop risk planning process
		3) Develop risk planning implementation approach
		4) Defvelop risk documentation approach
2) Risk identification	Manage interfaces and testing	1) Identify potential system risks
		2) Collect data if it exists on similar risks
		3) Document potential risks
3) Risk analysis	Perform risk analysis	1) Perform risk analysis, including contributing causes, outcomes and impacts of risk events
4) Risk handling	Develop risk handling approaches for identified risks	1) Develop risk handling approaches
5) Risk monitoring	Plan and perform risk monitoring	1) Develop risk monitoring approach
		2) Implement risk monitoring approach
		3) Mointor risks, and re-assess when needed

FIGURE 12.17   Risk management plan.

12.3.11 RISK MANAGEMENT PLAN

The Risk Management Plan is used to identify, assess, monitor, mitigate, and control system risks.

Risk management was initially discussed in Chapter 5. A sample Risk Management Plan is shown in Figure 12.17.

12.3.12 SPECIALTY ENGINEERING PLAN

The Specialty Engineering Plan is used to plan the work of the required specialty engineering disciplines. A sample of a high-level plan is shown in Figure 12.18.

12.3.13 SYSTEM DEVELOPMENT PLAN

The System Development Plan is used to plan the development of the system in the Phase 3 Detailed Design. A sample system development plan is shown in Figure 12.19.

Speciality Engineering Plan		
Topics	**Description**	**Activities**
1) Identify engineering speciality disciplines for the system	Identify engineering speciality disciplines	1) Develop engineering speciality disciplines based on system concepts and requirements.
		2) Update speciality engineering disciplines throughout lifecycle as needed
2) Plan and develop infrrasture and reseouces for special engineering disciplines	Develop specialty engineering disciplines plans	1) Develop "Functional" engineering (electrical, mechanical, structural, industrial, etc.) plans
		2) Develop software engineering plans
		3) Defvelop reliability engineering plans
		4) Develop maintainability engineering plans
		5) Develop human factors engineering plans
		6) Develop safety engineering plans
		7) Develeop security engineering plans
		8) Develop manufacturing and production engineering plans
		9) Develop logisitcs and supportabiity engineering plans
		10) Develop disposability engineering plans
		11) Develop quality engineering plans
		12) Develop environmental engineering plans
		13) Develop value / cost engineering plans
		14) Develop other engineering discipline plans (as appropriate)

FIGURE 12.18   Specialty engineering plan.

12.3.14 SYSTEM INTEGRATION PLAN

The System Integration Plan is used to plan how the system will be integrated to ensure verification and validation of the system. The system integration approach describes how the system will be built, integrated, and then tested. A system integration plan is shown in Figure 12.20.

System Development Plan		
Topics	**Description**	**Activities**
1) Define detailed design	Define detailed design	1) Define detailed design - physical architecture model -system elements - Quality Function Deployment
2) Perform systems analysis	Perform systems analysis with trade-off analysis, effectiveness analysis, cost analysis, technical analysis	1) Develop selection criteria for design elements 2) Perform trade - off analysis 3) Perform effectiveness analysis 4) Perform cost analysis 5) Perform technical risk analysis 6) Develop justification report

FIGURE 12.19   System development plan.

System Integration Plan		
Topics	**Description**	**Activities**
1) Define system integration requirements	Definesystem integration requjirements	1) Define system requjirements 2) Document system integration reqjuirements 4) Develop system integration approach 4) Develop test plans and test cases- for testing system integration
2) Deploy integration plans	Deploy system integration	1) Build and integrate system 2) Test system integration 3) Document test case results

FIGURE 12.20   System integration plan.

12.4 SUMMARY

This chapter covered the systems engineering planning and the multitude of plans that are associated with these planning efforts.

12.5 ACTIVE LEARNING EXERCISES

1) Tailor the plans that are part of the Systems Engineering Management Plan for your system. Define which plans should be part of your Systems Engineering Management Plan.
2) Develop each of the plans discussed in this chapter for your system:
   a. Assessment and Control Plan
   b. Configuration Management Plan
   c. Contractor Management Plan
   d. Deployment Plan
   e. Disposal and Retirement Plan
   f. Information Management Plan
   g. Interface Management Plan
   h. Maintainability Program Plan
   i. Measurement Plan
   j. Quality Management Plan
   k. Risk Management Plan
   l. Specialty Engineering Plan
   m. System Development Plan
   n. System Integration Plan

BIBLIOGRAPHY

Baldrige Excellence Framework, Baldrige Performance Excellence Program, National Institute of Standards and Technology (NIST), United States Department of Commerce, Gathersburg, MD, www.mist.gov/baldrige, 2019.

Burgess, T.F., Byrne K. and Kidd C., Making Project Status Visible in Complex Aerospace Projects, *International Journal of Project Management* 2003 Vol. 21 Issue 4, p. 251–259.

No Author, International Organization for Standardization, ISO Central Secretariat Chemin de Blandonnet 8 Case Postale 401 CH – 1214 Vernier, Geneva Switzerland, iso.org, 2015

No Author, The Certified Six Sigma Black Belt Primer, 4th Edition, Quality Council of Indiana, (2014)

SEBoK Editorial Board, *The Guide to the Systems Engineering Body of Knowledge (SEBoK)*, vol. 2.2, ed. R. J. Cloutier (Hoboken, NJ: Trustees of the Stevens Institute of Technology, 2020), accessed May 25, 2020, www.sebokwiki.org. BKCASE is managed and maintained by the Stevens Institute of Technology Systems Engineering Research Center, the International Council on Systems Engineering, and the Institute of Electrical and Electronics Engineers Computer Society.

Xu, Y., Malisetty, M. K., and Round, M., *Configuration management in aerospace industry*, 2nd International Through-life Engineering Services Conference, 2013.

13 Model-Based Systems Engineering (MBSE)

13.1 PURPOSE

The purpose of this chapter is to describe Model-Based Systems Engineering. We start first with providing a description and definition of MBSE and the benefits and challenges of implementing MBSE versus a traditional document-based orientation to systems engineering. We then address potential research opportunities in the MBSE space.

13.2 DESCRIPTION OF MODEL-BASED SYSTEMS ENGINEERING

Model-Based Systems Engineering is an approach to applying systems engineering principles, tools, and activities with a focus on developing electronic models that are architected to be integrated throughout the systems engineering life cycle as well as from high levels to more detailed levels of the models and from single elements to the integrated system, as illustrated in Figure 13.1.

Badiru identified two levels of integration, horizontal integration across the phases of the life cycle and vertical integration from the component to the system to the operational level of models (Badiru, 2019). We add a dimension for integrating elements to each other within the system. The three dimensions of integration in our view of MBSE are 1) between phases of the life cycle, 2) between levels of detail, and 3) integration between elements within the system. The models integrate elements between each phase of the life cycle: concepts, stakeholders, and risk in the concepts phase, to requirements and architecture in that phase, to specifications and integration and interfaces in detailed design, through test cases and results in implementation, verification, and validation, through to a deployed integrated system. The models integrate increasing detail from high-level functions through to decomposed functions providing more insight into details of supported system functions. Lastly, the models show integration and connectivity between elements, sub-systems, and systems.

MBSE requires a change in thinking from a modeling approach based on documents, such as requirements, specifications, drawings, test cases, to one based on integrated models that are architected to demonstrate traceability between the elements of the traditional documents used in systems engineering design and development. According to Madni and Sievers (2018), the value of MBSE comes from the system-related information being stored in a centralized and configuration managed repository. The software repository should enable the architected and interconnected capability of the system elements and models. It should also ensure alignment to the modeling language and standards and incorporate error identification as part of the information system.

DOI: 10.1201/9781003081258-15

3-Dimensional Nature of MBSE

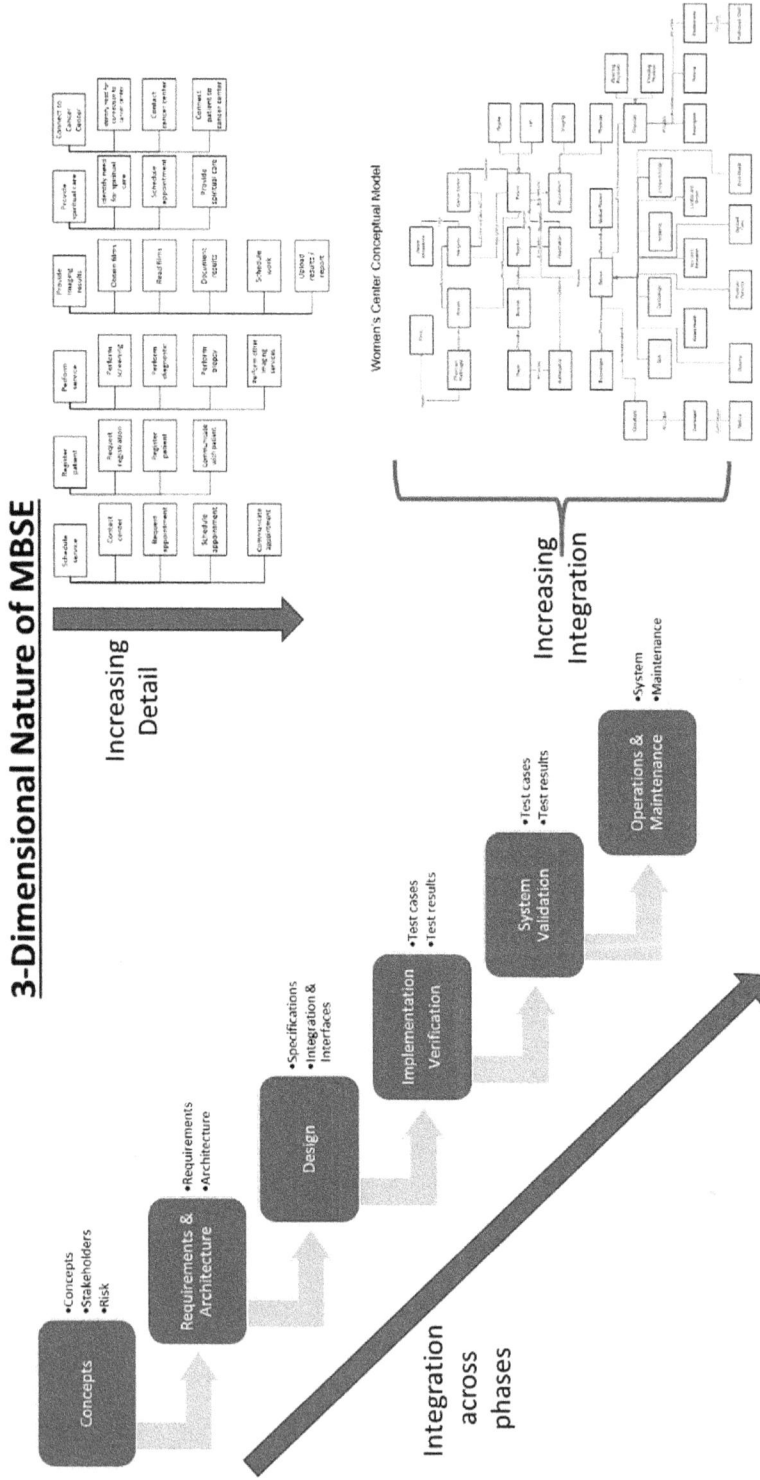

FIGURE 13.1   3-Dimensional nature of MBSE.

13.3 BENEFITS AND CHALLENGES OF IMPLEMENTING MBSE

13.3.1 BENEFITS

There are many benefits to using MBSE including, summarized in Figure 13.2 (Madni and Sievers, 2018), (Badiru, 2019):

1) Having a common modeling language with electronic architecture of models and model elements
2) Electronic repository, configuration management, and change control
3) Improved efficiency, productivity, and quality
4) Enhances communication and coordination amongst systems engineering stakeholders
5) Enables automation and optimization
6) Enables multiple views of the system and its elements, interfaces, and interaction points

Following is more detail related to each benefit.

1) Having a common modeling language with electronic architecture of models and model elements
   Using a MBSE software provides the ability to use a common modeling language that provides standard models and symbols. The system architecture is embedded within the software enabling consistency of the models within the system and across other systems projects. The use of the standard modeling language provides consistency, easier learning, and less variation in the use of the models across different users.
2) Electronic repository, with configuration management and change control
   The electronic repository of the models along with electronic configuration management and change control helps to manage changes to the models once they are baselined and version controlled. Automated distribution of the changes to affected stakeholders can be part of the configuration and change management processes.
3) Improved efficiency, productivity, and quality
   Because the models are electronic, standard, and include configuration and change control, this can improve efficiency and productivity of the development of the initial models as well as changes to the models during downstream phases of the systems engineering life cycle. The quality of the modeling and systems engineering processes can be enhanced due to controlling variation and standardizing the models, architecture, configuration, and change control. Embedded error checking can also ensure the quality of the models. According to Jorgensen (2019), a "traditional functional requirements decomposition approach is likely to capture

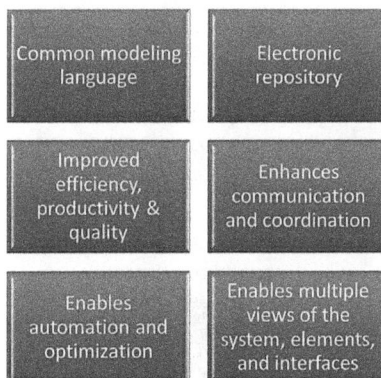

FIGURE 13.2 Benefits of MBSE.

50% of problem understanding." Incorporating use cases and scenarios can increase problem understanding to 90% or more the first time through. Up to 40% fewer requirement defects were found in systems developed with MBSE (Estefan, 2011).

4) Enhances communication and coordination amongst systems engineering stakeholders
When concurrent or simultaneous engineering was first instituted it led to huge improvements in communication and coordination amongst systems engineering teams and related stakeholders as teams worked together to develop requirements and engaged the entire team from marketing, design engineers, manufacturing engineers, process engineers, and so forth to be engaged together in a team versus having each discipline perform their activities in isolation and "throw" their deliverables over the wall to the next group. This has been further enhanced with MBSE through automation of much of the workflow and communication between stakeholders impacted by changes to the interconnected models. It doesn't rely on knowledge of the stakeholders for identifying who are impacted by changes to requirements and the cascading impacts to other related models, it is a part of the system functionality and architecture, leaving less to chance.

5) Enables automation and optimization
Incorporating electronic models for the systems engineering deliverables enable integration to other electronic automation and optimization software programs, such as simulation, statistical analysis, optimization, forecasting, and other decision analysis tools.

6) Enables multiple views of the system and its elements, interfaces, and interaction points
MBSE enables having multiple views, diagrams, reports, and documentation of the system and its elements, along with interfaces and interaction points. The systems engineer or other specialty engineers responsible for creating the system models can create them once, and the system enables presenting them as the stakeholder groups best understand and prefers viewing the models. Narratives and graphical views are all easily output to reports with the most up-to-date information, for review, use, and training purposes.

13.3.2 Challenges

There are many challenges that have hindered the adoption of MBSE, summarized in Figure 13.3, including (Madni and Sievers, 2018), (Badiru, 2019):

1) Lack of training and education of the MBSE tools, principles, and methods in both industry and academia, with a steep learning curve of tools and practices.
2) Need to change the culture and practice from document-centric to model-centric modeling.

FIGURE 13.3 Challenges of MBSE.

3) Limitations of preferred system architectures and models in existing MBSE software applications.
4) Lack of rigor in requirements analysis.
5) Activities that leverage MBSE has focused on requirements and architecture modeling versus design, verification, and validation phases of the life cycle.

Following is additional detail related to each of the challenges.

1) Lack of training and education of the MBSE tools, principles, and methods in both industry and academia, with a steep learning curve of tools and practices.
   There are many more traditional engineering programs that teach Computer-Aided Design for design drawings versus MBSE and systems engineering. Many specialty engineering disciplines learn CAD software but not necessarily MBSE and SySML. CAD applications have been around quite a bit longer than MBSE applications, so the maturity of CAD applications and use of them are more prevalent. There exists a steep learning curve related to the tools, language, and practices related to systems engineering and requirements analysis and elicitation techniques.
2) Need to change the culture and practice from document-centric to model-centric modeling.
   This challenge is probably one of the biggest challenges preventing the adoption of MBSE. Structures and processes embedded in organizations around traditional document-based applications, review, and configuration controls provide a strong inertia against migrating to MBSE. Not only do new applications, modeling approaches, and languages need to be learned but the ways that teams review models need to be learned as well. Reviewing graphical models with some narrative is much different than purely narrative requirements documents. There can be a substantial difference in review packages than what teams and customers are used to in the document-centric world (Madni and Sievers, 2018).
3) Limitations of preferred system architectures and models in existing MBSE software applications.
   MBSE applications have evolved out of Information Architecture and Enterprise Architecture modeling and tools. One of the seminal papers by John Zachman was published in 1987 (Zachman, 1987) that described the information architecture concepts which evolved into Enterprise Architecture. He developed what he called primitives, schemas, or an ontology that answered the fundamental architectural questions of what (data), how (process), where (location), who (roles), when (timing), and why (motivation) (Zachman, 1967-2011). So, the MBSE concepts have evolved only in the last 30 years, or so, and the tools even later. Some MBSE tools have embedded meta models that maintain the elements, their relationships, and behaviors, while other MBSE tools allow the user to create their own models, elements, and relationships, which can lead to lack of consistency and standardization. Additionally, integration with CAD and engineering drawings is limited, preventing traceability, configuration control between models and drawings, requirements, specifications, and test cases. It is also difficult to create integrated review packages automatically, requiring extensive manipulation.
4) Lack of rigor in requirements analysis.
   There are steep learning curves related to the MBSE models and language as well as the heavier qualitative requirements elicitation techniques based upon scenario building, process maps, and use case tools. Engineers typically do not learn and practice requirements analysis elicitation techniques in school, instead they are given requirements as an assumption. Interviewing, building user stories and scenarios requires extensive communication, listening, and facilitation skills that some engineers may not have been exposed to in their past educational and industry experience.
5) Activities that leverage MBSE has focused on requirements and architecture modeling versus design, verification, and validation phases of the life cycle.

Current MBSE tools are most applied in the upfront conceptual design, requirements, and architecture phases of the systems engineering life cycle versus the design, verification, and validation phases. There needs to be an integration across the life cycle so that requirements are traceable from requirements, the architecture, to design specifications all the way through to test cases for verification and validation activities (Glaessgen and Stargel, 2012). Design specifications and other artifacts are maintained outside of the MBSE database and information systems in CAD systems that enable design analysis tools to be used. There needs to be alignment between the design drawings and the MBSE models that doesn't usually exist in today's MBSE and CAD applications.

13.4   POTENTIAL FUTURE RESEARCH IN MBSE

There is so much richness for future research areas in MBSE. Some areas include:

- Incorporating quality management and measurement into MBSE to enhance consistency, standardization, measurement, and quality.
- Identification of resources needed, their skill sets, and experiences to be able to train and educate future systems engineers.
- Enhancing the culture change needed to go from document-centric to model-centric modeling.
- The importance of patterns in systems engineering of complex systems to assessing the technical feasibility of creating a more holistic MBSE approach (Blackburn, Cloutier, Hole, Bone, 2015).
- Enhancing model management, model transformation, and consistency checking (Qamar, Paredis and Wikander, 2012), (Schamai, Fritzson and Paredis, 2013), and (Herzig, Oamar and Paredis, 2014).
- There is ongoing research in interactive storytelling (Madni, 2015) and experiential design language (Madni, Spraragen, and Madni, 2014).
- There is the need for additional research in the areas of understanding human capabilities and limitation in the human models employed in MBSE (Orellana and Madni, 2014).
- Other researchers are focusing on incorporating human-system integration into the system engineering life cycle (Sharples, 2015), (HSI, 2013), (Madni, Sage and Madni, 2005), and (NATO Technical Report, 2010).
- Research is needed to incorporate MBSE into community-based systems engineering, similar to our cases studies throughout this book.
- Lastly, research is strongly needed to create and validate the MBSE value proposition, such as elimination of rework, cycle time reduction, risk reduction, and cost reduction (Madni and Sievers, 2018).

13.5   SUMMARY

In this chapter, we discussed the benefits, challenges, and future research in the areas of Model-Based Systems Engineering.

13.6   ACTIVE LEARNING EXERCISES

1) Select an off-the-shelf MBSE such as Sparx Systems Architecture or Cameo Systems Modeler No Magic and create systems models for each phase of the Vee Life Cycle. Some models to create can be:
   a. Use cases
   b. Functional Decomposition Model
   c. Process maps

d. Class diagrams
e. Physical architecture model
f. Test cases
g. Integration constraints or N-squared diagram

BIBLIOGRAPHY

Badiru, A., *Systems engineering models: theory, methods, and applications*, Taylor and Francis, Boca Raton, Florida, 2019.

Blackburn, M, Cloutier, R, Hole, E, Bone, M, and Witus, G. Transforming systems engineering through model-centric engineering. Technical Report SERC-2015-TR-044-3. 2015.

Estefan, J.A. Methodology and metrics activity: overview, update, and breakout agenda. INCOSE International Workshop, January 28–February 2, 2011; INCOSE MBSE Initiative, Phoenix, AZ.

Glaessgen, E.H., and Stargel, D.S. The digital twin paradigm for future NASA and U.S. air force vehicles. AIAA 53rd Structures, Structural Dynamics, and Materials Conference; 2012.

Herzig, S, Oamar, A, Paredis, C. An approach to identifying inconsistencies in model-based systems engineering. *Procedia Computer Science* 2014; 28:354–362.

Human System Integration (HSI). Interim DoDI Instruction 5000.2. November 25, 2013.

Jorgensen, R.W. Defining operational concepts using SysML: definition from the human perspective in rockwell collins. INCOSE International Symposium; 2011; Denver, CO.

Madni, A.M. Expanding stakeholder participation in upfront system engineering through storytelling in virtual worlds. *System Engineering* 2015; 18:16–27.

Madni, A.M., and Sievers, M. "Model-Based Systems Engineering: Motivation, Current Status, and Research Opportunities." *Systems Engineering*. vol. 21, no. 3, 8 May 2018, pp. 172–190. doi: 10.1002/sys.21438.

Madni, A.M., Sage, A.P., Madni, C.C. *Infusion of cognitive engineering into systems engineering processes and practices. IEEE International Conference on Systems, Man and Cybernetics*; 2005:960–965.

Madni, A.M., Spraragen, M., Madni, C.C. *Exploring and assessing complex system behavior through model-driven storytelling. IEEE Systems, Man and Cybernetics International Conference*; Invited special session "Frontiers of Model Based Systems Engineering," October 5–8, 2014; San Diego, CA.

NATO Technical Report. TR-HFM-155, HSI for NCW, February 2010.

Orellana, D.W., Madni, A.M. Human system integration ontology: enhancing model based systems engineering to evaluate human system performance. *Procedia Computer Science*. 2014; 28:19–25.

Qamar, A., Paredis, C., Wikander, J., During, C. Dependency modeling and model management in mechatronic design. *Journal of Computer Information Science Engineering*. 2012, doi: 10.1115/1.4007986.

Schamai, W, Fritzson, P, Paredis, C. Translation of UML state machines to Modelica: handling semantic issues. *Simulation*. 2013; 89:498–512.

Sharples, RA. Implementation of human system integration (HIS) and "non-functional characteristics" into the systems engineering lifecycle—a practical approach to airbus defence and space. Procedia Manuf. 2015; 3:1896–1902. 6th International Conference on Applied Human Factors and Ergonomics (AHFE 2015) and the Affiliated Conferences, AHFE 2015.

Zachman, J.A. A framework for information systems architecture, *IBM Systems Journal*, vol. 26, no. 3, pp. 276–292, 1987, doi: 10.1147/sj.263.0276.

Zachman, The Zachman Framework for Enterprise Architecture, https://www.zachman.com/images/ZI_PIcs/ZF3.0.jpg. Accessed 3/3/2021, 1967-2011.

14 Lean, Iterative, and Agile Life Cycles that Can Be Used to Streamline System Design and Development Timelines

14.1 PURPOSE

This chapter includes a discussion of lean, iterative, and agile life cycles that can reduce life cycle design and development timelines. We first discuss the traditional waterfall life cycle, also known as linear or predictive life cycle. Next, we discuss the agile or iterative life cycle, some lean concepts that can improve the efficiency of the life cycle, and finally ideas from the research literature for incorporating lean principles and tools to streamline the systems engineering life cycle.

14.2 WATERFALL, LINEAR, OR PREDICTIVE LIFE CYCLE

The waterfall life cycle is the traditional way that projects were performed until agile or iterative methods were created in the early 1990s. The waterfall cycle includes multiple phases that are followed sequentially, where the deliverables and knowledge from each phase flow into the next phase. The knowledge builds and becomes more detailed and also moves from conceptual to design-oriented or physical instantiations. The waterfall phases are:

1) Requirements' analysis and definition: The requirements are elicited and defined from the customer for the product, service, or process
2) Design: Definition of the design specifications that achieve the requirements
3) Implementation: Building the product
4) Testing: Verifying and validating the product
5) Installation: Installing the finished product
6) Maintenance: Maintaining the finished product
7) Disposal or retirement: Retiring or replacing the product and disposing products, services, or processes

14.3 EVOLUTIONARY OR INCREMENTAL LIFE CYCLE

The evolutionary or incremental life cycle model provides initial capability or functionality of the product, service, or process, which is then followed by successive releases of functionality until the final product is delivered. An evolutionary or incremental life cycle model is used when rapid exploration and implementation of part of the product is desired, the requirements are unclear at the beginning, funding is constrained, the customer wishes to remain flexible and allow new technology to be applied later, and experimentation is required to develop successive prototypes or versions (SeBoK, 2020).

DOI: 10.1201/9781003081258-16

14.4 ITERATIVE PROJECT MANAGEMENT, INCLUDING AGILE AND SCRUM

Agile and Scrum methods typically are considered to be provided using an iterative project management methodology (SeBoK, 2020). However, the PMI *PMBOK* describes leveraging both incremental and iterative approaches, as this allows for iterating on increments of the functionality (PMI, 2020). Iterative development processes provide several distinct advantages as described below (SeBok, 2020):

- This life cycle allows continuous integration and testing of the evolving product
- It allows the team to demonstrate progress through delivery of functionality more frequently than in a waterfall model
- Provides for early warnings of problems
- Provides for early delivery of subsets of capabilities or functionality

The Agile methodology is described next.

14.4.1 AGILE METHODOLOGY

The agile methodology is an iterative, adaptive, and interpersonal approach that focuses on delivering what the customer truly needs (Lappi, Karvonen, Lwakatare, Aaltonen, and Kuvaja, 2018), (Furterer and Wood, 2021). The agile method began in software development but has migrated to use in other industries, including financial services, consulting, traditional metal manufacturing, and defense industries as well as into systems engineering design and development (Conforto, Rebentisch, and Amaral, 2014).

Nerur, Mahapatra, and Mangalaraj (2005) contrasted the traditional software development approach to the agile development approach across several elements. First, in traditional development, the systems are considered fully specifiable and predictable and can be built through extensive planning. In contrast, in agile development, the systems have high-quality, adaptive software that can be continuously developed and improved in small teams, with rapid testing and feedback. The control mechanisms for the traditional methods are process centered, while agile control is performed by people. The management style for the traditional waterfall is command and control, while leadership and collaboration are key for agile methods. The knowledge management is explicit and well documented for traditional approaches but tacit and embedded within people in the agile method. The communication for a traditional approach is formal versus informal for the agile method. The customer is important in identifying and validating requirements for the traditional approach but considered critical in the agile method. In the traditional method, the project cycle is managed by tasks and activities, while in the agile method it is guided by the product features. The development model for the traditional method is some variation of the life cycle model, such as waterfall, spiral, or a variation of these, while an evolutionary model is applied for the agile method. The desired organizational structure is bureaucratic and formalized for the traditional method yet very flexible and organic in the agile method. The agile method favors object-oriented technology, while there is no technological restriction in the traditional approach (Nerur, Mahapatra, and Mangalaraj, 2005). The differences between the waterfall and the agile methods are illustrated in Table 14.1.

The Agile Manifesto states four principles that the authors value (Beck, Beedle, Bennekum, Cockburn, Cunningham, Fowler, …Thomas, 2001):

- Individuals and interactions over processes and tools
- Working software over comprehensive documentation
- Customer collaboration over contract negotiation
- Responding to change over following a plan

TABLE 14.1

Differences between the Waterfall and Agile Methods

Element	Waterfall / Traditional Life Cycle	Agile Methods
Final System	Fully specifiable and predictable, can be built through extensive planning	High-quality, adaptive that can be continuously developed and improved in small teams with rapid testing and feedback
Control mechanisms	Process centered	Performed by people
Management style	Command and control	Leadership and collaboration
Knowledge management	Explicit and well documented	Tacit and embedded within people
Communication	Formal	Informal
Customer	Customer is important in identifying and validating requirements	Customer is critical
Project cycle	Managed by tasks and activities	Guided by the product features
Development model	Waterfall, spiral, or a variation of these	Evolutionary model
Organizational structure	Bureaucratic and formalized	Flexible and organic
Technology	No technological restriction	Object-oriented

The authors of the Agile Manifesto defined 12 agile principles (Beck et al., 2001):

- Our highest priority is to satisfy the customer through early and continuous delivery of valuable software.
- Welcome changing requirements, even late in development. Agile processes harness change for the customer's competitive advantage.
- Deliver working software frequently, from a couple of weeks to a couple of months, with a preference to the shorter timescale.
- Businesspeople and developers must work together daily throughout the project.
- Build projects around motivated individuals. Give them the environment and support they need and trust them to get the job done.
- The most efficient and effective method of conveying information to and within a development team is face-to-face conversation.
- Working software is the primary measure of progress.
- Agile processes promote sustainable development. The sponsors, developers, and users should be able to maintain a constant pace indefinitely.
- Continuous attention to technical excellence and good design enhances agility.
- Simplicity – the art of maximizing the amount of work not done – is essential.
- The best architectures, requirements, and designs emerge from self-organizing teams.
- At regular intervals, the team reflects on how to become more effective, then tunes and adjusts its behavior accordingly.

A study by Lappi et al. identified important project governance elements in agile projects, categorized by goal setting, incentives, monitoring, coordination, roles and decision-making power, and capability-building practices.

The following lists the key agile practices within the governance categories (Lappi, date?): Lappi, 2018

Goal setting:

- Customer and team cooperation are critical for setting initial and flexible requirements and team-level goals
- The product backlog and vision guide the prioritization and iteration process
- The deliverable definition is based on user stories

Incentives:

- The agile philosophy is the best incentive for gaining team members' commitment
- Peer recognition, project-based organizational structure, decision-making authority, and customer contact supports team dedication
- Limited monetary incentives and risk and opportunity-sharing are addressed for motivation with the agile methods

Monitoring:

- Monitoring by teams using sprint and iteration reviews with customer feedback
- Standard and quantitative measures, with agile-specific qualitative measures
- Visual tracking for deliverables
- Testing is important to validate user stories, with emphasis on automated testing

Coordination:

- Real-time communication of information between empowered teams
- Iterative project planning with product vision and backlogs
- Providing infrastructure and practices for customer involvement and team autonomy
- Managing change through continuous prioritizations
- Facilitating frequent customer deliveries

Roles and decision-making power:

- Agile project teams with cross-functional roles and customer involvement
- Agile project team has total autonomy in decision-making
- Adaptive leadership by the project manager
- Agile coach oversees agile capabilities; Scrum master manages sprints and project team performance

Capability building:

- Clients' capabilities are key to success
- Optimal capability of agile team needs to address tensions between high- and low-skilled workers and specialization versus cross-functionality
- Key knowledge exchange within the team, with key stakeholders and permanent organization
- Agile practices and tools support routine and continuous learning

There are several methodologies that align with the statements and principles in the Agile Manifesto and include:

- Scrum: A framework to deliver high-quality products, quickly. Scrum is discussed further below.

- Lean (including Kanban): A methodology focused on reducing waste and improving the productivity and speed of processes and projects. Some of the lean tools applied to agile projects are Kanban (signals for production based on customer demand), visual control, pull, value stream mapping, value analysis, and waste analysis, to name a few.
- Extreme Programming (XP): An agile software development framework applying engineering practices for software development.
- Crystal: An agile software development approach focusing mainly on people and their interactions, rather than processes and tools.
- Dynamic systems development method (DSDM): An agile project delivery framework that is both iterative and incremental, incorporating agile methods and principles including continuous customer involvement.
- Feature-driven development (FDD): An agile software development methodology focusing on product features.
- Rapid application development (RAD): An agile software development methodology that focuses on rapid prototype releases and iterations.

14.4.2 Scrum

Scrum is defined by its creators as *a framework within which people can address complex adaptive problems while productively and creatively delivering products of the highest possible value* (Schwaber, and Sutherland, 2017). Scrum has been used for many different applications, such as software; hardware; embedded software; networks; autonomous vehicles; schools; and government, marketing, and managing organizations (Schwaber, and Sutherland, 2017).

Scrum is a simple yet incredibly powerful set of principles and practices that helps teams deliver products in short cycles, enabling fast feedback, continual improvement, and rapid adaptation to change (The Scrum Framework):

- A product owner creates a prioritized wish list called a *product backlog*.
- During *sprint planning*, the team pulls a small part of functionality from the top of that wish list, a *sprint backlog*, and decides how to implement those pieces.
- The team has a certain amount of time – a *sprint* (usually two to four weeks) – to complete its work but it meets each day to assess its progress (daily Scrum).
- Along the way, the *Scrum master* keeps the team focused on its goal.
- At the end of the sprint, the work should be potentially shippable: ready to hand to a customer, put on a store shelf, or show to a stakeholder.
- The sprint ends with a sprint review and retrospective.
- As the next sprint begins, the team chooses another part of the product backlog and begins working again.

14.4.3 Scrum Tools and Activities

User Stories. User stories are typically narrative descriptions of the functionality to be delivered. The user stories represent the customers' needs and requirements for the product, system, or service (Furterer and Wood, 2021).

Scrum Burndown Chart. The Scrum burndown chart is used to track the productivity of the team and to identify how much work remains and when it will be complete. The chart graphs the work remaining in hours (y-axis) versus the number of days worked (x-axis). The slope of the curve represents the burndown velocity or rate of productivity (hours per day). The days to complete is equal to work remaining (hours) divided by the rate of productivity (hours per day).

Product Backlog and Sprints. The product backlog is a prioritized list of ideas for the product. This helps to break the functionality into small increments or sprints. Sprints are typically for one to four weeks. At the end of the sprint, the functionality should be shippable to the customer, or complete, and meet the customer's requirements based on the items identified to be worked on for the sprint.

Sprint Planning. During the sprint planning, the team pulls a small group of the items from the highest-priority items on the product backlog. The team then determines how best to accomplish those items during the next sprint. A review and retrospective ends the sprint.

Daily Scrum. A daily meeting where the team meets to assess progress and make needed adjustments.

Scrum Master and Scrum Team. The Scrum master leads the Scrum team to the project's goal. The Scrum master removes barriers to successful project completion. The Scrum teams work due to the interpersonal nature of the team. The teams meet daily to assess progress and self-correct their direction. The teams are cross-functional and have all the competencies needed to accomplish the work. In large projects, multiple Scrum teams could work together to coordinate the work of the project. The product owner is responsible for the product vision and achieving the project's goals.

Scrum of Scrums. The Scrum of Scrum is a daily meeting for the Scrum teams to coordinate across multiple Scrum teams, typically for larger Scrum projects.

14.5 LEAN

Lean Enterprise is a methodology that focuses on reducing cycle time and waste in processes. Lean Enterprise originated from the Toyota Motor Corporation as the Toyota Production System and increased in popularity after the 1973 energy crisis. The term "Lean Thinking" is coined by James P. Womack and Daniel T. Jones in their book, *Lean Thinking* (Womack and Jones, 1996). The term "Lean Enterprise" is used to broaden the scope of a Lean program from manufacturing to embrace the enterprise or entire organization (Alukal, 2003). The Ford Production System was used to assemble cars, which was the basis for the Toyota Production System (TPS). Just-in-time (JIT) production philosophies joined with TPS which evolved into Lean.

The Lean tools that are most commonly used to eliminate waste and achieve flow and could be used to streamline the system development life cycle are discussed next:

- Value Stream Mapping
- Value analysis
- Waste analysis
- Why-why diagram
- 5S and visual management
- Kaizen
- Flow, Pull, and Kanban
- Mistake proofing
- Standard work
- Systems thinking and theory of constraints

14.5.1 Value Stream Mapping

The Value Stream Map is a Lean Enterprise tool that can help the organization understand the high-level overview of the processes performed within our system, with timelines for process times to determine how long it takes the customer (patient) to go through the system. The Value Stream Map for the Women's Healthcare Center is shown in Figure 14.1.

The main process steps are:

1) Patient calls and schedules an appointment for a screening mammogram.
2) The center verifies the insurance and authorizes the screening.
3) The screening is performed at the Women's Center.
4) The results are read by the radiologist.
5) The results are prepared in a letter and sent to the patient.
6) Twenty-eight and a half percent (28.5%) of the patients require a diagnostic appointment for further scans and views. The diagnostic appointment is made.
7) The center verifies the authorization for the diagnostic mammogram.
8) The diagnostic procedure is performed.
9) The results are read by the radiologist and sent to the patient.
10) About 3% of the patients require a biopsy. The patient calls to schedule the biopsy appointment.
11) The center verifies the authorization for the biopsy procedure.
12) The biopsy procedure is performed.
13) The results are provided to the patient.
14) The patient receives the results. Further surgical procedures would be scheduled at a surgical center if necessary.

For our current state system, it takes about 33 days for a patient to go through the entire system, if they need both a screening and diagnostic mammogram and a biopsy for suspected cancer. The value-added cycle time is only 185 minutes but included long wait times to get an appointment, authorize the services, and get the results of the tests.

14.5.2 VALUE ANALYSIS

A value analysis views activities from the perspective of how much these activities add value to the customer. While developing systems, the value analysis could be used to streamline the processes performed to design and develop the system.

The value analysis consists of the following steps:

1. Document the process using process maps.
2. Identify non-value-added activities and waste.
3. Consider eliminating non-value-added activities and waste.
4. Identify and validate (collect more data if necessary) root causes of non-value-added activities and waste.
5. Begin generating improvement opportunities to eliminate non-value-added activities and waste.

Value-added activities are those activities that the customer would pay for, that add value for the customer. Non-value-added activities are those that the customer would not want to pay-for or don't add value for the customer. Limited value activities are those activities that do not add value to the customer, but are necessary, such as for legal, financial reporting, documentation reasons. Defining the activities as value-added, limited value-added, or non-value-added provides a prioritization of where first to focus to eliminate or reduce these activities. Start first with the non-value-added activities, then the limited value activities, and then further improve the value-added activities. One can assess the percent of value-added activities as:

100% X (Number of value-added activities/Number of total activities),

where value-added activities include operations that add value for the customer, and non-value-added activities include delays, storage of materials, movement of materials, and inspections. The number of total activities include the value-added activities and the non-value-added activities.

One can also calculate the percent of value-added time as:

100% X (Total time spent in value-added activities/Total time for process).

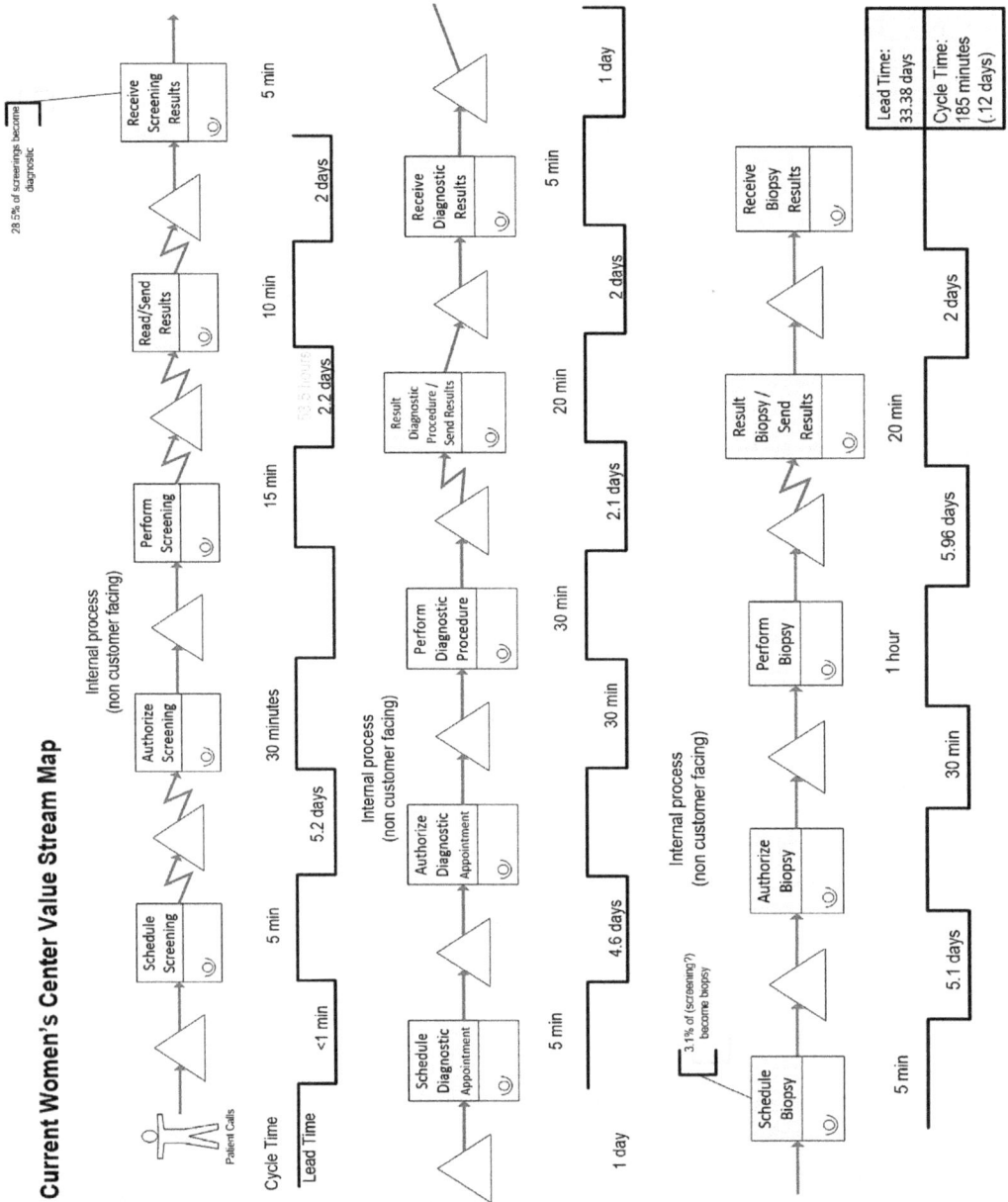

FIGURE 14.1 Women's Healthcare Center Value Stream Map.
Source: Furterer Class Notes, Management of Engineering Systems, 2018.

Typically, the percent of value-added time is about 1% to 5%, with total non-value-added time equal to 95% to 99%.

During the lean analysis, the team can focus on areas to identify inefficiencies in the following areas (Process Flow Analysis Training Manual, 1982):

- Can labor-intensive process be reduced, eliminated, or combined?
- Can delays be eliminated?
- Are all reviews and approvals necessary and value-added?
- Are decisions necessary?
- Why is rework required?
- Are all of the documentation, tracking, and reporting necessary?
- Duplicated process across the organization?
- What is slipping through the cracks causing customer dissatisfaction?
- Activities that require accessing multiple information systems.
- Travel – look at the layout requiring the travel.
- Is it necessary to store and retrieve all of that information, do we need that many copies?
- Are the inspections necessary?
- Is the sequence of activities or flow logical?
- Are standardization, training, and documentation needed?
- Are all of the inputs and outputs of a process necessary?
- How is the data and information stored and used?
- Are systems slow?
- Are systems usable?
- Are information systems user-friendly?
- Can you combine tasks?
- Is the responsible person at too high or too low a level?

14.5.3 Waste Analysis

Waste Analysis is a Lean tool that divides waste into eight different categories to help brainstorm and eliminate different types of wastes. The eight wastes are all considered limited value or non-value-added activities and should be reduced or eliminated when possible. The waste analysis aligns nicely to the value analysis above, while identifying limited value- or non-value-added activities, the wastes for these activities can be identified. Waste is defined as anything that adds cost to the product without adding value. This helps to identify root causes of the non-value-added and limited value-added activities and where to prioritize improvements.

The eight wastes are:

- Transportation: Moving people, equipment, materials, and tools.
- OverProduction: Producing more product or material than is necessary to satisfy the customers' orders or faster than is needed.
- Motion: Unnecessary motion, usually at a micro or workplace level.
- Defects: Any errors in not making the product or delivering the service correctly the first time.
- Delay: Wait or delay for equipment or people.
- Inventory: Storing product or materials.
- Processing: Effort which adds no value to a product or service. Incorporating requirements not requested by the customer.
- People: Not using people's skills, mental, creative, and physical abilities.

14.5.4 Why-Why Diagram and Five Whys

The Why-Why diagram is a powerful tool to generate root causes. It is used to identify root causes of problems and inefficiencies in the system development life cycle to reduce or eliminate them. It uses the concept of the five Whys, where you ask the question why several times until the root cause is revealed. Following are the steps to create a Why-Why diagram

1. Start on left with problem statement
2. State causes for the problem
3. State causes for each cause
4. Keep asking why five times
5. Try to substantiate the causes with data
6. Draw the diagram

Figure 14.2 shows a sample Why-Why diagram for why it takes a long time for patients to get results of women's center imaging results.

It is critical that once you brainstorm the potential root causes of the problems, you collect additional data to substantiate the causes.

14.5.5 5S and Visual Management

5S is a Lean tool that helps to organize a workplace in a visual manner. The 5Ss which are typically performed in the order identified are:

- Sort or Simplify: Clearly distinguish between what is necessary and what is unnecessary, disposing the unnecessary. A red tag is used to identify items that should be reviewed for disposal.
- Straighten: Organize the necessary items so that they can be used and returned easily.
- Scrub: Fix the root cause of the dirt or disorganization.
- Stabilize: Maintain and improve the standards of the first three Ss.
- Sustain: Achieving the discipline or habit of properly maintaining the correct 5S procedures.

14.5.6 Kaizen

Kaizen is a Lean tool that stands for "kai" (change) and Zen (for the good) or change for the good. It represents continuous incremental improvement of an activity to constantly create more value for the customer by eliminating waste. A Kaizen consists of short-term activities that focus on redesigning a particular process. A Kaizen event can be incorporated into any phase of the systems engineering project to help design and/or implement a focused improvement recommendation.

The Kaizen event follows the Plan-Do-Check-Act cycle, including the following steps:

PLAN:

1. Identify Need: Determine the purpose of the Kaizen.
2. Form Kaizen Team: Typically, six to eight team members.
3. Develop Kaizen objectives: To document the scope of the project. The objectives should be Specific, Measurable, Attainable, Realistic, and Time-based (SMART).

4. Collect current state baseline data: From Measure phase or additional data as needed.
5. Develop schedule and Kaizen event agenda: About one week or less.

DO:
6. Hold Kaizen event using DMAIC
   Sample Kaizen Event Agenda:
   • Review Kaizen event agenda
   • Review Kaizen objectives and approach
   • Develop Kaizen event ground rules with team
   • Present baseline measure and background information

   Hold Event:
   • Define: Problem (derived from objectives), agree on scope for the event
   • Measure: Review measure baseline collected
   • Analyze: Identify root causes, wastes, and inefficiencies
   • Improve: Create action item list and improvement recommendations
   • Control: Create standard operating procedures to document and sustain improvements. Prepare a summary report and present to sponsor.
Identify and assign action items
Document findings and results

Discuss next steps and close meeting

7. Implement: Implement recommendations, fine tune, and train

   CHECK/ACT:

   Summarize: Summarize results

   Kaizen Summary Report Items:
   • Team Members
   • Project Scope
   • Project Goals
   • Before Kaizen Description
   • Pictures (with captions)
   • Key Kaizen Breakthroughs
   • After Kaizen Description
   • Results
   • Summary
   • Lessons Learned
   • Kaizen Report Card with Follow-Up Date

8. Control: If targets are met, standardize the process. If targets are not met, or the process is not stabilized, restart Kaizen event Plan-Do-Check-Act (PDCA) cycle.

14.5.7 FLOW, PULL, AND KANBAN

The idea in creating flow in Lean is to deliver products and services just-in-time, in the right amounts, and at the right quality levels at the right place. This necessitates that products and services are produced and delivered only when a pull is exerted by the customer through a signal in the form of a purchase. A well-designed Lean system allows for an immediate and effective response to fluctuating customer demands and requirements.

(Cudney, Furterer, and Dietrich, 2014)

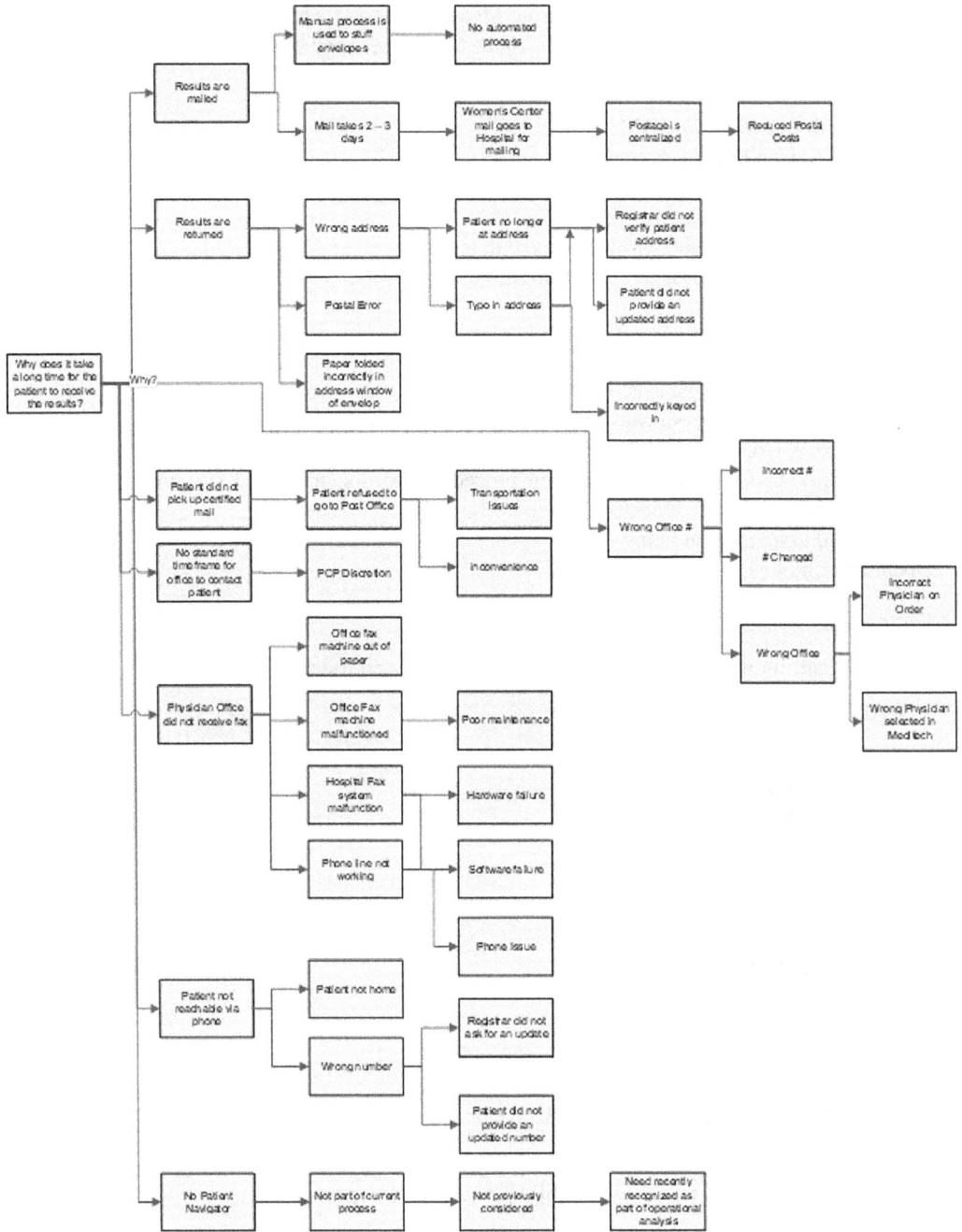

FIGURE 14.2   Why-why diagram for women's center – Why does it take a long time for the patient to receive the results?

The pull concept of manufacturing incorporates only producing product when there is an actual customer order. Kanbans or signals for production are used to notify the replenishment of product at each workstation, pulling product from the last step of the production process backward through the workflow. Kanbans and pull concepts control inventory based on the ability of the system to produce product. Production control should be visible, standard, and disciplined (Cudney, Furterer, and Dietrich, 2014).

14.5.8 Mistake Proofing

Mistake proofing is a tool that helps to prevent errors in your process. Errors are inadvertent, unintentional, accidental mistakes made by people because of the human sensitivity designed into our products and processes (Furterer, 2014).

Mistake proofing, also called Poka-Yoke, is the activity of awareness, detection, and prevention of errors which adversely affect our customers, our people, and result in Waste.

Some of the underlying mistake proofing concepts are:

- You should have to think to do it wrong, instead of right.
- Easy to perform inspection at the source.
- Reduces the need for rework and prevents further work (and cost) on a process step that is already defective.
- Simplifies prevention and repair of defects by placing responsibility on the responsible worker.

14.5.9 Standard Work

Standard work defines and documents the detailed work instructions of a process. It aids in training and ensuring that the process is performed in a consistent and standard manner. Standard work consists of three elements:

- Takt time: The rate at which a product or service must be created to meet customer demand.
- Takt Time = Available Time/Customer Demand
- Standard work sequence: The specific order in which the associate performs the process.
- Standard work in process: The minimum number of parts on the line that will allow the work to flow efficiently.

To implement standard work, the following characteristics must exist (Cudney, Furterer, Dietrich, 2014):

- The process must be able to be observed
- The process must be repetitive in nature
- The process must be based on human motion

14.6 SYSTEMS THINKING AND THEORY OF CONSTRAINTS

As we discussed earlier in this book, systems thinking helps us to understand how elements of the system connect to each other, interact, and interface with each other and behave. As systems become more complex, it becomes more critical to incorporate systems thinking in product, service, and process development. Theory of Constraints is a concept that incorporates systems thinking to understand the sequence of activities used to produce a product or service, understand where there are bottlenecks in the process, and to focus to eliminate these bottlenecks. There are typically constraints that limit the ability of the system to achieve its goals. Focusing on eliminating the weakest link or constraint is the underlying philosophy of Theory of Constraints. Eliyahu Goldratt developed a method with five steps to identify and eliminate the constraints (Goldratt, 1984).

- Step 1: identify the system constraint.
- Step 2: determine how to exploit this constraint.
- Step 3: subordinate the non-constraint components.
- Step 4: elevate the constraint.
- Step 5: return to Step 1 with the next constraint.

The five-step process is continued on other constraints in an effort for continuous improvement.

14.7   IDEAS FROM THE RESEARCH LITERATURE FOR INCORPORATING LEAN PRINCIPLES AND TOOLS INTO THE SYSTEMS ENGINEERING LIFE CYCLE

We first define Lean Systems Engineering:

> Lean Systems Engineering (LSE) is the application of Lean principles, practices, and tools to SE and to the related aspects of enterprise management (EM) in order to enhance the delivery of value (which is defined as flawless delivery of product or mission with satisfaction of all stakeholders) while reducing waste.
>
> (Oppenheim, Murman, and Secor, 2011)

An INCOSE working group developed lean enablers for systems engineering published by Oppenheim, Murman, and Secor in 2010. These enablers were developed by the working group consisting of Aerospace industry and Lean experts. Extensive rounds of surveys were distributed to subject matter experts in the areas of lean and systems engineering to prioritize the lean enablers. Following is a summary of the lean principles and enablers for systems engineering; however, the author encourages the reader to read the original journal article (Oppenheim, Murman, and Secor, 2011):

1) Value Principle – Establish the value of the end product or system to the customer
   a. Enablers:
      i.   Follow all practices for the requirements capture and development in the INCOSE Handbook.
      ii. Establish value of the end product or system to the customer. This includes developing a robust and effective process of capturing the customer value proposition, disseminating amongst the program team, involving appropriate customers and stakeholders, aligning the team, and developing the process to avoid future wastes.
      iii. Frequently involve the customer.
2) Map the Value Stream (Plan the Program) Principle – Map the planning value stream
   a. Enablers:
      i.   Develop the Value Stream for planning the value delivery process. The second principle includes a checklist of the practices for planning all end-to-end linked streamlined actions and processes necessary to realize value without waste. This principle encourages the integration of planning of Systems Engineering, Project Management, and other relevant enterprise planning activities.
      ii. Plan for frontloading the program by utilizing cross-functional teams with the most experienced and compatible people at the start of the project.
      iii. Plan to develop only what needs developing, promoting reuse, and sharing of program assets.
      iv. Plan to prevent potential conflicts with suppliers.
3) Flow Principle – Allows the value creation process to flow without waste
   a. Enablers:
      i.   Execute the Program according to the INCOSE handbook process
      ii. Clarify, derive, prioritize requirements early and often during execution, creative effective communication channels
      iii. Front load architectural design and implementation
      iv. Enable systems engineers to accept responsibility for coordination of product development activities.
      v.   Use efficient and effective communication and coordination
      vi. Promote smooth systems engineering flow through integrative events and programmatic reviews, align process flow to decision flow

      vii. Make program progress visible to all

      viii. Use lean tools to promote the flow of information, minimize software revision updates, adapt technology to fit people and process

4) Pull Principle – Tailor tasks to meet the needs of the team, cut tasks related to waste

   a. Enablers:

      i. Tailor for a given program according to the INCOSE Handbook process

      ii. Pull tasks and outputs based on need and reject others as waste

         1. Let information needs pull the needed work activities

         2. Engineers should pull knowledge as needed

5) Perfection Principle – Strive for excellence for continuous improvement for the SE process and related enterprise management

   a. Enablers:

      i. Pursue continuous improvement according to the INCOSE Handbook process

      ii. Strive for excellence of SE processes, building in robust quality, strive for perfection, apply PDCA, incorporate continuous improvement

      iii. Use lessons learned from past programs

      iv. Develop perfect communication, coordination, and collaboration

      v. Use a chief engineering role to lead and integrate the program from start to finish

      vi. Drive out waste through design standardization, process standardization, and skill-set standardization

      vii. Promote all three complementary continuous improvement methods (Lean, Kaizen, and Six Sigma).

6) Respect for People –Promote the best human relations based upon respect for people including trust, honesty, respect, empowerment, teamwork, stability, motivation, drive for excellence, and healthy hiring and promotion policies.

   a. Enablers:

      i. Pursue People Management according to the INCOSE Handbook process

      ii. Build an organization based upon respect for people

      iii. Expect and support engineers to strive for technical excellence

      iv. Nurture a learning environment

      v. Treat people as your most valued assets

14.8   SUMMARY

In this chapter we described more lean and agile life cycle approaches that can streamline more traditional design life cycles. We also discussed ideas for incorporating lean principles and tools into the systems engineering life cycle.

14.9   ACTIVE LEARNING EXERCISES

1) Find and read the following journal article

   a. Oppenheim, B., Murman, E., and Secor, D, *Lean Enablers for Systems Engineering*, Systems Engineering, Vo. 14, No. 1, 2011.

   b. Summarize the Lean principles and enablers in the article.

   c. Develop ideas for how you would implement these Lean principles and enablers into the Vee model life cycle.

2) After reviewing the Agile and Scrum methods, discuss how you might integrate Agile and Scrum principles and methods into the Systems Engineering Vee Life Cycle.

BIBLIOGRAPHY

Alukal, G. *Create a Lean, Mean Machine, Quality Progress*. ASQ, Milwaukee, WI, 2003.

Beck, K., Beedle, M., Van Bennekum, A., Cockburn, A., Cunningham, W., Fowler, M.,… Thomas, D. Agile Manifesto, 2001, http://agilemanifesto.org/.

Bott, M., & Mesmer, B. (2020). An Analysis of Theories Supporting Agile Scrum and the Use of Scrum in Systems Engineering. *Engineering Management Journal*, *32*(2), 76–85. https://doi-org.libproxy.udayton.edu/10.1080/10429247.2019.1659701

Conforto, E. C., Rebentisch, E., and Amaral, D. C. *Project Management Agility Global Survey*, Massachusetts Institute of Technology, Consortium for Engineering Program Excellence, Cambridge, MA: 2014).

Cudney, E., Furterer, S., and Dietrich, D., *Lean Systems: Applications and Case Studies in Manufacturing, Service, and Healthcare*, CRC Press, Boca Raton, FL, 2014.

Duffy, G. L. and Furterer, S.L. *The ASQ Certified Quality Improvement Associate Handbook*, 4th ed. (Milwaukee: ASQ Quality Press, 2020).

Fleming, W., and Koppelman, J. M., *Earned Value Project Management*, 2nd ed (Newtown Square, PA: Project Management Institute, 2000).

Furterer, S., *Lean Six Sigma Case Studies in the Healthcare Enterprise*, Springer, London, 2014.

Furterer, S. and Wood, D., *The ASQ Certified Manager of Quality / Organizational Excellent Handbook*, ASQ Quality Press, Milwaukee, WI, 2021.

Goldratt, E. *The Goal: A Process of Ongoing Improvement*, North River Press, 1984.

Kerzner, H. *Project Management: A Systems Approach to Planning, Scheduling, and Controlling* 12th Edition, John Wiley & Sons, Hoboken, NJ, 2017.

Lane J.A., and Valerdi R. *Accelerating system of systems engineering understanding and optimization through lean enterprise principles. 2010 IEEE International Systems Conference, Systems Conference, 2010 4th Annual IEEE*. April 2010:196–201. doi:10.1109/SYSTEMS.2010.5482339

Lappi, T., Karvonen, T., Lwakatare, L.E., Aaltonen, K., and Kuvaja, P. "Toward an Improved Understanding of Agile Project Governance: A Systematic Literature Review," *Project Management Journal* 49, no. (2018): 39–63, 2018. doi:10.1177/8756972818803482.

Nerur, S., Mahapatra, R., and Mangalaraj, G., "Challenges of Migrating to Agile Methodologies," *Communications of the ACM* 48, no. 5 (May 2005).

Oppenheim, B., Murman, E., and Secor, D, Lean Enablers for Systems Engineering, *Systems Engineering*, 14, no. 1, 2011.

PMI Global Standard, *Project Management Body of Knowledge*. 2020.

Porter, M. E. *Competitive Strategy: Techniques for Analyzing Industries and Competitors*, New York, Free Press, 1980.

Process Flow Analysis Training Manual, Control Data, 1982.

Santos, A. C. O., da Silva, C. E. S., Braga, R. A. da S., Corrêa, J. É., and de Almeida, F. A. Customer value in lean product development: Conceptual model for incremental innovations. *Systems Engineering*, 23(3), 281–293, 2020.

Schwaber, K., and Sutherland, J. "The Scrum Guide, The Definitive Guide to Scrum: The Rules of the Game," 2017, Scrum.org.

SEBoK Editorial Board, *The Guide to the Systems Engineering Body of Knowledge (SEBoK)*, vol. 2.2, ed. R. J. Cloutier (Hoboken, NJ: Trustees of the Stevens Institute of Technology, 2020), accessed May 25, 2020, www.sebokwiki.org. BKCASE is managed and maintained by the Stevens Institute of Technology Systems Engineering Research Center, the International Council on Systems Engineering, and the Institute of Electrical and Electronics Engineers Computer Society.

Sreeram, TR and Thondiyath, A. Combining Lean and Six Sigma in the context of Systems Engineering design. *International Journal of Lean Six Sigma*, 6(4), 290–312. 2015. https://doi-org.libproxy.udayton.edu/10.1108/IJLSS-07-2014-0022

Stage-gate (go/no-go) decision process is discussed in R. G. Cooper. *Winning at New Products: Accelerating the Process from Idea to Launch*, 3rd ed. (Cambridge, MA: Perseus, 2001).

"The Scrum Framework," accessed May 25, 2020, https://www.scrumalliance.org/why-scrum#why-its-called-scrum.

Womack, J.P., and Jones, D.T. *Lean Thinking: Banish Waste and Create Wealth in Your Corporation*. New York, NY: Simon & Shuster, 1996.

15 Holistic Culture and Change Management Concepts and Skills for Enhancing Systems Engineering Practices

15.1 PURPOSE

This last chapter includes a discussion of the impacts of culture and change management concepts and skills needed by systems engineers to enhance systems engineering practices. We first discuss change and change management, culture and culture change, structures, and people skills that are needed to successfully apply systems engineering tools and methods. We then discuss some techniques that can be used in each phase of the Vee Life Cycle to enhance the communication with customers and stakeholders.

15.2 CHANGE AND CHANGE MANAGEMENT

Change is necessary to continue to refresh the practices, technology, and skills required to successfully enact systems engineering methods, tools, and principles. As we discussed in our chapter on Model-Based Systems Engineering, change is needed to move from a technology and practice of document-based product design to one of model-based. Change must be implemented within the structures, culture, and with the people working in the organization. We first discuss a definition of change and different approaches to change from the literature. We then describe culture and culture change. Finally, we provide ideas, tools, and skills that can be applied within each phase of the Vee Model Life Cycle that are given less emphasis in engineering and design curriculum and within the INCOSE Handbook as well.

15.2.1 What Is Change?

A definition of change is to "making things different" (Robbins and Judge, 2019).

15.2.2 Making Change Happen

Different approaches to change (Smith and Graetz, 2011):

There are several different approaches to change found in the literature:

- Rational approaches: Organizational change being led by managers and leaders based upon strategies and plans within the organization.
- Biological approaches: Managers and leaders focus on environmental factors progressing through an evolutionary life cycle.
- Institutional approaches: Change is determined by industry pressure, where managers and leaders focus upon achieving industry standards and benchmarks.
- Resource-based approaches: Change depends upon access to resources and core competencies.

- Psychological-based approaches: Change is embedded in the minds of those affected.
- Systems-based approaches: Focus on the interconnected nature of organizing, through considering all stakeholders and components of the organization.
- Cultural-based approaches: Change is determined by the culture and their values, which encourages managers and leaders to focus on rites, rituals, and values.
- Critical approaches: Change is realized through focus on conflict, power, and the rejection of universal rules, incorporating empowerment and emancipatory practices.

There seems to be little consensus in the literature as to the best approaches to change management. Additionally, research methodology is somewhat lacking to be able to provide definitive, repeatable research methods that point us to the best change management approaches.

(Hughes, 2019)

Diffusion of Innovations, a book by Everett Rogers, originally published in 1962, presented how innovations can be successfully implemented. He described the different types of adopters in an organization as follows:

1) Innovators: Venturesome types enjoy being at the forefront. They align to the innovation and its possibilities. They represent about 2.5% of the adopters.
2) Early adopters: They look to the data of the innovators and the opinion leaders to make their adoption decision. Most of the opinion leaders reside in this group, they represent about 13.5% of the adopters.
3) Early majority and late majority make up about 34% of the adopters. They will listen to the opinions of the opinion leaders and can be swayed by them. Once the early and late majority embrace the innovation, it becomes a tipping point for a successful innovation.
4) Laggards represent about 16% of the adopters. They can be very traditional or isolates in their social systems. Traditional laggards can be suspicious of innovations. The isolates tend to be less aware of an innovation's benefits and it takes them much longer to adopt innovations, if at all.

Rogers encouraged leveraging opinion leaders for change since they can be powerful change agents to affect the diffusion of an innovation. Communication exchanges with peers and opinion leaders also help develop attitudes about innovations. The types of opinion leaders that should be targeted for dissemination of communication depends upon the type of social system that exists. There are two types of social systems (Rogers, 1962):

1) Heterophilous: These types of social systems tend to encourage change from system norms. There is more interaction between people from different backgrounds with different ideas. The opinion leadership tends to be more innovative, due to the system wanting innovation. Change and innovation is much easier in a heterophilous organization where new ideas and innovations are more widely accepted.
2) Homophilous: These social systems tend toward system norms. Most of the interaction within these social systems tend to be people from similar backgrounds and ideas. People and ideas that differ from the norm are seen as strange and undesirable. These systems have opinion leadership that is not very innovative because these systems are averse to innovation. Encouraging the diffusion of innovation can be much more difficult, since opinion leaders will not risk suggesting ideas that are far outside of the norms of the system, because they don't want to lose their role as an opinion leader. Change agents must communicate a convincing argument aligned to the system's norms so that the opinion leaders can use this argument with the others in the system.

The culture, structures, people skills, and technology must change to move from a documents-based product development practice to a Model-Based Systems Engineering practice. The culture is probably the most difficult of the three elements to change. The culture must embrace the idea of moving to MBSE, so that they are willing to support the changing structures, people skills, and technology.

Kotter's *Leading Change* is the most cited book in the field of organizational change studies. His change model includes eight steps for leading change (Kotter, 1995):

1) Acting with Urgency:
   - Gaining top management support and visibly communicating this support for the projects, emphasizing the importance of the project to the organization.
2) Develop the Guiding Coalition:
   - Form the project team and properly scope the projects. Throughout the project we engage appropriate subject matter experts who help develop the new process and generate improvement ideas.
3) Developing a Change Vision:
   - Develop the new processes, creating innovative ways to transfer this knowledge (social media, intranet, videos, online training, etc.).
   - Develop the implementation plan including a detailed work plan to stage the improvements.
4) Communicate the Vision Buy-in
   - Develop and deploy a communication plan to communicate the changes and status of the improvement efforts.
   - Management must model the behavior of the new process, assess understanding, and compliance to the new process.
5) Empowering Broad-Based Action:
   - Identify and remove obstacles to change. Implement suggestion systems to get ideas from any and all of the staff.
   - Incorporate accountability to adhere to the new processes.
6) Generating Short-Term Wins:
   - Implement a reward system to encourage change and improvement.
   - Communicate the wins and significant improvement to all employees.
   - Celebrate success and share best practices and lessons learned.
7) Don't Let Up:
   - Incorporate the new metrics dashboards into the daily metrics monitoring and control.
   - Incorporate the process competencies into the performance evaluation system.
8) Make Change Stick:
   - Provide change management and Lean Six Sigma training to staff and management to ensure the succession of the change in thinking to a fact-based decision process and one of continuous improvement.

15.2.3 What Is Culture?

Culture is a system of shared meaning held by members that distinguishes the organization from other organizations (Robbins and Judge, 2019).

The shared meaning of culture includes values, beliefs, and assumptions. Culture can be informed by observed behaviors when people interact, group norms, espoused values, formal philosophy, rules of the game and climate, embedded skills, habits of thinking, mental models, linguistic paradigms, shared meanings, symbols, rituals, and celebrations (Schein, 2010). Many times, stories about the organization, people that tripped up in the organization, and even the founder of the company help to convey the culture. I was part of an organization that the stories of the company founder representing his values of how to treat people, which had passed 15 years prior, were so strong that one could almost feel his ghost walking through the hallways of the home office headquarters.

15.2.4   What Is Culture Change?

Culture change is the act of changing some of the values, beliefs, or assumptions embedded in the organization. It is important to understand which of the values, beliefs, or assumptions that need to be changed, in our case to implement MBSE. Following are some questions to help understand the culture through the artifacts that represent the culture (Johnson, 1992):

> Answers to the questions are provided by the author from an organization where she worked.

15.2.4.1   Symbols: Organizational Logos, Office Layouts, Parking Lots, Artwork in the Hallways

Questions to ask to understand symbol artifacts:

- What language and jargon are used? Community is critical.
- What status symbols exist? No preference for parking spots between faculty and staff.
- Which symbols denote the purpose of the organization exist? The chapel on campus.

15.2.4.2   Power Structures: Managerial Groupings and the Most Powerful Functional Areas

- How is power distributed in the organization? Faculty senate governs faculty policies.
- What are the main blockages to change? It takes a long time to make change happen due to needing to get input from "everyone."
- How is the organization led? President leads with a leadership team from the business side and the academic side.

15.2.4.3   Organizational Structures: Organizational Hierarchy and Informal Structures

- Is the organization hierarchical? There is no formal organization chart.
- Do structures encourage collaboration or cooperation? When the new president came into the organization, he spent a year "listening" to the organization about what they needed.
- Top-down/bottom-up decision-making? Some top-down, some unit based.

15.2.4.4   Control Systems: Performance Measurement and Reward Systems

- How closely monitored are employees? Academic freedom, annual reviews, periodic reviews for promotion and tenure.
- Are there too many/not enough controls? Many financial controls, human resource controls.
- How are good behaviors rewarded? Rewards and recognition throughout the university.

15.2.4.5   Rituals and Routines: The Christmas Party and Meetings around the Water Cooler

- Which regular routines are valued? Faculty meetings monthly, Christmas parties, Industry Advisory Committee (IAC) meetings.
- What are the key rituals? Annual fund-raising campaign, community service.
- What behaviors do the routines encourage? Community, collaboration, networking.

15.2.4.6   Stories and Myths: Stories about Innovative Curriculum and Experiential Learning

- What core beliefs do the stories reflect? Supporting students through experiential learning and innovative curriculum

- Who are the heroes and villains? Faculty and staff are the heroes
- What norms do the mavericks deviate from? Innovative curriculum

15.2.4.7 The Paradigm: Student Satisfaction and Excellence in Teaching

- What are the taken-for-granted assumptions and beliefs? Everyone is treated with kindness
- What paradigm do the cultural artifacts suggest? Community is critical
- What is the dominant culture? Collaborative and caring

One can answer these questions related to the culture in the systems engineering organization related to openness to change, controls, beliefs, artifacts, etc., that are important to understand to affect culture change.

15.3 STRUCTURES

There are several structural elements that can help create a Model-Based Systems Engineering (MBSE) practice. These are derived from the author's experience with the initial attempt at creating a Business Architecture practice at a large international retailer and from best practices from the literature. John Kotter's eight-step change model was used as a framework, incorporating and adapting many of the steps of Kotter's change model to the MBSE success factors (Kotter, 1996). Other references that provided some of the success factors are (Furterer, 2008); (Hoffman and Mera, 1999); (Newman, 1994); (Sureshchandar, Chandrasekharan, and Anantharaman, 2001).

- Establish and Communicate the MBSE Practice Mission, Vision, and Core Values.
- Gain Top Management Support Across the Organization.
- Develop and deploy the practice Infrastructure and governance structure Integrated with the product development and systems engineering life cycles.
- Define and gain buy-in to the MBSE Meta Models.
- Train engineers and modelers in conceptual modeling and MBSE tools and methods.
- Align the MBSE Vision with the Culture.
- Assess the Culture and Readiness for Change and Incorporate Change Management tools.
- Engage Subject Matter Experts.
- Incorporate modeling with the MBSE into the product development and systems engineering life cycles.
- Create and communicate MBSE practice status and successes.
- Remove obstacles (including people if necessary) to change.
- Empower the MBSE modelers.

Each element is discussed in more detail.

1) **Establish and Communicate the MBSE Practice Mission, Vision, and Core Values:**
   The most critical first step of any change program is for top management to define the purpose of the program, develop the vision, in our case, the systems engineering program vision and develop the core values of the systems engineering organization. For many organizations that haven't had a formal MBSE practice in the past, it is important for management to set the vision and gain buy-in for this vision across the organization. It is important that all of the stakeholders related to the MBSE practice are aligned with the MBSE vision. All stakeholders, including contractors, must understand the purpose and benefits of the MBSE practice and how the modelers and systems engineers will engage with the stakeholders and customers to understand the customer and system requirements. The mission, vision, and core values of the MBSE practice should be developed in strategic planning sessions with product development, systems engineering, and senior management.

A sample mission for the MBSE practice is:

Provide MBSE models, deliverables, best practices, standards, guidance, and coordination to:

- Ensure requirements are elicited and validated from the customers.
- Provide a system view of the system through modeling, documentation, and usage.
- Provide modeling integration with specialty engineering disciplines.
- Guide MBSE defined artifacts/models.
- Provide best practices, standards, and guidance.
- Guide and enhance modeling skills maturity and practice.

A sample vision for the MBSE practice is:

To create MBSE system models that enhance a system's holistic view, incorporate requirements elicited from the customers that are traceable throughout the systems engineering life cycle.

The core values of the MBSE practicing organization could be:

- Innovation: Provide innovative system solutions that support the business.
- Integrity: Create MBSE with the highest integrity.
- Respect: Treat all associates and stakeholders with respect.
- Knowledge: Share, create, and learn knowledge.
- Teamwork: Associates working together in teams.

The mission, vision, and core values should be shared with the organization. There are many methods for communicating this information, including company newsletters, staff meetings, email, townhalls, one-to-one meetings, company intranets, hard copy, and electronic bulletin boards.

2) **Gain Top Management Support across the Organization**:
Once the mission, vision, and core values are established, the systems engineering organization should gain top management support across the organization. The governance structure, methodology, and roles and responsibilities of the systems and specialty engineers should be communicated to these organizations. If the MBSE program is new, the organization may already have been performing some of the roles and responsibilities that the new practice will be planning to do. There needs to be clear explanations of the expectations of all stakeholders, as roles and business knowledge transfers to the systems engineers. It will need to be determined who will create, revise, and maintain artifacts and the repository in which they will reside.

3) **Develop and Deploy the MBSE Infrastructure and Governance Structure Integrated with the Systems Engineering Life Cycle**:
It is important to develop the program infrastructure and governance structure. There are many components of the infrastructure and governance structure, including the following elements:
1) Systems Engineers' roles and responsibilities would include:
   a. Engage the customer to elicit the customers' requirement through developing process scenarios and use cases.
   b. Create and revise the systems engineering models.
   c. Create the technical systems engineering management plans (SEMPs) based on the systems engineering models.
   d. Share the systems engineering models with the specialty engineering disciplines to integrate with the additional architecture and requirements layers.
   e. Share the systems engineering models with the software development teams, mainly the business analysts, to integrate and kick off the software development activities, as applicable.

2) Integration of systems engineering activities within the systems engineering life cycle:
   a. A sample integration is shown in Figure 15.1. The systems engineering activities would start the development life cycle for the defined architecture engagement. The business would be engaged to develop business scenarios that guide the systems engineering modeling. The systems engineering models are inputs into the business requirements elicitation and analysis. They provide the high-level guidance for the more detailed requirements extraction. Typical hand-off points are between the systems engineers and the specialty engineers from the scenarios, conceptual models, value chains, functional decompositions, and process maps to the use case models and the requirements document. There should also be feedback to ensure that in post implementation the proposed system requirements were realized.
3) How the systems engineers will engage the customers for planning, elicitation of requirements, gathering process and system knowledge, building process scenarios, identifying improvement areas, process measurement, etc.
4) How the systems engineers will engage with the engineering teams, how they will integrate with and handoff to the teams.
5) The systems engineering governance structure:
   a. This will include the reporting structure of the systems engineers.
6) Systems engineer skills sets. This will describe the technical, change management, and relationship building skills of the systems engineers. The skills would include:
   a. Process management, process mapping, documentation, analysis, improvement, and control skills.
   b. Conceptual modeling, SysML, and/or UML within the software engineering body of knowledge.
   c. Systems engineering modeling tools and repository management and control.
   d. Business analysis skills including requirements elicitation, scenario building, use case modeling, requirements analysis.
   e. Change management and facilitation skills.
   f. Relationship building and "people" skills.
7) The systems engineering methodology.
   a. The systems engineering life cycle methodology discussed in the prior chapters is used to build the systems engineering modeling artifacts.
8) Define the modeling standards and governance process. The governance process should include how the models will be assessed and reviewed to ensure that they are meeting the goals of the tools, the users' requirements, and adhering to the modeling standards. Consider maturity levels and absorption rate to determine a phased roll-out approach for standards adherence.

4) **Define and Gain Buy-in to the Systems Engineering Meta Models**
After the roles and responsibilities and governance structure are defined, it is critical to create consistent and standard Systems Engineering Meta Models. The models and their relationships will set the stage, framework, and rules for the instance models. There needs to be consensus across the modeling organization, but the skilled systems engineers must be empowered to break any ties and make decisions when impasses occur. The systems engineers must work with the stakeholders to understand the modeling meta models and use this knowledge to create the meta models. The meta models developed in this book can be used as the foundation for any organization's Systems Engineering models and then adapted for the industry and company specific terminology. This should take no more than three to six months, depending upon the size of the organization to create the initial draft of the meta models. The meta models can always be enhanced in the future, but the analysis paralysis can occur in this stage and totally derail the effort to get the systems engineering practice off the ground. This is where the "obstacles" can totally block progress and put the success of the practice at great risk.

5) **Train Systems Engineers in Conceptual Modeling and Systems Engineering Tools and Methods**
Appropriate skill sets would include:
- Business process mapping, analysis, documentation, improvement, control, and optimization
- Lean Six Sigma

Traceability of Systems Engineering Model Artifacts

1 Concept of Operations
- Mission strategy
- Concepts
- Analyses
- Risks
- Stakeholder analysis
- CTS

2 Requirements & Architecture
- SIPOC
- Functional Decomp.
- Use cases
- Process Arch. Map
- Requirements
- Class diagram
- Quality management plan

3 Detailed Design
- Physical arch. Model
- System elements
- QFD
- System analysis

4 Implementation, Integration, Test & Verification
- Integration constraints
- Implementation strategy
- System elements supplied
- Operator training
- Verification criteria
- Verification test cases & results
- N-squared diagram

5 System Verification & Validation
- Verification & validation criteria
- Verification & validation test cases & results

6 Operations & Maintenance
- Training plan & materials
- Certification plan & materials
- Operations manuals
- Performance reports
- Improvement & maintenance plans
- Disposal & retirement plans

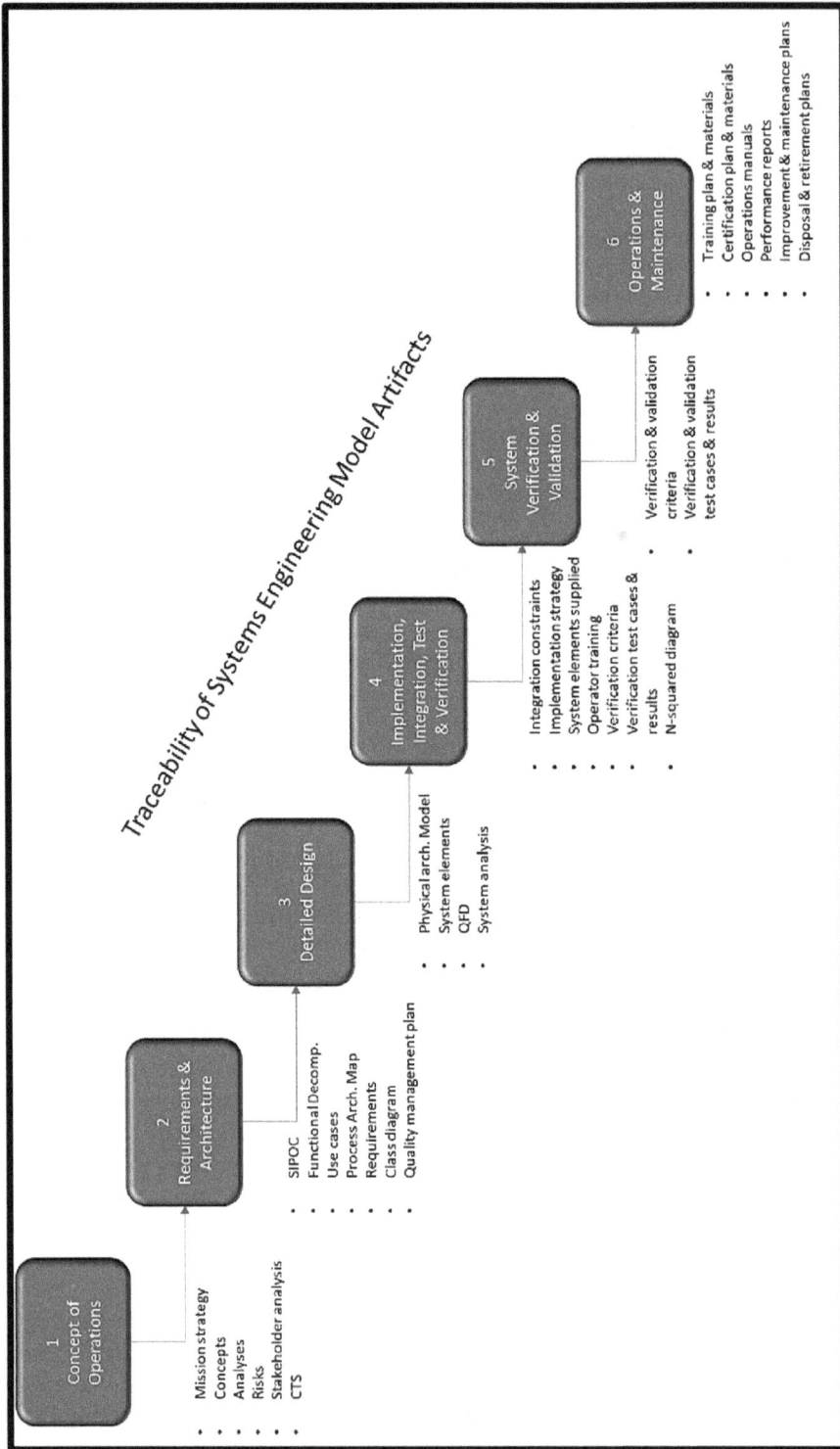

FIGURE 15.1 Traceability of systems engineering model artifacts.

- Process metrics and measurement
- Facilitation and elicitation
- Communication
- People skills
- Relationship building
- Change management
- Conceptual modeling, SysML, Unified Modeling Language (UML)
- Business analysis: requirements elicitation, use case modeling
- Modeling and architecture standards and governance process

6) **Align the Systems Engineering Vision with the Culture**

The modeling vision should be aligned with the culture. The core values of the modeling organization can be compared to the core values of the company to assess how well they are aligned. For instance, an autocratic organization that doesn't empower its people would not be aligned with empowering the systems engineers to elicit strategic information from the customers. Also, an organization that will not train associates in the new skills required to perform systems engineering modeling or bring in and support new associates with those different skill sets, it would also be extremely difficult to create successful systems engineering program. Intense soul-searching to understand the existing cultural barriers for a successful modeling program needs to be performed, so that these barriers can be removed.

7) **Assess the Culture and Readiness for Change and Incorporate Change Management tools**

The stakeholders of the systems engineering life cycle should be identified and their readiness for or resistance to change should be assessed. Change management tools should be incorporated to help enhance the acceptance of the systems engineering organization. Figure 15.2 Stakeholder Commitment Scale can be used to assess how receptive the various stakeholder groups are to change and where they need to be by the end of the systems modeling effort or by key milestones of the project. A plan for working with each stakeholder group can be developed to make the required change happen. These critical success factors can be used as a guide for changing the organization to embrace the new modeling program. Additionally, a stakeholder analysis discussed in the Concept of Operations phase can be used to identify the impacted stakeholder groups of systems modeling and projects, their concerns or how they are impacted, and their receptivity, combined together.

8) **Engage Subject Matter Experts**

How the systems engineers will engage with the subject matter experts should be defined, communicated, and monitored. In most cases, the systems engineers should work directly with the customers and specialty engineers to share and elicit information.

9) **Systems Engineering Life Cycle Methodology**

It is critical that the organization develop the modeling and systems engineering life cycle methodology that they will follow to plan and develop the models. The systems engineering life cycle prescribes a detailed methodology that aligns to the meta models described in this book.

10) **Create and Communicate Modeling Status and Successes**

A formal program management structure should be applied to plan, track, and ensure that the systems engineering engagements are progressing and showing success. Project management tools should be applied, including developing project engagement charters, project plans (work breakdown structures, milestones, time and duration estimates, resources, etc.), and tracking progress status.

Stakeholder Commitment Scale					
Stakeholders	Strongly Against	Moderate Against	Neutral	Moderate Support	Strongly Support
Customers			X -------		▸ O
Contractors		X -------			▸ O
Systems Engineers					XO
Specialty Engineers			X -------		▸ O
Leaders				X -------	▸ O
X = Receptivity at beginning of project				O = Receptivity at end of project	

FIGURE 15.2   Stakeholder commitment scale.

A formal communication plan should be developed for communicating the modeling prac-tice project in progress, completed, and pending. The success, especially at the beginning of the program, should be shared within the stakeholders and leadership. Short-term wins early in the practice should be held up as examples of what is possible if the practice is allowed to develop and grow to fruition.

The communication plan should include the following elements:

1) Desired audience to communicate to (who to communicate)
2) Communication objective/purpose (why and what to communicate)
3) Communication media or mechanism (how to communicate)
4) Frequency of communication (how often)
5) Responsible parties to communicate (who communicates the message)

A communication plan template and example are shown in Figure 15.3.

11) **Remove Obstacles (Including People as Necessary) to Change**
As the new modeling practice is created, there could be and most probably will be power strug-gles as "old-time" engineers abdicate some of their power and knowledge to the new systems engineers. In traditional-cultured organizations, the longer-term more seasoned engineers may have assumed system engineer roles and could be incredibly resistant to the new role of the

Stakeholder Receptivity Communication Plan

A communication plan is a road map for getting the message across to the desired audience. Identify how to communicate with stakeholders during the project and help move receptivity in a positive direction.

Audience	Objective/Why	Key Message	Media	Owner	When/ Frequency
Leadership	Status of system project	Current status from status reports	Email and dashboards	Program Manager or Systems Engineering	Monthly, or as requested
Specialty Engineers	Assessment & review feedback	Detailed feedback on appropriate artifacts	Review meetings and summary review documents	System Engineer	End of each reviews
Systems Engineer modelers	Assessment & review feedback	Detailed feedback on appropriate artifacts	Review meetings and summary review documents	Lead Systems Engineer	End of each reviews

FIGURE 15.3   Communication plan template and example.

formal systems engineer. They may try to block development of the models, creation of the new roles, responsibilities, and governance structure, and most importantly engagement with the customers. These old-time engineers may even sabotage the stature of the systems engineers within the organization and on the teams. This behavior must not be allowed to happen and the obstacles to change must be removed, even if it means removing the people who are obstacles to change. This may be difficult, since one of the reasons that the old-time engineers are so powerful is in the tribal knowledge that they harbor, which was never before formalized or written down, due to the lack of a formal systems engineering practice. Even at the risk of losing this knowledge and having to collect it again from the subject matter experts, these obstacles (people) must be removed. This will send a message that resistance to change will not be tolerated and that the new systems engineering organization and practice will be supported and will be successful.

12) **Empower the Systems Engineers**

The systems engineers should be completely empowered to engage with the customers and the specialty engineering stakeholders who create and revise the models. Their roles and responsibilities must be clearly defined and communicated, so that everyone understands how they are the integrating role to develop a holistic system and modeling practice. Senior management must nurture and stay connected with the systems engineers to assess their success, barriers to change, and be the champions for the systems engineering practice. The systems engineers must not be allowed to flounder without the support of leadership.

Conclusions Regarding Structure:

The structural elements, tools, and best practices described above set the stage for a successful system engineering modeling practice. The new modeling practice faces many challenges but addressing these challenges and incorporating these practices will greatly enhance the probability of success and remove the obstacles that could derail the systems engineering practice.

15.4 PEOPLE SKILLS

As discussed earlier, there are critical people, communication, facilitation, and leadership skills beyond the technical systems engineering skills that are important for the systems engineer to acquire to be successful, as listed:

a. Since our recommended approach for requirements elicitation is process oriented, it is important to know the following process skills: process management, process mapping, documentation, analysis, improvement, and control skills.

b. Conceptual modeling, SysML, and/or Unified Modifying Language (UML) drawn from the software engineering body of knowledge but adapted to systems engineering modeling related to case diagrams, some of the physical systems models, use case diagrams, and use cases.

c. Systems engineering modeling tools and repository management and control, including many of the tools that have been discussed in this book.

d. Business analysis skills including requirements elicitation, scenario building, use case modeling, requirements analysis for eliciting, and documentation customers and systems requirements.

e. Change management skills to ensure change is embraced and managed. There are many change management tools that also should be known including communication plans, change plans, and stakeholder analysis with receptivity analysis.

f. Facilitation skills are important as the systems engineer will likely facilitate meetings to focus on improving culture change, change management, requirements elicitation, and assessment reviews for deliverables and artifacts.

g. Relationship-building and "people" skills are some of the most important skills, as none of the system design and development would be possible without the teams of specialists that are required to design and develop systems.

h. Teamwork and teambuilding skills, which are discussed next.

15.4.1 Teamwork

This section describes what a team needs and teamwork skills that can be enhanced and acquired by the team members to be a more effective team member and leader.

What is a team?

A team is a group of people working together to achieve a common purpose and the following elements (Scholtes, Joiner and Streibel, 2003):

- Members have a shared work product
- Tasks are interdependent
- There is shared responsibility for output and results
- There is a commitment to a common approach to working together

Teams with one or more multicultural members were found to outperform teams devoid of multicultural individuals (HRD, 2018).

Teams outperform individuals when:

- The task is complex
- Creativity is needed
- The path forward is unclear
- More efficient use of resources is required
- Fast learning is necessary
- High commitment is desirable
- Cooperation is essential for implementation
- Members have a stake in the outcome
- The task or process involved is cross-functional
- No individual has sufficient knowledge to solve the problem

An often overlooked yet key aspect of systems engineering projects is identifying and managing teams. Effective teams are a critical element of team efforts and can significantly impact the results. We discuss a few key topics that can enhance teamwork, including:

1) Understanding what teams need
2) Team stages for growth
3) Lencioni Model of Team Dysfunction
   - Psychological safety
   - Building team trust
4) Establishing rules of engagement
5) Understanding diversity and how it impacts teams, strategies for leveraging team diversity

15.4.2 Understanding What Teams Need

Table 15.1 lists what teams need and some tools that can provide those elements for the team.

TABLE 15.1

What Teams Need and Tools for Team Building

What Teams Need	Tools for Team Building
• Understanding what a team and the power of teams is	• Stages of team growth • What is a team?
• Clearly defined purposes and goals • Clear project scope	• Project charter
• Well-defined roles and responsibilities • Well-defined and agreed upon ground rules/rules of engagement	• Rules of engagement
• Common set of values and ethics • Trust with team members	• Trust building
• Effective communication • Ability to reach consensus	• Brainstorming • Communication tips • Consensus building
• Way to handle issues	• Items for resolution • Conflict management
• Project plan	• Work plan, Work Breakdown Structure (WBS)
• Measures to evaluate their work	• Team evaluation

Source: Furterer, 2020.

15.4.3 TEAM STAGES FOR GROWTH

Teams go through fairly predictable stages according to the Tuckman model (Scholtes, Joiner, and Streibel, 2003).

Knowing what the stages are and how to meet the needs of the group at each stage can help to prevent many of the pitfall's teams come up against. The stages are:

- Forming – where the team forms and begins their activities
- Storming – unrest within the team due to many factors
- Norming – understanding each other and reaching norms of behavior
- Performing – the team gets the work done and is high performing

For each team stage, the figures show the characteristics that describe the stage, how teams feel being in this stage, the behaviors that team members show within the stage, and how the teams can work through this stage to move to the next stage. One important element of this model is that teams typically go through each of the stages, but depending upon the team, the length of time they remain in the stage varies and the teams can return to prior stages and return to forming when additional members are added.

Team Stage: Forming (Table 15.2)
Team Stage: Storming (Table 15.3)
Team Stage: Norming (Table 15.4)
Team Stage: Performing (Table 15.5).

15.4.4 LENCIONI MODEL OF TEAM DYSFUNCTION

The Lencioni Model of Team Dysfunction describes five types of dysfunctions that contribute to team's not working well together. The foundational type is the absence of trust. This is when the

TABLE 15.2
Forming Team Stage Characteristics, Feelings, Behaviors, and Leading through this Stage

Characteristics	Feelings that team members could have	Behaviors that team members can exhibit
• Members are cautious and explore appropriate group behavior • The members' transition from being an individual to a team member	• There is pride in being chosen to be a part of the team • There is excitement and optimism • There is initially tentative attachment to the team • Suspicion, fear, anxiety	• Attempting to design the purpose and how it will be carried out • Deciding what information is needed to complete tasks • Establishing team rules • Waiting to be told what to do and directing most communication to the team leader • Complaining about the organization and difficulties in completing the tasks

Leading the team through this stage

- Help members to get to know each other
- Provide clear direction and purpose
- Involve members in defining roles, activities, ways to work together
- Provide information on structure
- Define expectations
- Define why the team members are on the team, what is in it for me (WIIFM)

TABLE 15.3
Storming Team Stage Characteristics and Behaviors

Characteristics	Feelings that team members could have	Behaviors that team members can exhibit
• Most difficult stage for the team • Tasks are difficult • Team members become testy, anxious, or overzealous • Team can fall apart in this stage • Should encourage collaboration	• Frustration and resistance to tasks • Fluctuation in attitudes about the team's probability of success • Anxiety • Withdrawal from conflict	• Arguing between members, even when they agree upon the issues • Defensive, competitive, splitting into factions • Questioning the wisdom of those who select the project and appointed the team members • Establish unrealistic goals • Increased tension and jealousy • Withdrawing or dropping out of the team

Leading the team through this stage

- Resolve issues of power and authority
- Develop and implement agreements about how decisions are made and who makes them
- Allow the team to become more independent
- Use established ground rules
- Develop and/or clarify roles and responsibilities
- Update the project charter with the project goals and scope

team members fear being vulnerable with other team members and prevents them from developing trust with their team members. The second type of dysfunction is the fear of conflict. Team members want to avoid conflict and preserve an artificial harmony. This artificial harmony prevents productive ideological conflict. The third dysfunction is lack of commitment. This includes lack of buy-in which prevents team members from making decisions that they can commit to. Dysfunction four is avoidance of accountability, where team members do not hold each other accountable to complete

TABLE 15.4
Norming Team Stage Characteristics and Behaviors

Characteristics	Feelings that team members could have	Behaviors that team members can exhibit
• Members reconcile competing loyalties and responsibilities • They accept the team ground rules and norms • Emotional conflict is reduced	• A sense of team cohesion • Acceptance of membership in the team • Relief that it seems that everything will work out	• Laughing, joking, and attempts to achieve harmony • Experimenting with ways to raise and discuss differences in opinion effectively • Confiding in each other and sharing personal problems, becoming more friendly • Expressing criticism constructively and maintaining team ground rules

Leading the team through this stage

- Fully utilize team members' skills, knowledge, and experience
- Encourage and acknowledge members' respect for each other
- Refer the team to established ground rules

TABLE 15.5
Performing Team Stage Characteristics and Behaviors

Characteristics	Feelings that team members could have	Behaviors that team members can exhibit
• Team has settled into its relationships and expectations. They can perform consistently, diagnosing and solving problems, and choosing and implementing changes	• Insights into personal and group processes and a better understanding of each other's strengths and weaknesses • Satisfaction at the team's progress • Close attachment to the team	• Creating constructive self-change • Preventing or working through group problems • You get a lot work done

Leading the team through this stage

- Update the team's methods and procedures to support cooperation
- Help the team to manage change
- Advocate for the team to others

tasks or complete them at a required quality level. The last dysfunction is inattention to results, where the focus on the individual's own goals or personal status could erode the team's success. The Team Dysfunction model is shown in Figure 15.4. Building trust relates to psychological safety. Psychological safety is the belief that you won't be punished for making a mistake.

15.4.5 PSYCHOLOGICAL SAFETY (EDMONDSON, 2021)

- Psychological safety, the belief that you won't be punished when you make a mistake.
- Studies show that psychological safety allows for moderate risk-taking, speaking your mind, creativity, and sticking your neck out without fear of having it cut off — just the types of behavior that lead to market breakthroughs.
- We become more open-minded, resilient, motivated, and persistent when we feel safe. Humor increases, as does solution-finding and divergent thinking — the cognitive process underlying creativity.

FIGURE 15.4 The five dysfunctions of a team by Patrick M. Lencioni.

- Represents the extent to which the team views the social climate as conducive to interpersonal risk; it is a measure of people's willingness to trust others not to attempt to gain personal advantage at their expense.
- When someone makes a mistake in this team, it is often held against him or her.
- In this team, it is easy to discuss difficult issues and problems.
- In this team, people are sometimes rejected for being different.
- It is completely safe to take a risk on this team.
- It is difficult to ask other members of this team for help.
- Members of this team value and respect each other's contributions.
- To increase psychological safety:
  - Approach conflict as a collaborator, not an adversary.
  - Speak human to human.
  - Anticipate reactions and plan countermoves.
  - Replace blame with curiosity.
  - Ask for feedback on delivery.
  - Measure psychological safety.

15.4.6 RULES OF ENGAGEMENT

Rules of engagement are tools to develop the norms and expectations of how the team will work with each other and behave. It is important to collaboratively develop the rules of engagement with the team members and create rules that everyone agrees to abide by.

- Following is a list of the rules of how the team will work with each other, which can include the categories:
  - Team logistics, scheduling meetings, due dates.
  - Treatment of each other.
  - Roles and responsibilities.

- How you'll work together to get tasks done?
- How you'll ask for help?
- How will you communicate with team members, project sponsors, and instructor?
- How you'll delegate tasks?
- How to define quality communication?
- How assess quality of work products?
- Other.

15.4.7 TEAM DIVERSITY

- In the McKinsey Study, "Diversity Matters," there was a correlation between diversity and financial performance.
- The companies in the top quartile for gender diversity were 15% more likely to have financial returns that were above their national industry median.
- The companies in the top quartile for racial/ethnic diversity were 35% more likely to have financial returns above their industry median.
- Companies that commit to diverse leadership are more successful.
- More diverse companies are better able to:
  - Win top talent,
  - Improve their customer orientation,
  - Employee satisfaction,
  - Decision-making, Increasing returns and competitive advantage.

15.5 TECHNIQUES THAT CAN BE USED IN EACH PHASE OF THE VEE LIFE CYCLE TO ENHANCE THE COMMUNICATION WITH CUSTOMERS AND STAKEHOLDERS

15.5.1 PHASE 1 – CONCEPTS OF OPERATIONS

15.5.1.1 Concept Generation Applying the IDA Method (Furterer, 2020)

- The IDA (Ideation, Disruption, Aha) method, shown in Figure 15.5, is used at the University of Dayton's Institute of Applied Creativity for Transformation (IACT) to generate innovative and creative design concepts.
- IDA will be applied to develop system concepts in the Concept of Operations phase.

Ideation – Step 1:

- In the Ideation stage of the methodology, Step (1), the participants on the original concept generation teams that having been working on them in the first phase, use content from the analyses within the phase to generate concepts.
- The team can use free form brainstorming where they all throw out ideas verbally or on pieces of paper to everyone at the table.
- The team can also use a more structured brainstorming technique that is part of the Nominal Group Technique.
  - Define the goal of the brainstorming: generate system concepts
  - Give everyone 5 to 10 minutes to silently generate and write down potential concepts
  - In a round-robin fashion, each team member shares their concept idea with the team, and someone writes it on a white board, flip chart paper, or Smart-Sheet®.
  - There should be no evaluation of the ideas during the brainstorming activity.

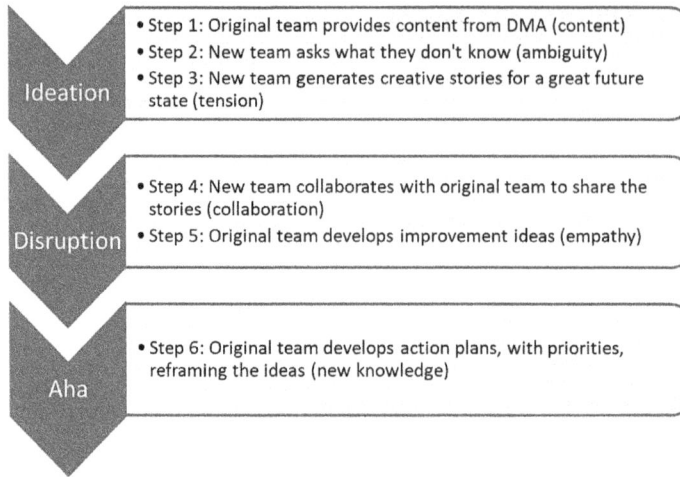

FIGURE 15.5   IDA process for concept generation.

Example: Concepts from Team 1:

- Community meetings
- Website/form
- Corner stores
- Farms
- Delivery service

Ideation – Step 2:

- The list of concepts is left on the table. One original team member stays at the original concepts table, and the other team members move as a group to a new table.
- The new team writes questions that they don't know or understand about the existing content/concepts. The original team member answers the questions.

Example:

The teams had an open discussion and asked questions to better understand the proposed concepts.

Ideation – Step 3:

- The new team creates a story or narrative of a great future state that could include the concepts.

Typically, most narratives work along these lines in order to build the story's arc:

- **Trigger:** what is the problem, challenge, need, or desire that your story/video will address?
- **Action:** what happens to resolve this issue, what is the journey like?
- **Reward:** what is the outcome or reward, what are the benefits that the action delivered?

The trick is to create a story that resonates with your target audience and engages them, so they want to engage with your organization (storyboarding).

Ideation – Step 3:

Example story:

There is no way to easily connect local farmers to places for them to sell their produce. We envision an application where we can connect farmers to the corner store via a website form. The farmers would enter the produce that they have available to sell. The buyers could go into the application and select the produce that they need. The farmers and buyers could be part of community meetings where farmers can share what produce they have, and buyers can select to buy the produce. A delivery service could also pick up the produce from several local farms and take it to a community market as needed, based upon the orders from the website form. There is a perfect match between what the farmers produce and can sell and what the market and community want to buy.

Disruption – Step 4:

• The new team collaborates with the original team and shares the stories with the class.

Disruption – Step 5:

• The original team uses the new story to develop additional concepts and ideas around these stories and selects categories organizing these ideas into concepts.

Examples:

Disruption – Step 4:

• The new team collaborates with the original team and shares the stories with the class.

Disruption – Step 5:

• See the concepts listed in the Pugh matrices in chapter 5 for each team.

Aha– Step 6:

• In the Aha phase, in Step 6, the original team develops descriptions of the concepts reframing the ideas and creating the new knowledge and realizing the Aha experience.
• Example: See the concepts in the Pugh matrices in Chapter 5.

15.5.2 Phase 2 – Requirements and Architecture

15.5.2.1 Process Scenarios

It is suggested that the Process Architecture Maps are used to elicit and capture the requirements in a storyboarding manner. The IDA method discussed earlier could be applied to develop the stories and the different stakeholders expand on them in a gallery walking manner, where each team moves from table to table, contributing to the stories. Following is a reprint of the list of elements and factors to consider when generating the stories. Remember that for a product system, the functions and ways that the system would be used would be developed in the PAMs. For a service system, the processes for how the system would work would be developed within the scenarios and documented in the PAMs.

Following is a process scenario use case checklist that can be used to generate process scenarios and use case documents (TOGAF8.0).

Process Scenario Use Case Checklist

- What is the overall objective of the scenario?
- What are any preconditions that must exist to start the scenario?
- Who is the "actor" of the scenario (who is performing the scenario)?
- What are the inputs (information, etc.) used to start the scenario?
- Walk through the "happy path" steps to get to the end result?
- Walk through alternate steps to get to end result and "not" to get to end result (failure)?
- Think of who the person would interact with along the way? Who are the suppliers of inputs to the process and which customers receives outputs from the process?
- Think of what systems, information systems, or technology the person might interact with (generally)?
- Think of assumptions or needs a person might expect to require during the scenario?
- What is the final outcome, result, or output? What is transformed?
- What are the post conditions, what is the state of the business after the scenario is finished?
- What are the resources, people, and tools needed?
- What is the information, materials, information systems, policies, and procedures used and/or produced?
- How is the process measured today?
- What are the failure points?
- What decisions are made?
- How is work distributed?

15.5.3   PHASE 3 – DESIGN

15.5.3.1   Design Idea Generation

The Theory of Inventive Problem Solving (TRIZ) method is developed by Genrich S. Altshuller and his colleagues. According to TRIZ, universal principles of creativity form the basis of innovation. The following principles could be applied to generating design ideas for a service system. Additional principles could be applied for a product system.

1. Think of the ideal vision, process, or system
2. Think of ways to improve the process or function
3. Think of ways to eliminate or reduce undesired functions
4. Think of ways to segment the process
5. Think of ways to copy existing ideas or processes
6. Think of a disposable concept

15.5.4   PHASES 4 AND 5 – VERIFICATION AND VALIDATION

15.5.4.1   Test Cases

For developing different test cases, the use cases can be used almost verbatim to develop the test cases. However, additional test cases may need to be brainstormed to consider integration points found in the N-squared diagrams developed earlier.

Some different types of brainstorming not already discussed include:

- Assumption busting: Instead of asking why, ask "why not?"
- Brainwriting: Write down an idea; pass it to the person next to them who then builds on the idea or concept.
- Anti-solution brainstorming: Generate ideas of how they could make the process or system even worse, punching holes in your own argument.

Standard Work Instruction Sheet
Identifies whether the operators are trained in each key process.

Activity	Time	Notes

FIGURE 15.6   Standard work form.

15.5.5   Phase 6 – Operations and Maintenance

15.5.5.1   Standard Work

- Standardized work is one of the most powerful but least used lean tools. By documenting the current best practice, standardized work forms the baseline for kaizen or continuous improvement. As the standard is improved, the new standard becomes the baseline for further improvements and so on. Improving standardized work is a never-ending process.
- Standardized work consists of three elements:
  - Takt time, which is the rate at which products must be made in a process to meet customer demand.
  - The precise work sequence in which an operator performs tasks within takt time.
  - The standard volume and resources required to keep the process operating smoothly.
- Establishing standardized work relies on collecting and recording data on forms. These forms are used by engineers and front-line supervisors to design the process and by operators to make improvements in their own jobs.

A sample standard work instruction sheet is shown in Figure 15.6. It includes the following elements:

- Identify activities
- Time for each activity
- Notes

It could also include the following elements, more common in manufacturing for product systems:

- Takt time
- Cycle time
- Volume in inventory

15.6   SUMMARY

In this chapter, we discussed change management, culture and culture change, structures that can help build a Model-Based Systems Engineering practice, and ideas for enhancing communication in each phase of the Vee Life Cycle.

15.7 ACTIVE LEARNING EXERCISES

1) Teambuilding: team experience
   - Think of the best team experience that you had
   - What made it the best team experience?
   - Develop an affinity diagram, identifying themes for what makes a great team experience
   - Reflect with your classmates
2) Silently generate different strategies for forming teams
   - On a Smart-Sheet® or white board create an affinity diagram of these strategies, grouping similar ideas into themes
3) As you discuss "trust" as a basis of teamwork, pass out color-coated M&Ms. Ask everyone to take several, eat as many as they wish, but save one (1) in sight on top of their desk.
   - Discuss findings from Google about psychological safety.
   - Explain that each person is going to tell a two (2) minute story about themselves as a way to get to know each other better.
   - List colors of M&Ms and type of story on board.
     a. Blue = proud
     b. Orange = failure
     c. Yellow = scared
     d. Green = embarrassed
     e. Brown = success/accomplishment
     f. Red = aspiration
   - Explain that the types of personal experiences each person tells in their true stories must correspond to the color of M&M left on their desks.
   - Tell a sample personal story to the class. Choose a self-deprecating story such as failure or embarrassment to model.
   - Allow two (2) minutes to prepare stories.
   - Work around room telling stories. (Can be adjusted to group size).
4) Develop rules of engagement for your team considering the following topics:
   - Describe how you will deal with scheduling meetings.
   - State time and condition for how long before a deadline our individual work will be completed.
   - Describe the process we will use to make decisions.
   - How will we deal with a team member who is not holding up his/her "part of the bargain" in terms of quality amount of work, timing, and/or communication?
   - How will we deal with conflict and disagreement?
   - What specific rules of engagement (communication) do we pledge to follow in working with each other on this project? (Corresponding to working outside of meetings, quality of communication, providing status.)
   - Consider any other challenges that your teams have faced in the past, include those topics.
   - Share with the class.

BIBLIOGRAPHY

Edmondson, A. *Team Learning and Psychological Safety Survey.* https://www.midss.org/content/team-learning-and-psychological-safety-survey. Accessed 3/15/2021

Furterer, S., Lean Six Sigma for Engineers, Graduate Program Course Material, 2020.

Furterer. Sandra, Lean Six Sigma program success factors in a retail application. International Conference on Industry, Engineering and Management Systems. Cocoa Beach, FL 2008.

HRD, Do Diverse Teams Perform Better, Human Resources Director, Australia, January 15, 2018.

Hoffman, J. and Mera S. Management leadership and productivity improvement program, *International Journal of Applied Quality Management*, 2(2), pp 221–232, 1999.

Hughes, M. *Managing and Leading Organizational Change*, Routledge, New York, 2019.

Johnson, G. Managing Strategic Change Strategy, Culture and Action, *Long Range Planning*, Vol. 25, No. 1, pp. 28–36, 1992.

Kotter, J.P. Leading Change: Why Transformation Efforts Fail. *Harvard Business Review* 73(2): 59–67, 1995.

Kotter, J. *Leading Change*, Harvard Business School Press, 1996.

Lencioni, P. *The Five Dysfunctions of a Team: A Leadership Fable*, Jossey-Bass, San Francisco, CA, 2002.

Newman, J. Beyond the vision: Cultural change in the public sector. *Public Money & Management*. April-June 1994.

Robbins, S. and Judge, A. *Organizational Behavior*, Pearson, New York, 2019.

Rogers, E. *Diffusion of Innovations*, The Free Press, New York, 1962.

Schein, E. *Organizational Culture and Leadership*, John Wiley & Sons, San Francisco, CA, 2010.

Scholtes, P., Joiner, B., Streibel, B. *The Team Handbook*, 3rd Edition, Oriel Inc., Madision, WI, 2003

https://www.sightline.co.uk/storyboarding-the-key-to-telling-your-story-right/, Accessed 3/14/2021.

Smith, A.C.T., and Graetz, F.M. *Philosophies of Organizational Change*. Cheltenham, Edward Elgar, 2011.

Sureshchandar, G., Chandrasekharan, R., and Anantharaman, R. A holistic model for total quality service. *International Journal of Service Industry Management* 12 (4), pp. 378–412, 2001.

Index

Taylor & Francis Group
an **informa** business

Taylor & Francis eBooks

www.taylorfrancis.com

A single destination for eBooks from Taylor & Francis
with increased functionality and an improved user
experience to meet the needs of our customers.

90,000+ eBooks of award-winning academic content in
Humanities, Social Science, Science, Technology, Engineering,
and Medical written by a global network of editors and authors.

TAYLOR & FRANCIS EBOOKS OFFERS:

A streamlined
experience for
our library
customers

A single point
of discovery
for all of our
eBook content

Improved
search and
discovery of
content at both
book and
chapter level

REQUEST A FREE TRIAL
support@taylorfrancis.com

Routledge
Taylor & Francis Group

CRC Press
Taylor & Francis Group

For Product Safety Concerns and Information please contact our EU
representative GPSR@taylorandfrancis.com
Taylor & Francis Verlag GmbH, Kaufingerstraße 24, 80331 München, Germany

www.ingramcontent.com/pod-product-compliance
Lightning Source LLC
Chambersburg PA
CBHW080700220326
41598CB00033B/5272